168 Topics in Current Chemistry

W0107622

Photoinduced Electron Transfer V

Editor: J. Mattay

With contributions by
A. Albini, R. A. Bissell, E. Fasani, P. K. Freeman,
H. Q. N. Gunaratne, S. A. Hatlevig, P. L. M. Lynch,
G. E. M. Maguire, P. Maslak, C. P. McCoy, M. Mella,
G. Pandey, K. R. A. S. Sandanayake, A. P. de Silva,
J.-P. Soumillion

With 33 Figures and 11 Tables

Springer-Verlag
Berlin Heidelberg GmbH

This series presents critical reviews of the present position and future trends in modern chemical research. It is addressed to all research and industrial chemists who wish to keep abreast of advances in their subject.

As a rule, contributions are specially commissioned. The editors and publishers will, however, always be pleased to receive suggestions and supplementary information. Papers are accepted for "Topics in Current Chemistry" in English.

ISBN 978-3-662-14951-5   ISBN 978-3-540-47646-7 (eBook)
DOI 10.1007/978-3-540-47646-7

Library of Congress Catalog Card Number 74-644622

© Springer-Verlag Berlin Heidelberg 1993
Originally published by Springer-Verlag Berlin Heidelberg New York in 1993
Softcover reprint of the hardcover 1st edition 1993

Typesetting: Macmillan India Ltd., Bangalore-25

51/3020-5 4 3 2 1 0 – Printed on acid-free paper

Guest Editor

Prof. Dr. *Jochen Mattay*
Organisch-Chemisches Institut,
Westfälische Wilhelms-Universität Münster,
Orléansring 23, 48149 Münster, FRG

Editorial Board

Preface to the Series
on Photoinduced Electron Transfer

The exchange of an electron from a donor molecule to an acceptor molecule belongs to the most fundamental processes in artificial and natural systems, although, at the primary stage, bonds are neither broken nor formed. However, the transfer of an electron determines the chemical fate of the molecular entities to a great extent. Nature has made use of this principle since the early beginnings of life by converting light energy into chemical energy via charge separation. In recent years man has learnt, e.g. from X-ray analyses performed by Huber, Michel and Deisenhofer, how elaborately the molecular entities are constructed within the supermolecular framework of proteins. The light energy is transferred along cascades of donor and acceptor substrates in order to prevent back electron transfer as an energy wasting step and chemical changes are thus induced in the desired manner.

Today we are still far away from a complete understanding of light-driven electron transfer processes in natural systems. It is not without reason that the Pimentel Report emphasizes the necessity of future efforts in this field, since to understand and "to replicate photosynthesis in the laboratory would clearly be a major triumph with dramatic implications". Despite the fact that we are at the very beginning of knowledge about these fundamental natural processes, we have made much progress in understanding electron transfer reactions in "simple" molecular systems. For example most recently a unified view of organic and inorganic reaction mechanisms has been discussed by Kochi. In this context photochemistry plays a crucial role not only for the reasons mentioned above, but also as a tool to achieve electron transfer reactions. The literature contains a host of examples, inorganic as well as organic, homogeneous as well as heterogeneous. Not surprisingly, most of them were published within the last decade, although early examples have been known since the beginning of photochemistry (cf. Roth's article in Vol.156). A reason is certainly the rapid development of analytical methods, which makes possible the study of chemical processes at very short time ranges. Eberson in his monograph, printed by this publishing company two years ago, nicely pointed out that "electron transfer theories come in cycles". Though electron transfer has been known to inorganic chemists for a relatively long time, organic chemist have still to make up for missing concepts (cf. Eberson).

A major challenge for research in future, the "control of chemical reactions" as stated by the Pimentel Report, can be approached by various methods; light-driven processes are among the most important ones. Without interaction of the diverse scientific disciplines, recent progress in photochemistry, as well as future developments would scarcely be possible. This is particularly true for the study of electron transfer processes. Herein lies a challenge for science and economy and the special facsination of this topic – at least for the guest editor.

The scope of photochemistry and the knowledge about the fundamentals of photoinduced electron transfer reactions have tremendously broadened within the last decade, as have their applications. Therefore I deeply appreciate that the Springer-Verlag has shown interest in this important development and is introducing a series of volumes on new trends in this field. It is clear that not all aspects of this rapidly developing topic can be exhaustively compiled. I have therefore tried to select some papers which most representatively reflect the current state of research. Several important contributions might be considered missing by those readers who are currently involved in this field, however, these scientists are referred to other monographs and periodical review series which have been published recently. These volumes are meant to give an impression of this newly discovered reaction type, its potential and on the other hand to complement other series.

The guest editor deeply appreciates that well-known experts have decided to contribute to this series. Their effort was substantial and I am thankful to all of them. Finally, I wish to express my appreciation to Dr. Stumpe and his coworkers at the Springer-Verlag for helping me with all the problems which arose during the process of bringing the manuscript together.

Münster, December 1989   Jochen Mattay

Preface to Volume V

Included in this volume are six articles devoted to a process which is basic not only in chemistry, but in biology and physics. The intent is to provide a critical review of some topics in photo-induced electron transfer (PET) in order to complement existing works. All articles are concerned with experimental and theoretical aspects of PET in organic chemistry. The interdisciplinary character of PET is already reflected in the preceding volumes of this series as will be so in the following volumes for electron transfer in general.

PET V starts with a survey of fragmentation reactions which, as one of the most important events after the initial transfer of an electron, competes with back-electron transfer. The factors controlling this latter process as well as the thermodynamics, kinetics and the stereochemistry of the fragmentations are comprehensively discussed. The second article deals with dehydrohalogenation of polyhalocompounds emphasizing both the nature of excimers in the phototransformation and the nature of exciplexes formed in the presence of electron transfer reagents. The impact photochemistry has on toxic waste disposal concludes this chapter. The special features of organic anions in PET-reactions are critically discussed in the following article. Despite the advantages such as the increased electron donating capacity, the reactivity is strongly influenced by the experimental conditions. S_{RN}^1 reactions, alkylations of carbanions and reductions initiated by oxyanions are discussed in detail as well as the quenching of excited states by anions in their ground state.

Aromatic compounds have been used as electron acceptors and electron donors in PET-processes since the early beginnings of photochemistry. Chapter 4 of this volume is devoted to this still fascinating area mainly from the synthetic point of view. Special emphasis is given to recombination reactions of radical ions, electrophilic, nucleophilic and radicalic additions. The following article extends this topic to recent advances of PET in organic synthesis. Several new reactions are discussed covering photo-sensitized electron transfer processes and donor–acceptor reactions. The final chapter turns to new developments in the field of fluorescent PET sensors. In this article the design of an efficient model system containing both a selective receptor and a fluorophore is vividly described. A variety of sensor systems with an eventually high potential for application is discussed.

The guest editor deeply acknowledges that colleagues well-known as experts for years have agreed to contribute to this series. Their effort was substantial just as the support of Dr. Stumpe and his coworkers at Springer-Verlag in preparing the volume.

Münster, May 1993   Jochen Mattay

Attention all "Topics in Current Chemistry" readers:

A file with the complete volume indexes Vols. 22 (1972) through 167 (1993) in delimited ASCII format is available for downloading at no charge from the Springer EARN mailbox. Delimited ASCII format can be imported into most databanks.

The file has been compressed using the popular shareware program "PKZIP" (Trademark of PKware Inc., PKZIP is available from most BBS and shareware distributors).

This file is distributed without any expressed or implied warranty.

To receive this file send an e-mail message to:
SVSERV@DHDSPRI6.BITNET.
The message must be: "GET/CHEMISTRY/TCC_CONT.ZIP".

SVSERV is an automatic data distribution system. It responds to your message. The following commands are available:

HELP	returns a detailed instruction set for the use of SVSERV,
DIR (*name*)	returns a list of files available in the directory "name",
INDEX (*name*)	same as "DIR",
CD ⟨*name*⟩	changes to directory "name",
SEND ⟨*filename*⟩	invokes a message with the file "filename",
GET ⟨*filename*⟩	same as "SEND".

For more information send a message to:
INTERNET: STUMPE@ SPINT. COMPUSERVE. COM

Table of Contents

Table of Contents of Volume 156

Table of Contents of Volume 158

Table of Contents of Volume 159

Table of Contents of Volume 163

Fragmentations by Photoinduced Electron Transfer. Fundamentals and Practical Aspects

Przemyslaw Maslak

Department of Chemistry, The Pennsylvania State University, University Park, PA 16802, USA

Table of Contents

Topics in Current Chemistry, Vol. 168
© Springer-Verlag Berlin Heidelberg 1993

Removal of an electron from a bonding orbital or addition of an electron to an antibonding orbital of a diamagnetic molecule activates the resulting radical ion for fragmentation. Such reactive radical ions may be generated by photoinduced electron transfer (PET). There are two alternative ways to accomplish the transfer of an electron: (1) the local excitation of a donor or an acceptor which is well described by the empirical Weller equation and (2) the excitation of the charge-transfer complexes according to the Mulliken theory. The fragmentation reaction competes with back-electron transfer (BET) within the photogenerated radical ion pairs. The back electron transfer is well described by the Marcus theory. In most PET systems the rate of BET decreases with the increasing exergonicity and the rate is faster within contact ion pairs than with solvent separated ion pairs. The exergonicity of BET as well as ion pair solvation and spin multiplicity are predetermined by the method of ion-pair generation. These factors, in addition to the rate of cleavage, are critical in determining the overall efficiency of the PET fragmentation.

The thermodynamics of the unassisted fragmentation reaction is determined by the homolytic bond strengths and the difference in redox potentials of the radical ion and the ionic fragment. The overlap between the scissile-bond orbital and the orbital bearing the unpaired electron is the critical stereoelectronic factor affecting the cleavage. For endergonic fragmentations, the "intrinsic" kinetic barriers are low, i.e. the reverse reactions (radical/ion coupling) have low activation energies. For exergonic scissions the reactions are very rapid. The fragmentation reactions may be used to rapidly generate reactive intermediates and expeditiously fragment homolytically strong bonds at ambient or low temperatures.

1 Introduction

Photon energy absorbed by a chemical system may be used to transfer an electron between a donor (D) and an acceptor (A), thus generating a pair of high-energy intermediates [1–6]. A conceptually simple reaction that often follows is a cleavage of a single bond within one of the intermediates (usually a radical ion) to yield two fragments [7–10]. The overall process utilizes light energy to carry out a chemical reaction and may serve as a simple model for energy conversion as well as a means to generate reactive intermediates (ions and radicals) for the purpose of a scientific inquiry or even commercial applications.

This review presents the theoretical and practical aspects of the three elementary steps involved in the overall process: 1) photochemically induced electron transfer (PET), 2) energy-wasting back-electron transfer (BET) within the pair of reactive intermediates, and 3) fragmentation of single bonds within the reactive intermediates and its consequences. All these steps contribute to the overall efficiency of the process. The first step serves to generate the high energy intermediates. Its efficiency not only directly provides the impetus for the process, but indirectly, by producing the high energy intermediates in specific solvation and spin states, predetermines the efficiency of the back-electron transfer step. The competition between the back-electron transfer and a net chemical reaction (bond cleavage) is, in turn, the major factor determining the usefulness of the overall process. That competition is, however, not straightforward. The high energy intermediates (radical ions) involved are in a state of constant change: the ions are being solvated, the ions or radicals separate by diffusion, the radical pairs change their spin status, etc. All that dynamic behavior influences the thermodynamics and kinetics of BET and, to a smaller degree, that of the fragmentation reaction. Similarly to the rate of BET, the fragmentation rate is primarily controlled by the thermodynamic factors, but, as in any bond making or breaking process, the stereoelectronic considerations are also important.

The purpose of this review is to highlight the above-mentioned aspects of the photoinduced bond cleavage reactions. The aim is to summarize our current understanding of these processes from a subjective point of view of a practicing physical organic chemist, and to provide illustrative examples, rather than to give an exhaustive review of the literature. To keep the subject manageable, the review will be mostly limited to organic examples consisting of "true" fragmentation reactions, i.e. those in which two unconnected fragments are produced. Thus, cleavages of cyclic bonds [8,11–13] or rearrangements [14] are outside the scope of this review. The proton transfer reactions which are often found to accompany the electron transfer are also excluded [15]. Such processes, where the bond breaking is accompanied by a significant bond making in the transition state, are clearly more complicated and require separate analysis.

2 Photoinduced Electron Transfer

2.1 The Marcus Theory of Electron Transfer

A complete theory of any elementary reaction must include a functional dependence of the reaction rate (kinetics) on the driving force (thermodynamics) which should embody all pertinent experimental variables, such as medium effects or stereoelectronic factors. Such a theory must have a predictive power. In the case of electron transfer (ET) between organic donors and acceptors, the Marcus theory comes very close to satisfying these requirements [16–18]. Within the classical [17] (simplified) framework the activation energy ($\Delta G_{et}^{\ddagger}$) of electron transfer depends on the driving force ($-\Delta G_{et}$) and the reorganization energy, λ:

$$\Delta G_{et}^{\ddagger} = \Delta G_{es} + \lambda/4[1 + \Delta G_{et}/\lambda]^2 \tag{1}$$

where ΔG_{es} is a change in electrostatic free-energy as the reactants form an encounter complex. If the donor or acceptor is neutral the electrostatic term is neglected ($\Delta G_{es} = 0$). The relationship of Eq. 1 is quite unlike other free-energy relationships such as the linear functions of Hammett or Brønsted. The rate of endergonic ET is predicted to increase in a parabolic fashion as the driving force increases, reaching a maximum at $-\Delta G_{et} = \lambda$ (Fig. 1). Further increases in the driving force lead to decreases in the reaction rate. This unique driving force region is commonly termed an "inverted region".

The electron is transferred essentially instantaneously from the donor to the acceptor. The nuclear motions of the participants (the reactants and the solvent) are, however, much slower. Thus, for a successful transfer, both the solvent and the substrates must "reorganize" first to bring both reacting systems to a transition state configuration where the transferred electron will have similar energy (within $k_b T$) at the initial and final sites. This Franck-Condon factor is responsible for the fact that the observed rates of electron transfer can vary over many orders of magnitude, even if they have similar driving forces [17]. The reorganization energy is usually considered as a sum of the internal reorganization energy, λ_i, associated with changes in bond lengths and angles of the substrates, and the solvent reorganization energy, λ_s, related to solvent polarizability and realignment of solvent dipoles.

In the semi-classical approach [18], the golden-rule type expression is used (Eq. 2) in which the rate is a product of an electronic matrix element squared

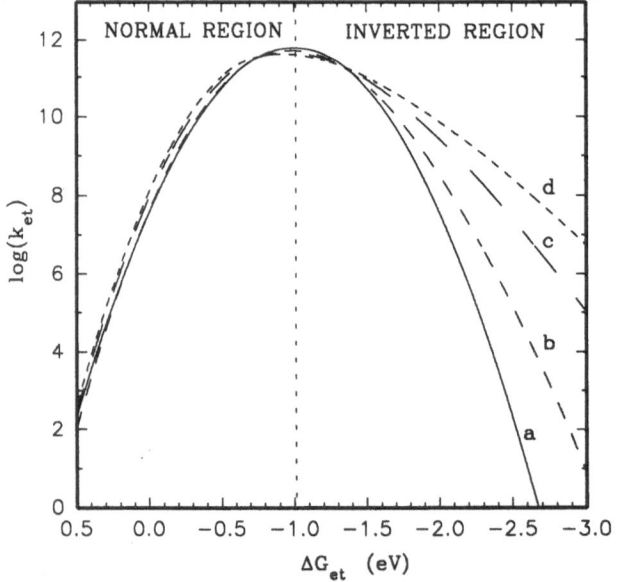

Fig. 1. The classical (*a*) and the semi-classical (*b–d*) representations of the Marcus theory for $\lambda = 1.0$ eV at $T = 300$ K. In the classical expression (Eq. 1), λ determines both the position of the maximum and the breadth of the parabola. The maximum k_{et} is determined by the frequency factor (Z, here taken as 6×10^{11} s^{-1}) in the Eyring expression ($k_{et} = \kappa Z \exp(-\Delta G_{et}^{\ddagger}/k_b T)$ where κ is the transmission coefficient, usually taken to be unity). In the semi-classical approach the reorganization energy is explicitly divided into λ_i (here equal 0.2 eV) and λ_s (0.8 eV). The value of V is chosen to match Z (V = 48.2 cm^{-1}) and three different vibrational frequencies (ν) are used: 750 cm^{-1} (*b*), 1500 cm^{-1} (*c*), and 2250 cm^{-1} (*d*) for illustration purposes

(V^2) and a Franck-Condon weighted density (FCWD) of states. The FCWD term contains the dependence of the rate on ΔG_{et}. In this approach the solvent reorganization energy is treated classically and the internal reorganization energy quantum mechanically, with the vibrational frequencies represented by an average frequency, v. The rate–driving force relationship is usually written as:

$$k_{et} = (2\pi/h) \cdot V^2 \cdot \text{FCWD} \tag{2a}$$

or

$$k_{et} = (\pi/\hbar^2 \lambda_s k_b T)^{1/2} \cdot |V|^2 \sum_{w=0}^{\infty} (e^{-S} S^w / w!) \cdot \exp\{-[(\lambda_s + \Delta G_{et} + whv)^2 / 4\lambda_s k_b T]\} \tag{2b}$$

where $S = \lambda_i / hv$, h is the Planck constant, $\hbar = h/2\pi$, k_b is the Boltzman constant, T is temperature and w is an integer [19]. The average vibrational frequency (v) is commonly taken as 1500 cm^{-1}, a value typical for carbon–carbon skeletal vibrations [18].

Equation 2 is functionally very similar to Eq. 1 (Fig. 1) in the endergonic region (normal driving force region), but differs significantly in the inverted region. Both, the simplified version (Eq. 1) and the more cumbersome variant (Eq. 2b) provide a convenient framework for at least qualitative discussion of the factors affecting the forward and backward electron-transfer rates in organic systems. It has to be understood, however, that variation of a single experimental quantity, such as solvent polarity [20], or even an apparently minor structural feature of the donor (or acceptor) usually simultaneously effects several variables within the Marcus theory.

For example (Fig. 2), in the case of ET within a geminate pair of ions of opposite charge, a change in solvent may affect ΔG_{et}, λ_s and V. A decrease in the dielectric constant of the medium will increase the Coulombic attraction of the ions, but also change their redox potentials [20]. Very often even if the change in redox potentials of the components is significant, the difference between them (which determines ΔG_{et}) remains relatively constant in different solvents. For the purpose of this example, no change in the driving force may be assumed, although the net effect on ΔG_{et} is usually difficult to quantify. The distance between the ions will also be affected, in the extreme shifting the equilibrium from solvent separated (SSIP) to contact (CIP) ion pairs. The shortening of the distance between the ions may significantly increase the electronic coupling (V) between them by increasing the overlap of orbitals participating in the electron transfer. It is usually assumed that V is inversely related to the separation distance (r) and decreases exponentially from its maximum value (V°) when the donor and acceptor are in contact (r_0):

$$V = V^\circ \cdot \exp[-\beta(r - r_0)/2] \tag{3}$$

where the value of β is usually between 0.9 and 2.0 Å^{-1}, although the precise physical meaning of this parameter is unclear [21]. Finally, the change of solvent is going to affect the solvent reorganization energy (λ_s). The usual

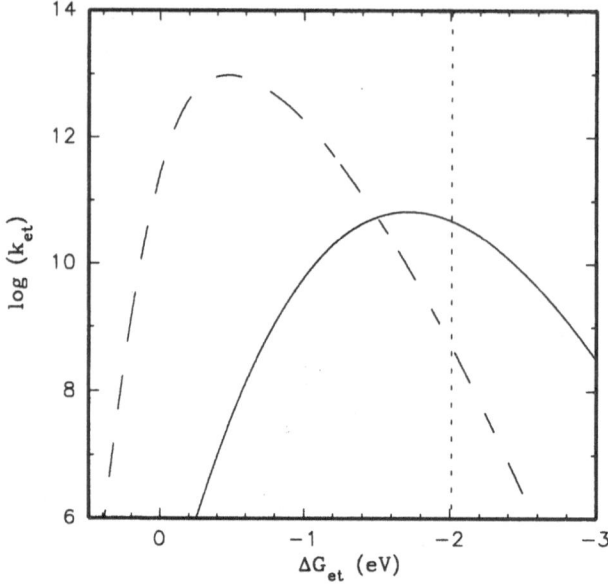

Fig. 2. Hypothetical effect of solvent change on the rate of BET within geminate ion pairs. The *solid line* represents BET in a polar solvent with a relatively large solvent reorganization energy ($\lambda_s = 1.5$ eV) and a small electronic coupling (V = 20 cm^{-1}). The *broken line* depicts BET in a non-polar solvent with a small reorganization energy ($\lambda_s = 0.3$ eV) and a shorter interionic separation (V = 200 cm^{-1}). In both cases $\lambda_i = 0.25$ eV and v = 1500 cm^{-1}. Assuming that ΔG_{et} remains relatively constant and that the reaction takes place in the inverted region ($\Delta G_{et} < -2.0$ eV), the rate of BET would diminish by at least two orders of magnitude upon change from the polar to the non-polar solvent

approximation [16, 18] used to calculate λ_s is based on spherical donors and acceptors (Eq. 4):

$$\lambda_s = (\Delta e)^2/4\pi\varepsilon_0[(1/2a_D + 1/2a_A - 1/r)(1/\varepsilon_{op} - 1/\varepsilon_s) \qquad (4)$$

where Δe is the charge transferred in the reaction, a_D and a_A are ionic radii of the reactants, ε_{op} is the optical dielectric constant (square of the refractive index) and ε_s is the static dielectric constant. Generally, a less polar solvent will lower the value of λ_s. Assuming that BET takes place in the inverted region (a typical situation for photogenerated ion pairs), the lowering of solvent polarity will have a cumulative effect of decelerating the process, but many different factors have contributed to that outcome (Fig. 2).

2.2 Forward Electron Transfer

There are two basic alternative ways to use the photon energy to accomplish electron transfer between donors and acceptors. The first strategy involves deposition of energy into one of the components of the electron transfer system, i.e. it is equivalent to "intramolecular" (localized) excitation. The second approach is based on electron promotion within charge-transfer complexes formed by the ground-state acceptors and donors, i.e. it involves an "intermolecular" (delocalized) transfer.

2.2.1 Local Excitation

As the result of localized excitation, an electron from the highest occupied orbital (HOMO) is promoted to the lowest unoccupied orbital (LUMO) in the same molecule [1, 2, 6], or on much rarer occasions, if the photon energy is sufficient, an electron can be ejected into the medium [23]. Either process generates both a very strong reductant and a strong oxidant. The promoted (or ejected) electron can be easily transferred to a lower-laying LUMO of the other component (acceptor) or an electron from the other component (donor) can fill the vacancy created by the electron promotion (or ejection). Thus, in principle, the excited component can serve both as the donor and acceptor, and its ultimate role is determined by the electronic properties of the other component.

The feasibility of electron transfer between the excited component and the ground state reagent is determined largely by thermodynamic considerations. As pointed out by Weller [24], the free energy for electron transfer can be estimated from Eq. 5:

$$\Delta G_{et} = E_D - E_A - \Delta G_s - E_{00} \tag{5}$$

where E_D and E_A are the redox potentials of the donor and acceptor, ΔG_s represents any extra stabilization due to Coulombic interactions or exciplex formation and E_{00} represents the energy difference between the ground state and excited state of the light-absorbing component. The redox potentials are usually obtained by electrochemical measurements [22] in an electrolyte-containing medium and represent free ion conditions. In actual photochemical system ion pairs are typically involved, and therefore, any Coulombic and charge-transfer interactions (exciplex formation) have to be accounted for. In polar solvents these corrections are usually minimal. The energy of the excited state is commonly taken to correspond to the crossing point of the absorption and emission (normalized) spectra of the component being excited [2, 24].

With the estimate of ΔG_{et} available one can predict the rate of forward electron transfer using the Marcus theory. However, experimental studies of Rehm and Weller [25] have indicated absence of the inverted region in the case of bimolecular ET fluorescence quenching (Fig. 3). Instead, an empirical relation was developed [25] which has a hyperbolic shape (Eq. 6) and fits the experimental data rather well:

$$\Delta G_{et}^{\ddagger} = \Delta G_{et}/2 + [(\Delta G_{et}/2)^2 + (\lambda/4)^2]^{1/2} \tag{6}$$

It has to be noted that bimolecular forward electron transfer is the only situation where the presence of the inverted region has not been confirmed experimentally [26]. Several plausible explanations have been proposed to account for the

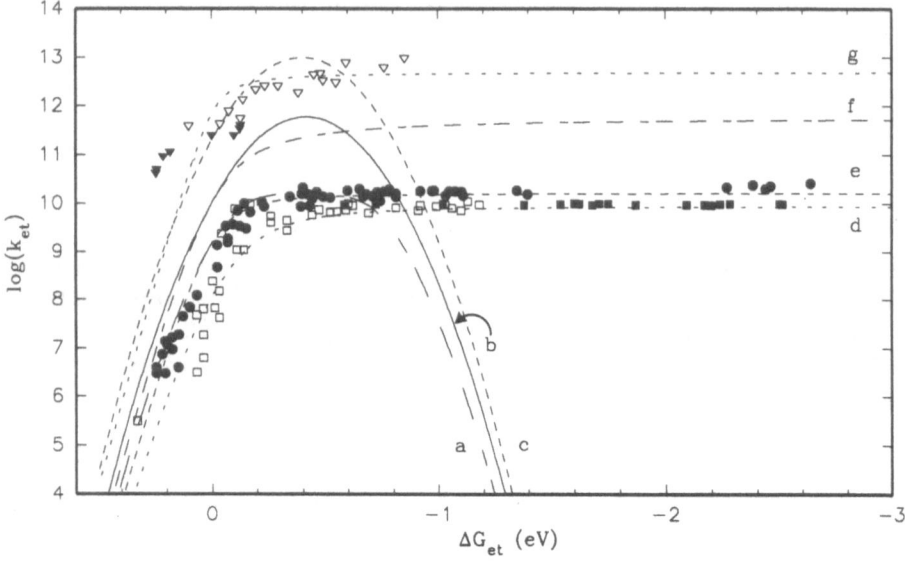

Fig. 3. Bimolecular electron transfer of photoexcited donors and acceptors. The parabolas (*a*) and (*b*) represent Eq. 1 with $\lambda = 0.42$ eV. Curve (*a*) is flattened at the top due to the diffusional limit [27]. The Rehm-Weller equations (Eq. 6) with identical λ is presented by lines (*e*) and (*f*), where the former includes a diffusional limit of $k_d = 2 \times 10^{10}$ M^{-1} s^{-1}. The *filled circles* represent the Rehm-Weller data for neutral aromatic donors and acceptors in acetonitrile [25]. The *squares* denote similar data for inorganic (charged) acceptors and organic (neutral) donors also in acetonitrile [28]. The data are fitted into Eq. 6 with $\lambda = 0.69$ eV and $k_d = 9 \times 10^9$ M^{-1} s^{-1} (*line d*). The *open squares* represent the forward ET to the excited inorganic complexes; the *filled squares* depict the bimolecular BET within the photogenerated ion pairs of the same systems [28]. The *triangles* represent forward electron transfer between organic borates [29] and cyanines (*filled triangles*) or pyrylium cations (*open triangles*) within contact ion pairs in benzene. Even in this case, without diffusional interference, the data seem to fit better the Rehm-Weller equation (*g*) than the Marcus equation (*c*)

apparent failure of the theory in this case [8, 18], but a generally satisfactory solution is not yet available. A part of the problem has to do with diffusion. As can be observed in Fig. 3 electron transfer with $\Delta G_{et} < -0.4$ eV is predicted to be diffusion controlled. The diffusion limit obscures the rapid rate region, and may mask the top of the parabola. However, the descending part of the parabola is not experimentally observed at very negative ΔG_{et}.

The requirement for diffusional collision between the excited state molecule and the ground state species has also practical consequences for photoinduced electron transfer. The lifetimes of singlet excited states are usually no longer than several nanoseconds [2]. Even with the diffusion limited rate ($\Delta G_{et} < -0.4$ eV) the concentration of the intercepting species has to be about tenth molar to give 90% quenching. An additional complication is a possibility of intersystem crossing from the singlet to the triplet manifold [2]. For many kinds of excited state molecules the process is facile. Thus, the incomplete quenching of the singlet state molecules results in ion pairs of different multiplicity. The triplet states are usually longer lived [2] (microsecond time scale) and are quantitatively quenched by millimolar concentration of quencher, provided there is a sufficient driving force for ET.

The microscopic picture of the diffusional ET quenching consists of formation of the encounter complex in which the ground state molecule may collide up to several hundred times with the excited state molecule. If the number of collisions is sufficient to overcome the reorganization barrier, the electron transfer takes place. As illustrated in Fig. 3, λ's are quite small for aromatic compounds, even in a polar solvent such as acetonitrile. The result of forward ET is a pair of reactive intermediates. If the components of the pair are radical ions, there is a Coulombic attractive force which slows their separation. Depending on the solvation status these pairs [30] can be divided into: 1) contact ion pairs (CIP) wherein the ions are in direct (but not necessarily spatially defined) contact without any solvent penetration, 2) solvent separated ion pairs (SSIP) which are sometimes further divided into solvent shared [31] ion pairs, wherein a solvent molecule or two are shared between the ions and solvent separated ion pairs in which the ions touch through their respective solvation spheres, and 3) free ions (FI) with no Coulombic interactions between specific ions of opposite charge. Although, the solvent polarity will determine the solvation status of the incipient ion pair and the equilibrium population of various ion pairs as well as the rate of their interconversion, the rate of forward ET for freely diffusing systems is only weakly solvent dependent, including the region where the reaction is slower than the diffusion limit [32].

If ΔG_{et} is insufficient for electron transfer, the components of the ET system (one of them in the excited state) may have a spatially specific charge-transfer interaction. Such interactions require good orbital overlap and, therefore, are limited to species in direct contact. These interactions may cover a wide spectrum of stabilization energies. On one end, the interactions will be weak, not unlike in weak ground state charge-transfer (CT) complexes, with little charge separation. On the other end of the spectrum, the products will be highly ionic. The first situation, often arbitrarily referred to as formation of exciplexes [33a], is typical for non-polar solvents and weak acceptors or donors. Conversely, after the electron transfer step, the CT interactions between the ions produced may diminish the degree of charge separation. Thus, the distinction between "exciplexes" and "contact ion pairs" is rather ill defined [33b].

Of course, appropriately modified considerations also apply to situations where no charge separation takes place. For organic substrates, which are rarely multicharged, it is convenient to divide ET reactions into three groups: 1) charge separation where both A and D are neutral, 2) charge shift where only one of the components is charged, and 3) charge annihilation where D and A have opposite charge. The Coulombic term will be important only when both partners are charged; however, CT interactions are charge impartial. As will be discussed below, the charge may have important implications to both the forward electron transfer and BET. That importance is reflected in different perturbations in driving forces ($-\Delta G_{et}$) and solvent reorganization energies (λ_s) with different charge characteristics.

In addition to the charge attribute, very often the intermediates generated via ET will have unpaired spins, thus constituting radical pairs. For two freely floating radicals the energy difference between the singlet and triplet states will be small and usually of no consequence for ΔG_{et} [34]. However, the electron transfer can only take place within singlet pairs. Thus, the barrier of intersystem crossing may control the rate of electron transfer. The spin dynamics [35] is a complicating factor, but also provides extra control of ET reactions by appropriate selection of triplet excited state donors or acceptors, or by using magnetic field [34, 36] effects to perturb the intersystem crossing rates.

It has to be noted that various solvation states of radical-ion pairs rarely can be distinguished by spectral means [30], and even then, the structural information is very indirect. Most often the distinction between various pairs is kinetic, by definition representing averages of many spatial orientations, perhaps around quite a shallow energy minimum. The ion pairs can interconvert rapidly (Sect. 4), with rates comparable to ET or the follow-up reaction (such as bond cleavage). To a smaller degree, similar considerations apply to CT interactions, where even for π-donors or acceptors several spatial orientations [37] are possible. Under such circumstances any conclusions about orbital overlap (and V), the ion separation distance, the solvation and spin status have to be treated with care. However, not entirely surprisingly, the delocalized organic ET systems behave very much alike, and even semi-quantitative predictions can be made about new systems based on results of several existing studies.

From the point of view of the efficiency of the follow-up reaction, the localized excitation mode has the drawback of a diffusional barrier (see above) which has to be overcome in order to bring the excited state component within reaction distance of the ground state component. Under usual conditions, within a practical concentration range of reactants, a fraction of the excitation energy will be lost by radiative or thermal means before the encounter. The longer-lived triplet excited states offer here a clear advantage [38]. An alternative solution is to eliminate the diffusional step altogether. This can be accomplished, for example, by 1) chemically linking the donor and acceptor component [39], 2) introducing an attractive force between the ground state A and D (see below), or 3) placing donors and acceptors in an organized environment [40, 41].

All these solutions have their problems. A chemical link will always present a synthetic challenge and may hamper optimal orbital overlap for the forward electron transfer [39]. The possible non-covalent forces to bring A and D together may involve hydrophobic binding, CT interactions or Coulombic forces. The design of appropriate photochemically active artificial binding cavities is just beginning [41], and practical systems for PET fragmentations have not been yet demonstrated. The CT interactions will be discussed in the next section. The approach using Coulombic attractions has been most promising, and has even found commercial applications (Sect. 5.3). The donor and acceptor are both charged and ET takes place within the contact ion pairs. Since the ions are in contact, the diffusional limit does not apply in this case (Fig. 3).

The presence of the inverted region, however, has not yet been demonstrated [29]. With the appropriate design (see below) some of these ions can be considered as a new type of an ion pair [42], a penetrated ion pair (PIP) in which one of the ions is buried inside the other. Research into PET in organized media is very active [40–41] and encompasses a variety of topics from semiconductors [43] and zeolites [44], through various photoconductive polymers [45] to PET-initiated polymerizations and depolymerizations [46] that are generally outside of the scope of this review.

The non-diffusional methods of bringing D and A together may also result in payment of an ultimate price of diminished overall efficiency since BET within contact ion pairs is usually more rapid than in solvent separated ion pairs (see below). Indeed, the most important aspect of the forward electron transfer is that it presets the conditions for the competition between BET and the fragmentation reaction. The reactive intermediates (ion pairs) are generated in specific solvation and spin states. That state can be controlled or at least influenced by a selection of the excited state component, the ground state component and solvent [20], as well as by magnetic and electric fields [36].

2.2.2 Irradiation of Charge-Transfer Complexes

Within the locally excited mode of PET, the forward electron transfer takes place between a different set of orbitals than does the back-electron transfer [29]. For example, if the acceptor is excited, the forward electron transfer is between the HOMO of the donor and an orbital that was the HOMO in the ground-state acceptor, while BET is between the orbital that was the LUMO of the ground-state acceptor and the HOMO of the donor. A different situation is encountered when photon energy is deposited into a charge-transfer band of the ground-state donor–acceptor complex [47–51]. In this case the forward and backward ET take place between the same set of orbitals: the HOMO of the donor and the LUMO of the acceptor engaged in the CT interaction [47–49]. Such ground-state complexes form when the gap between the frontier orbitals of the donor and acceptor is sufficiently small and the orbitals have adequate overlap. In most instances, the complexes are rather weak, i.e. only a small fraction of electron charge is formally transferred between the donor and acceptor. Thus, the net stabilization is small, and consequently the formation constants are small. The complexes form with diffusion limited rates and also dissociate rapidly.

The CT complexes are characterized by a new absorption band which is usually red-shifted as compared to local excitation bands [47–49]. According to the Mulliken formulation the CT-excitation corresponds to an electronic transition from the HOMO of the donor to the LUMO of the acceptor, i.e. it accomplishes full electron transfer [47]. The transition is instantaneous, producing two intermediates (ions) in a direct contact but in a non-equilibrium, Franck-Condon state. The relaxation of the pair competes with BET, diminishing the quantum yield for ion generation [49]. This process is believed to take no longer than a few picoseconds, and is not yet well understood in terms of solvent dynamics [49–51]. The equilibrium contact ion pairs are qualitatively no different than those produced by ET quenching methods. The contact ion pairs undergo rapid BET by radiative (exciplex fluorescence, see above) or thermal pathways [50]. In the case of radiative BET, the fluorescence spectrum contains the kinetic and thermodynamic information about the electron transfer [50]. Specifically, the width of the fluorescence band yields reorganization energy and the crossing point of absorbance (CT) and fluorescence bands provides ΔG_{et}.

The photoexcitation of CT bands leads, at least initially, to contact ion pairs in the singlet state [49]. Since the ion separation is small, the electronic coupling, V, is relatively large and the reorganization energy is small. BET within these pairs is expected to be very efficient (see below). In competition with BET, in polar solvents, these pairs may undergo separation into SSIP or intersystem crossing, in addition to the follow up reaction. Thus, the dynamic behavior of these ion pairs is identical to that of those produced by local excitation, but the starting state is better defined. That property of CIP produced by CT excitation has been successfully used to probe the dynamics of ion pair solvation (Sect. 4) [52, 53].

The formation of a detectable quantity of weak CT complexes usually requires a large excess of one component, or high concentrations of both components. Such a situation may lead to mechanistic complications. Thus, 1:2 complexes may form in the ground state [54], or the photogenerated ion pairs may be rapidly (subnanosecond or nanosecond time scale) intercepted to form "triplexes" [55] or ionic dimers [56]. It may be expected that the chemical behavior of these aggregates will crudely resemble the reactivity of "normal" ion pairs.

2.3 Backward Electron Transfer

For practical reasons, BET between photogenerated intermediates (ion pairs) falls most often into the inverted region. The photochemical systems are chosen to be thermally stable. The thermally activated electron transfer between the HOMO of the donor and the LUMO of the acceptor (with or without intervening CT complexation) will have a reasonable rate at room temperature if the activation energy is on the order of 0.5–1.0 eV. Thus, D and A of the photochemical system are often selected in such a way that ΔG_{et} for the thermal process is ca. 1 eV or more. This condition is equivalent to ΔG_{bet} (within the photogenerated pair of intermediates) being -1 eV or less, since BET is microscopically the reverse of the thermally activated forward ET. Typical reorganization energies of organic systems are ca. 1–2 eV. As discussed above, the inverted region starts at $-\Delta G_{bet} = \lambda$. Thus, the need to eliminate thermal ET between A and D leads to selection of systems with driving forces which yield the near-maximum rates for BET, unless the system is shifted to even larger driving forces (more negative ΔG_{bet}) as is commonly observed. Shifts into the smaller driving-force region, although workable, may bring the possible complication of thermal ET.

Regardless of the method of generation, BET within the intermediates (ion pairs) will take place between the same orbitals: the LUMO of A and the HOMO of D. Only singlet pairs, however, are able to undergo BET; triplet pairs must first undergo intersystem crossing. The mechanism of intersystem crossing in freely-floating radical ion pairs is analogous to that observed in uncharged radicals [34]. Due to the Coulombic attraction between the oppositely charged ions, the lifetimes of geminate pairs may be significantly longer than that of uncharged radicals, increasing the probability of intersystem crossing. The intersystem crossing may be rate limiting, or have a rate comparable to BET. Under these conditions, the Marcus equation is not expected to hold. If the intersystem crossing is rapid compared to BET the golden-rule expression (Eq. 2) may still be applicable, as it is for singlet pairs. Experimental results from many laboratories have confirmed the general parabolic shape of free-energy relationship and the existence of the inverted region and also have shown that the simplified (classical) version of Eq. 1 is inadequate [57].

One dramatic demonstration involves the temperature dependence of the ET rate. In the normal region, an Arrhenius-type relationship holds; however, there is almost no temperature effect on the rate in the inverted region [58]. The lack of temperature effect on rates, however, may also be accounted for by changes in the dielectric constant of the medium [58, 59]. The lowering of temperature leads to an increase in the dielectric constant and to an increase in the rate of ET. This trend is opposite to the "normal" lowering of the rate with the lowering of temperature. For reactions with small activation energies these effects may cancel out [59]. The change in dielectric constant with temperature may also lead to a change in the driving force ($-\Delta G_{bet}$), complicating the situation even more. Regardless of the explanation, experimentally the rates of BET in the inverted region do not vary strongly with temperature [18h, 58, 59].

The experimental data on BET within photogenerated ion pairs fit the theory very well. Figures 4 and 5 present data obtained by four different research groups [60–63] for charge annihilation within solvent separated radical ion pairs derived from aromatic systems. The qualitative agreement between the data obtained by different methods for different systems is outstanding [64]. The parabolas drawn correspond to Eq. 2. There are four adjustable parameters in the golden-rule expression (λ_i, λ_s, V, and ν). In all cases ν has been set to 1500 cm^{-1}; however, the remaining three parameters allow for some flexibility. With data only on one "side" of the parabola, many similar solution sets can be selected and, therefore, some caution is required in the analysis.

In spite of those difficulties, generalizations can be easily made. The onset of the inverted region starts around 1.5–2 eV of driving force (corresponds to $\lambda_i + \lambda_s$). The value of V is small (10–20 cm^{-1}), and the solvent reorganization energy is the dominant component of the total reorganization energy. An increase in size of the aromatic system leads to a decrease in V and λ_s, but the

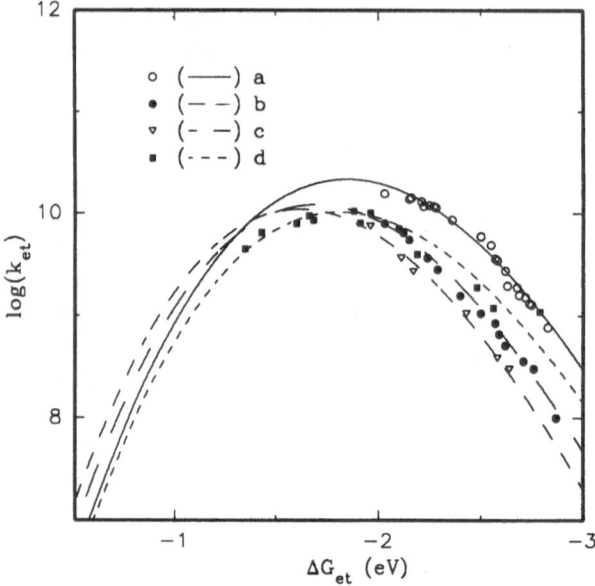

Fig. 4. Back electron transfer rates in photogenerated radical ion pairs in acetonitrile. In all cases cyanoanthracenes served as electron acceptors in their excited states. (*a*) Methylated benzene derivatives [60b] as donors (one-ring compounds), $V = 11.5\ cm^{-1}$, $\lambda_s = 1.63\ eV$. (*b*) Methylated biphenyls or naphthalenes [60b] as donors (two-ring compounds), $V = 8.5\ cm^{-1}$, $\lambda_s = 1.48\ eV$. (c) Methylated phenanthrenes [60b] as donors (three-ring compounds) $V = 8.0\ cm^{-1}$, $\lambda_s = 1.40\ eV$. (*d*) Diphenylbutadienes [61] as donors, $V = 8\ cm^{-1}$, $\lambda_s = 1.55\ eV$. In all cases $\lambda_i = 0.25\ eV$ and $V = 1500\ cm^{-1}$

changes in k_{et} are within a factor of 10 for a given value of the driving force. Similarly, a change from a charge annihilation reaction to a charge shift process (by using a charged acceptor) has only a minor influence [65] on the observed rate (at given ΔG_{et}).

The data in Fig. 5 are a little more scattered, but they involve donors and acceptors of various sizes that are quite distinct electronically [64]. Taking into account the uncertainties in the determination of the driving force for BET from (sometimes irreversible) cyclic voltammetry experiments, the data of Figs. 4 and 5 are an excellent starting point for semi-quantitative predictions of BET rates within photoinduced ion pairs, and may be used to estimate efficiencies of follow up reactions, including fragmentation reactions.

The analogous data for contact ion pairs are presented in Fig. 6. The data of Mataga [66] (filled symbols) do not seem to follow the parabolic relationship. The data of Farid and Gould [67] (although fitted into the gold-rule expression) cover too small a range of driving forces to make conclusions about the parabolic functional dependence of k_{bet}. It may be concluded, however, that the rates of BET within CIP follow an inverted relationship with the driving force,

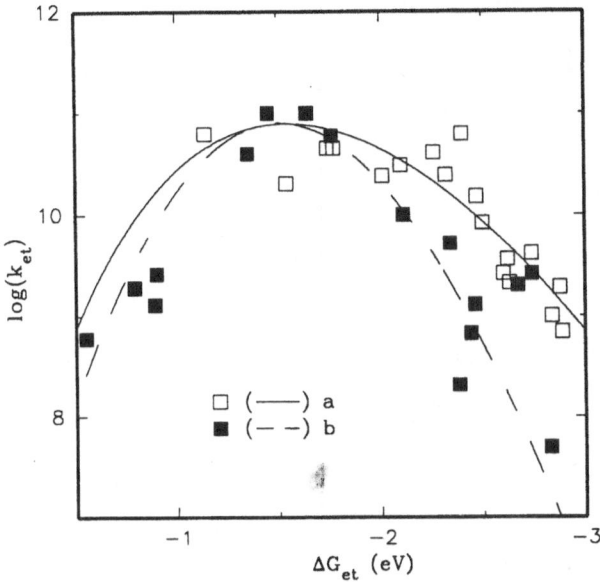

Fig. 5. Back electron transfer rates in photogenerated radical ion pairs in acetonitrile: (*a*) 9,10-Dicyanoanthracene in its excited state served as the acceptor. Aryl, alkyl, methoxy and amino benzene derivatives as well as aliphatic amines served as donors [62] ($V = 23$ cm^{-1}, $\lambda_s = 0.97$ eV, $\lambda_i = 0.64$ eV). (*b*) Perylene, pyrene, benzperylene, and aromatic amines served as donors. Tetracyanoethylene (TCNE), pyromellitic dianhydride (PMDA), phthalic anhydride (PA), maleic anhydride, pyrene and perylene served as electron acceptors [63]. Various combinations of donors or acceptors were excited ($V = 20$ cm^{-1}, $\lambda_s = 1.45$ eV, $\lambda_i = 0.07$ eV). The parabolas drawn are different from those offered in the original analysis. The parameters that were used were selected to emphasize the similarity to Fig. 4 (in all cases $v = 1500$ cm^{-1})

and are up to two orders of magnitude faster than BET rates within SSIP. The comparison of systems where both SSIP and CIP may be generated using the same acceptors and donors (lines (a) of Figs. 4 and 6) suggest that at sufficiently large driving forces ($\Delta G_{et} = -2.0$ eV) the rates of BET within CIP may in fact be slower than those within SSIP [67]. Mataga's data [66] obtained for different, but not too dissimilar systems, do not agree with that conclusion, however. Again, the differences between the data sets are small compared to the uncertainties of ΔG_{et} estimates and the data can be successfully used to estimate efficiencies of overall reactions within CIP. It is noteworthy that BET rates measured for borate salts in benzene [29] (also CIP) have comparable values to those in Fig. 6, despite the fact that those reactions 1) are run in solvents of drastically different polarity, 2) belong to the charge separation category, and 3) do not involve any CT interactions in the ground state. These similarities between different CIP suggest that the proximity of the components (large V, small λ_s) is the dominant factor in their ET rates.

Most of the studies of BET within geminate ion pairs have been carried out in acetonitrile. As discussed in Sect. 2.1, change of solvent may simultaneously

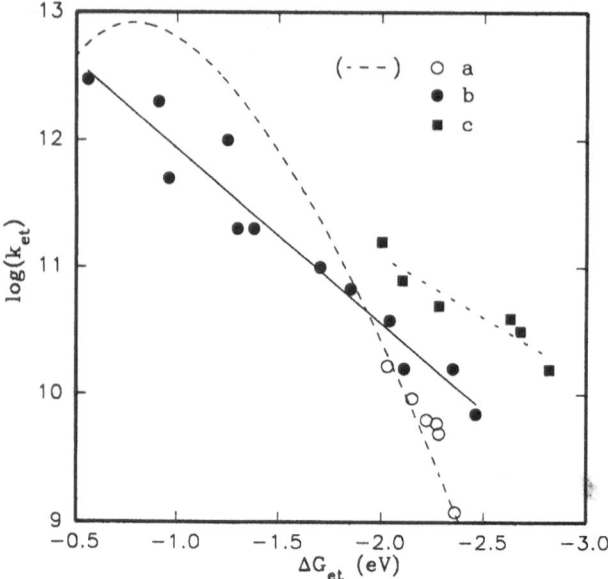

Fig. 6. Electron transfer within CIP formed by irradiation of (*a*) CT complexes between tetra-cyanoanthracene and alkyl-substituted benzenes [67], (*b*) CT complexes of TCNE, tetracyanoquinoldimethane (TCNQ), PMDA or PA with aromatic polycyclic hydrocarbons [66], and (*c*) CT complexes of PMDA with alkyl-substituted benzenes [66], all in acetonitrile. The parabola (*a*) drawn with a *broken line* represents one of possible sets of parameters fitting the data, and illustrates (by comparison with Fig. 4) a possible functional dependence of k_{bet} on ΔG_{bet} within ions in direct contact ($V = 200$ cm^{-1}, $\lambda_s = 0.55$ eV, $\lambda_i = 0.3$ eV, $\nu = 1500$ cm^{-1}). Mataga's data [66] are presented with least-squares lines. These data are obtained for aromatic compounds identical or similar to those shown in Fig. 5(b)

affect several parameters within the Marcus equation. The qualitative predictions are, however, almost always correct. Thus, in less polar solvents electron transfer is usually slower than in polar solvents and the rates of BET correlate well with E_T of the solvent [53]. A complication may exist in the case of ET with a small reorganization barrier. The Marcus theory implicitly assumed infinitely fast solvent motions. This assumption, however, may not always be valid. In cases with small λ, and small driving force ($\Delta G_{et} \approx 0$) the rate of electron transfer is predicted to correlate with solvent relaxation times [68]. Thus, under these circumstances, the solvent polarity will not affect the rate in a predictable way.

3 Fragmentation Reactions of Radical Ions

3.1 Reactive Intermediates in PET Reactions

Single bond fragmentations within high energy intermediates generated via PET take place almost exclusively at the radical ion stage. Although in principle fragmentation reactions may involve diamagnetic ions or uncharged radicals, in practice these are rarely, if ever, the bond cleaving intermediates. In the case of organic substrates, the donors and acceptors are almost always diamagnetic and neutral or, at most, monocharged. After the electron transfer step the components become radicals with or without charge. Commonly one of three situations is encountered. If the ET step involves charge separation (donor and acceptor are neutral) the intermediates generated constitute a pair of radical ions. If one component is neutral and the other charged (charge shift ET), one of the intermediates generated is a radical ion and the other beomes a radical. In the third case, when the two components have opposite charges (charge annihilation ET), the intermediates form a pair of radicals. In the cases where fragmentations are observed, these radicals are (at least formally) equivalent to radical ions (see Sect. 5.3). Thus, most organic PET systems produce at least one radical ion and – with the exception of rare instances of dissociative electron transfer (see below) – a well defined radical ion is the species undergoing the fragmentation reaction.

Well documented dissociative electron transfer cases are rare [69], especially in photochemical systems [70]. The dissociative case requires that the species losing or gaining an electron does not last for even one full vibration of the scissile bond. Such reactions are characterized by relatively large internal reorganization energies, since a significant bond elongation must be a part of the ET reaction coordinate [71]. Savéant has proposed a modification to the Marcus equation to account for such cases [71]. The best examples of dissociative fragmentations are found in reductive cleavages of aliphatic carbon–halogen bonds where the electron is injected directly into the elongated σ^* bond [69].

Much more common are cases where the radical ions are kinetically or structurally well defined. These radical ions undergo thermally activated fragmentations. In principle, the cleavage reactions of photochemically (PET) or thermally generated radical ions are identical. One possible difference is the presence of *a high-energy* geminate partner (a radical, an ion or a radical ion) in the photochemical systems. This presence is of potential importance for several reasons. Charge-transfer interactions between partners may perturb the thermodynamics of the cleavage reaction, the counterion may act as a nucleophile or base, assisting in the cleavage and, most importantly, BET within the pair of the reactive intermediates will compete with the bond scission, determining the overall efficiency of the reaction.

Radical ions may be divided into two groups [72]: 1) σ radical ions, wherein the unpaired electron resides formally in a σ orbital (one-electron bond radical cation) or a σ^* orbital (three-electron bond radical anion) of the scissile bond, and 2) π radical ions, where the unpaired electron is delocalized over a π-system (removal of a non-bonding electron leads to a π-like radical cation). The π radical ions are most frequently encountered in PET systems. In such radical ions the unpaired electron is initially highly localized on one side of the scissile bond and there are two formal ways in which electrons can be apportioned to

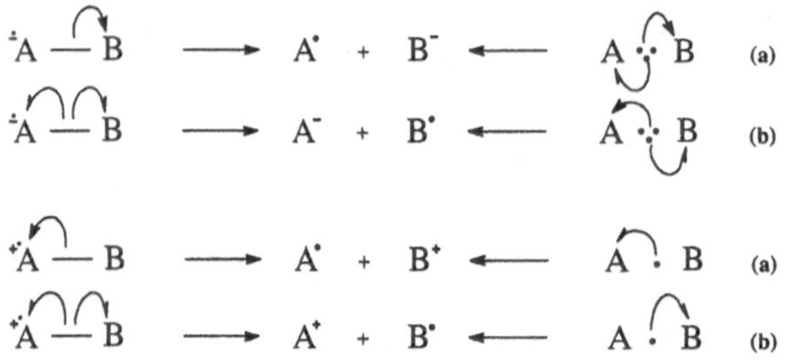

Scheme 1.

the fragments [73] (Scheme 1). In the heterolytic mode (a) the charge is transferred across the scissile bond; in the homolytic mode (b) the spin is transferred across the bond being cleaved. The demarcation line between the homolytic and heterolytic modes is less clear in σ radical ions, but these radical ions may generate both ions and both radicals. Indeed, the term *mesolysis* was proposed [74] to describe the unimolecular (unassisted) fragmentation of radical ions. In this context, the majority of cleavage reactions in PET systems are mesolytic fragmentations.

3.2 Thermodynamics of Mesolytic Fragmentations

In order to successfully compete with BET the cleavage reactions must have small activation barriers. A prerequisite for small activation energies is a small positive (for endergonic processes) or negative free energy (ΔG_m). The free energy of radical ion cleavage (mesolysis) may be conveniently estimated from simple thermodynamic cycles [74, 75] (Scheme 2).

Regardless of the mode of cleavage, the free energy of mesolysis (ΔG_m) is a function of the free energy of homolysis (ΔG_h) of the radical ion precursor, and the difference in redox potentials of the radical ion precursor and the ionic fragment produced:

$$\Delta G_m = \Delta G_h - \Delta E \tag{7}$$

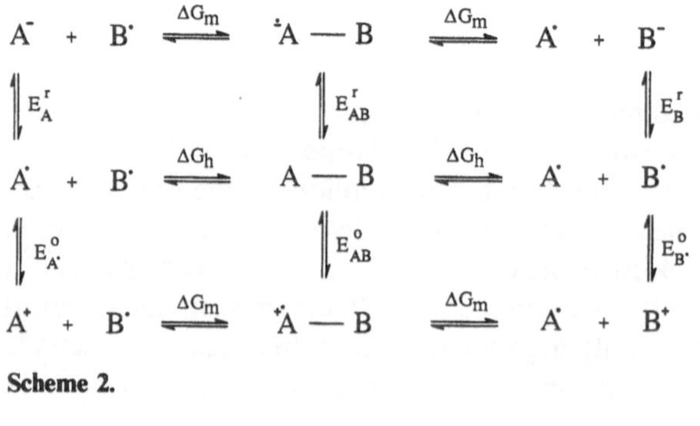

Scheme 2.

where, $\Delta E = E^0_{AB} - E^0_A$. (or E^0_B.) for radical cation cleavages, and $\Delta E = E^r_A$. (or E^r_B.) $- E^r_{AB}$ for radical anion fragmentations, and all quantities are in eV (1 eV is equivalent to 23.06 kcal/mol). The thermodynamic factors determine where the unpaired electron is initially localized (A or B) and which of the cleavage modes (homolytic or heterolytic) takes place. Of course in closely balanced cases, the populations of different radical ions (with the unpaired electron on A or B) and different cleavage modes may compete with each other.

The radicals, wherein the electron is added to, or removed from, a formally nonbonding orbital, are easier to reduce or to oxidize than the corresponding neutrals, where the electron is added to an antibonding orbital (radical anion), or removed from a bonding orbital (radical cation). The ΔE term is, therefore, always positive in such cases. As an outcome, a *weakening* of the scissile bond is observed (as compared to homolysis of the neutral). The redox activation of many reactions may be traced to this simple thermodynamic factor [72a]. It may be added that the removal of electrons from non-bonding orbitals does not have to weaken the bond; it may in certain cases (for example in hydrazines) lead to bond strengthening [76].

The homolytic bond strengths (ΔG_h) in organic compounds vary from ca. 5 eV to just under 1.5 eV. Removal (or addition) of an electron from (or to) organic compounds can, in principle, activate even the strongest bonds. In PET systems high energy radical ions with redox potentials ranging from -2.5 V to $+2.5$ V vs SCE can be easily produced. If the ionic fragment is particularly stable (i.e. the cation is difficult to reduce, or the anion is difficult to oxidize) the value of ΔE may be large enough to make ΔG_m low, or even negative. The purposeful design of a system capable of undergoing efficient (rapid) fragmentation within a PET system, should include a donor (or acceptor) having a homolytically weak bond, high reduction (or oxidation) potential, and ability to generate a stable closed-shell ion (one with very positive E^0_A. or E^0_B., or one with very negative E^r_A. or E^r_B) upon fragmentation. With these guidelines, radical ions with large driving forces (large negative ΔG_m) for mesolysis have been designed [75, 77, 78].

The data necessary for thermodynamic estimates are available from experimental as well as computational methods. In many systems ΔG_h can be approximated by experimentally accessible $\Delta G^{\ddagger}_h$. The approximation is valid (to within 0.05–0.15 eV) if the radical coupling has no barrier (is diffusion limited) and the thermolysis is carried out under conditions selected to minimize the cage recombination [79]. The homolytic bond strengths can also be obtained in many cases from the Benson group-additivity tables [80] or semiempirical quantum or molecular mechanics calculations [81]. With appropriate entropy corrections [75f], relatively accurate ΔG_h values can be obtained in that way.

The redox potentials for many organic ions and radicals are available in the literature [22, 82–86], or they can be estimated (usually to within 0.1 eV) by several simple electrochemical methods. The redox potentials of radical ion precursors are similarly available by direct measurement or by comparison with the redox data of appropriate electronically equivalent, but kinetically stable, model compounds [87]. The common problem of electrochemical measurements is their irreversibility due to the rapid follow up reaction. The data are obtained under nonequilibrium conditions and are thermodynamically ill defined, but the errors are usually small [82–86]. These problems may be circumvented in some cases by the use of kinetic techniques based on Marcus or Rehm-Weller equations [17, 88]. It has to be noted that the redox data are commonly obtained in a different medium than that used for a photochemical reaction. However, for thermodynamic estimates only the difference in redox potentials (ΔE) is required. Quite often, especially for aromatic systems, that difference is relatively solvent insensitive despite the fact that individual redox potentials may vary significantly from solvent to solvent. This simplification breaks down for small ions such as halides where the solvation of the organic substrate is quite different than that of small "inorganic" ions in different solvents. The electrochemical data are free energy quantities and should not be mixed with the enthalpies occasionally used to estimate homolytic bond energies. However, entropy contributions (ΔS) to ΔE for many electrochemical systems have been found to be rather small [84d].

The above considerations apply to mesolytic fragmentations only. When the radical ion fragmentation is assisted by a bond-forming nucleophile or by a base the thermodynamic cycles must be modified accordingly. Also any Coulombic or CT stabilization due to the counterion within the photogenerated ion pair will modify the thermodynamics of mesolysis. Analogously to ET processes (Sect. 2.2) such perturbations are small in polar solvents.

3.3 Stereoelectronic Considerations

For the cleavage to take place it is essential that the electron density is added to, or removed from, the space between the atoms forming the scissile bond [89]. In σ radical ions the unpaired electron is already "localized" within the scissile bond. A similar situation exists in dissociative electron transfer cases, where the stretching bond accepts (or loses) an electron. In contrast, within π radical ions the unpaired electron has to be transferred from (or to) the π system into (or from) the scissile bond. Two extreme situations may be visualized [72a, 89, 90]; one is termed a π–π fragmentation, the other a π–σ fragmentation.

In the first case the scissile bond overlaps with the π system containing the unpaired electron. This situation may be illustrated by radical ions of benzylic substrates (Fig. 7a). The best overlap is obtained if the dihedral angle between the plane of the π system and the plane defined by the scissile bond and the atom of the π system to which that bond is connected is 90°. With this geometry the charge and spin within the radical ion can delocalize freely throughout the molecule as the bond elongates. Dihedral angles less than 90° should lead to less efficient (slower) cleavage, although no systematic studies of the angle dependence have been yet carried out [91]. In the extreme, the dihedral angle is 0°, and there is no overlap between the scissile bond and the π system. This is an

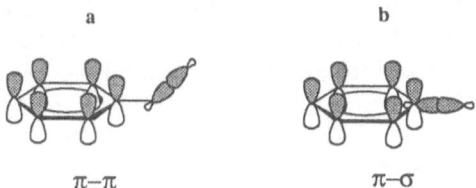

a   b

π–π π–σ

Fig. 7. Schematic representation of the stereoelectronic factors in the cleavage of radical ions. Case (*a*) involves a benzylic system, where the scissile bond can strongly overlap with the (generic) π orbitals of the aromatic system bearing the unpaired electron. Case (*b*) shows an aromatic system which is orthogonal to the scissile bond

example of π–σ fragmentation, an extreme case of which may be illustrated by a radical ion of aromatic or vinylic system (Fig. 7b). In these cases the scissile bond must elongate first (and distort to introduce some electronic coupling between the orthogonal orbitals), until the energy of the scissile bond is close (within $k_b T$) to the energy of orbital holding the unpaired electron. What follows is equivalent to dissociative *intramolecular* electron transfer.

The cases of π–σ fragmentations are limited to polar bonds where the scissile-bond orbitals have energies not too far removed from the energies of the π orbitals bearing the unpaired electron. For example, carbon–halogen [92], carbon–oxygen [93] or carbon–sulfur [94] bonds may participate in reductive cleavages of this type, and carbon–silicon, or carbon–boron [8, 29] bonds are potential candidates in the oxidative fragmentations. In principle, the scissile bond does not have to be directly attached to the π system [95]. Increased distance will, however, lower the electronic coupling between the orbitals participating in the electron transfer.

In the case of π–π fragmentations, the coupling between the relevant orbitals may also diminish due to the rotation of the scissile bond out of the perfect dihedral angle, or due to a change of symmetry of the π system. For example, a change in the position of a strongly conjugating substituent from *para* to *meta* in the system shown in Fig. 7a will dramatically reduce the charge and spin density at the carbon to which the scissile bond is attached, thus diminishing the effective orbital overlap [96, 97]. There is, of course, a continuum of fragmentation modes from the "strongly coupled" π–π cases to the orthogonal π–σ examples.

Another aspect of stereoelectronic control is related to the heterolytic and homolytic alternatives of mesolytic fragmentation (Scheme 1). In the case of π–π fragmentations, the extent of charge delocalization in the transition state may be greater for the heterolytic than for the homolytic mode, since the "departing" ionic fragment is predisposed to accommodate more charge than the "leaving" radical fragment [73, 98, 99]. If such charge delocalization leads to the lowering of transition state energy the heterolytic mode may have inherently lower activation energy than the homolytic mode. This effect may be especially accentuated for polar bonds [73]. Additional and complicating effects may be related to counterion movement accompanying the cleavage. In situations where ion pairing or CT interactions are favored (for example in non-polar solvents) the homolytic mode may have the advantage of little charge movement in the cleavage process, thus counterbalancing potential delocalization effects.

These effects are minimized for $\pi-\sigma$ fragmentations, especially those where the intramolecular electron transfer takes place between spatially adjacent orbitals. Somewhat related stereoelectronic considerations apply to nucleophile or base assisted fragmentations where certain three-dimensional dispositions of existing bonds may favor the assistance, in a way related to, for example, the anti stereochemistry of elimination reactions (Sect. 5.2).

3.4 Kinetics of Radical Ion Fragmentation

The interplay of stereoelectronic and thermodynamic factors determines the kinetics of the fragmentation process. Although, a quantitative theory of the fragmentation reaction of radical ions that ties these aspects together has not yet been formulated [100], the analysis of experimental data indicates that the kinetic barrier is small. For endergonic ($\Delta G_m > 0$) $\pi-\pi$ cleavages with optimal overlap, the activation energies are only slightly higher than the thermodynamic barriers [74, 78, 99]. The exergonic processes usually have very fast rates of cleavage ($>10^9 \text{ s}^{-1}$) that are difficult to measure directly. Thus, activation parameters are only rarely available in these instances. In general, only a relatively small number of kinetic data have been obtained for radical ion cleavages, and most of these for systems other than PET reactions. Table 1 collects rate constants for selected rapid ($k_m > 10^5 \text{ s}^{-1}$) fragmentation reactions measured by various techniques for several kinds of radical ions.

Both the thermodynamic and stereoelectronic factors discussed in the previous sections are reflected in the data of Table 1. For example, radical anions of bromo compounds undergo fragmentation faster than the corresponding chloro derivatives [92, 96]. That trend has thermodynamic origin. The homolytic bond strengths of C–Cl bonds are higher than those of analogous C–Br bonds [75f] by ca. 0.65 eV, and although Cl$^-$ is more difficult to oxidize than Br$^-$, that difference in redox potentials (ca. 0.38 eV [87]) is not sufficient to compensate for the bond-strength effect. The high oxidation potentials of halide ions [87] (in excess of 1 V vs SCE, except I$^-$) also explain why halo derivatives of aromatic compounds with negative reduction potentials (let's say $E^r_{AB} < -1.5$ V vs SCE) fragment so rapidly. The bond-weakening due to the ΔE term (Eq. 7) can be much larger than 2.5 eV, making ΔG_m for fragmentation small or negative.

Within the group of radical anions with the same kind of scissile bond, the rate of cleavage correlates with the redox potential of the precursor [92]. This is another manifestation of the thermodynamic effect. If the homolytic bond strength and the oxidation potential of the anionic fragment remain largely unchanged, the driving force should correlate with the energy of the LUMO of the precursor. Indeed, a crude correlation between $2.3RT \log(k_m)$ and E^r_{AB} with a slope of ca. 0.4–0.6 was observed for the $\pi-\sigma$ cleavages of aromatic halides [92].

Figure 8 presents a free-energy relationship for the case of $\pi-\pi$ fragmentations of C–C bonds in bicumene radical ions [78, 99, 102]. The thermodynamics of

Table 1. Rate constants[a] for fragmentation of radical ions

Radical ion precursor	log (k_m)	Solvent[b]	Reference
Radical anions:			
2-Cl--Naphthalene	8.0	DMF/e	[92]
2-Cl--Naphtalene	7.2	DMF/e	[92]
2-Cl--Quinoline	5.8	DMF/e	[92]
4-Cl--Quinoline	5.8	DMF/e	[92]
1-Br--Naphthalene	7.8	DMF/e	[92]
9-Br--Anthracene	6.4	DMF/e	[92]
m-Br--C$_6$H$_4$COMe	5.3	DMF/e	[92]
o-Cl--C$_6$H$_4$COMe	5.48	ACN	[97]
	5.5	DMF/e	[92]
p-Br--C$_6$H$_4$COMe	7.51	ACN	[97]
	7.5	DMF/e	[92]
o-Br--C$_6$H$_4$COMe	9.71	ACN	[97]
m-I--C$_6$H$_4$COMe	8.28	ACN	[97]
p-I--C$_6$H$_4$COMe	9.54	ACN	[97]
p-Br--C$_6$H$_4$COPh	5.0	DMF/e	[92]
m-I--C$_6$H$_4$COPh	6.40	ACN	[97]
PhCOCH$_2$--F	9.72	ACN	[97]
PhCOCH$_2$--OCOPh	9.80	ACN	[97]
PhCOCH$_2$--OCOMe	8.98	ACN	[97]
PhCOCH$_2$--OPh	6.98	ACN	[97]
PhCOCH$_2$--SPh	6.97	ACN	[97]
PhCOCH$_2$--SO$_2$$p$-Tol	8.72	ACN	[97]
PhCO-(CH$_2$)$_2$--Cl	6.92	HMPA	[95]
PhCO-(CH$_2$)$_3$--Cl	5.46	HMPA	[95]
PhCO-(CH$_2$)$_4$--Cl	5.40	HMPA	[95]
PhCO-(CH$_2$)$_2$--Br	7.26	HMPA	[95]
PhCO-(CH$_2$)$_3$--Br	6.61	HMPA	[95]
PhCO-(CH$_2$)$_4$--Br	6.46	HMPA	[95]
PhCO-(CH$_2$)$_3$--I	7.93	HMPA	[95]
PhCO-(CH$_2$)$_4$--I	7.78	HMPA	[95]
PhCOO--C(Me)$_3$	5.36	Hex	[101]
PhCOO--C(Me)$_2$C $\equiv$ CH	5.04	Hex	[101]
PhCOO--CH$_2$Ph	5.34	Hex	[101]
PhCOO--CH$_2$C$_6$H$_4$-p-OMe	5.70	Hex	[101]
PhCOO--CH(Me)Ph	6.15	Hex	[101]
	5.18	EtOH	[101]
PhCOO--CHPh$_2$	6.70	Hex	[101]
	5.66	EtOH	[101]
PhCOO--COPh	5.51	Hex	[101]
MeCOO--CH$_2$Ph	6.30	Hex	[101]
	6.15	EtOH	[101]
MeCOO--CHPh$_2$	6.70	Hex	[101]
p-NO$_2$-C$_6$H$_4$CH$_2$--Br	5.23	W/ROH	[96f]
p-NO$_2$-C$_6$H$_4$CH(Me)--Br	6.54	W/ROH	[96f]
p-NO$_2$-C$_6$H$_4$CH=CHCH$_2$--Br	5.00	W/ROH	[96c]
p-NO$_2$-C$_6$H$_4$CH$_2$--Br	5.23	W/ROH	[96c]
p-NO$_2$-C$_6$H$_4$CH$_2$--I	5.76	W/ROH	[96b]
p-NO$_2$-C$_6$H$_4$CH$_2$--Cl	6.60	DMF/e	[69d]
m-CN-C$_6$H$_4$CH$_2$--Br	7.11	W/ROH	[96d]
m-CN-C$_6$H$_4$--Br	6.90	W/ROH	[96d]
o-CN-C$_6$H$_4$--Cl	6.95	W/ROH	[96d]
p-CN-C$_6$H$_4$--Cl	6.70	W/ROH	[96d]
	8.2	DMF/e	[92]

Table 1. Continued

Radical ion precursor	log (k_m)	Solvent[b]	Reference
p-CN-C$_6$H$_4$--F	5.81	W/ROH	[96d]
p-CN-C$_6$H$_4$C(Me)$_2$--C(Me) (CN)C$_6$H$_4$-p-OMe	5.90	ACN	[78]
p-CN-C$_6$H$_4$C(Me)$_2$--C(Me)(CN)Ph	6.95	ACN	[78]
p-CN-C$_6$H$_4$C(Me)$_2$--C(Me) (CN)C$_6$H$_4$-p-CF$_3$	8.60	ACN	[78]
p-CN-C$_6$H$_4$C(Me)$_2$--C(CN)$_2$C$_6$H$_4$-p-OMe	9.60	ACN	[78]
p-CN-C$_6$H$_4$C(Me)$_2$--C(CN)$_2$Ph	$\leq$10.5	ACN	[78]
C(NO$_2$)$_4$	$\leq$12.5	MeCl$_2$	[49f]
Radical cations			
p-Me$_2$N-C$_6$H$_4$C(Et)$_2$--C(Et)$_2$Ph	5.20	MeCl$_2$/MeOH	[102]
p-Me$_2$N-C$_6$H$_4$C(Pr)$_2$--C(Pr)$_2$Ph	5.64	MeCl$_2$/MeOH	[102]
p-Me$_2$N-C$_6$H$_4$C(Bu)$_2$--C(Bu)$_2$Ph	5.73	MeCl$_2$/MeOH	[102]
p-MeO-C$_6$H$_4$C(Me)$_2$--C(Me)$_2$C$_6$H$_4$-p-CN	6.42	MeCl$_2$/MeOH	[102]
p-MeO-C$_6$H$_4$C(Me)$_2$--C(Me)$_2$C$_6$H$_4$-p-CF$_3$	6.20	MeCl$_2$/MeOH	[102]
p-MeO-C$_6$H$_4$C(Me)$_2$--C(Me)$_2$Ph	7.39	MeCl$_2$/MeOH	[102]
	> 7.70	MeCl$_2$	[103]
p-MeO-C$_6$H$_4$C(Me)$_2$--C(Me)$_2$C$_6$H$_4$-p-Me	8.04	MeCl$_2$/MeOH	[102]
p-MeO-C$_6$H$_4$C(Me)$_2$--C(Me)$_2$C$_6$H$_4$-p-OMe	8.78	MeCl$_2$/MeOH	[102]
p-MeO-C$_6$H$_4$C(Me)(Et)--C(Me)(Et)C$_6$H$_4$-p-OMe	9.08	MeCl$_2$/MeOH	[102]
p-MeO-C$_6$H$_4$C(Et)$_2$--C(Et)$_2$C$_6$H$_4$-p-OME	9.64	MeCl$_2$/MeOH	[102]
(p-MeO-C$_6$H$_4$)$_2$C(OH)--C(OH)(C$_6$H$_4$-p-OMe)$_2$	10.18	MeCL$_2$	[103]
p-MeO-C$_6$H$_4$C(Me) (OSiMe$_3$)--C(Me) (OSiMe$_3$)C$_6$H$_4$-p-OMe	9.74	MeCl$_2$	[103]
p-MeO-C$_6$H$_4$CH(OH)--CH(OH)C$_6$H$_4$-p-OMe	9.10	MeCl$_2$	[103]
p-MeO-C$_6$H$_4$CH (OSiMe$_3$)--CH(OSiMe$_3$)C$_6$H$_4$-p-OMe	8.62	MeCl$_2$	[103]
p-MeO-C$_6$H$_4$CH$_2$--SiMe$_3$	6.36[c]	ACN	[104]
p-MeO-C$_6$H$_4$CH$_2$--SiEt$_3$	6.11[c]	ACN	[104]

[a] The scissile bonds are indicated by *bold broken* lines. Arbitrarily, only rate constants larger than 10^5 s^{-1} (at room temperature) are presented here. The rates were obtained by various techniques involving direct observations of radical ions or by indirect methods. Only some of the radical ions presented were produced in PET systems.
[b] ACN = acetonitrile, DMF/e = dimethylformamide with a supporting electrolyte, W/ROH = water containing t-butanol or isopropanol, HMPA = hexamethylphosphoramide, Hex = hexane, EtOH = ethanol, MeCl$_2$ = dichloromethane, MeCl$_2$/MeOH = dichloromethane containing methanol.
[c] Pseudo-first-order rate constants for solvent-assisted cleavage.

the fragmentation was evaluated based on Eq. 7, substituting $\Delta G_h^{\ddagger}$ for ΔG_h. In these compounds the overlap between the π system and the scissile bond is nearly perfect [105]. The "intrinsic" barrier is small, and for endergonic reactions the kinetic "overhead" ($\Delta G_m^{\ddagger} - \Delta G_m$) is minimal. In other words, the barrier to the coupling of a radical and an ion to yield a radical ion is small.

The stereoelectronic effects discussed in Sect. 3.3 are also illustrated by the data in Table 1. Thus, the *meta* derivatives fragment slower than *para* isomers; an increase in the spatial separation between the π system and the scissile bond lowers the rate of fragmentation; and substituents increasing the extent of charge delocalization in the transition state accelerate the cleavage [99, 106].

The limited available data suggest that solvent has relatively small effect on the fragmentation rate, although the change in solvent may affect the relative magnitudes of entropy and enthalpy of activation [102]. For those reasons the

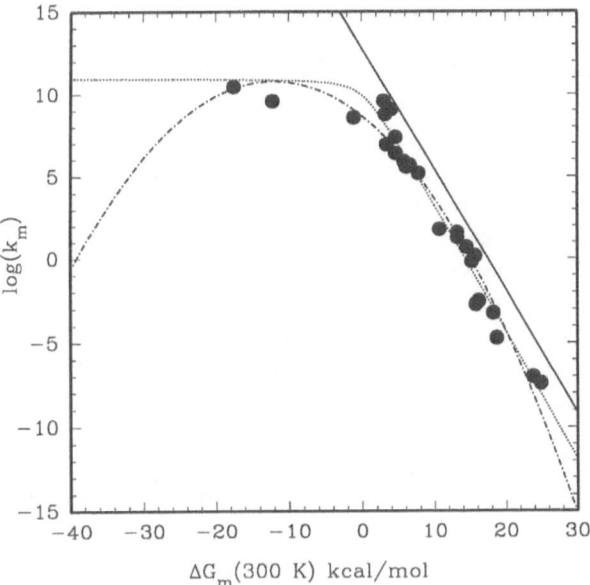

Fig. 8. A free-energy relationship for mesolytic cleavage of C–C bonds in π-radical ions of bicumene derivatives [78, 99, 102]. The solid line represents a hypothetical reaction with no "overhead" ($k_m = (k_b T/h)\exp(-\Delta G_m/(RT))$), i.e. one with only the thermodynamic barrier). The broken lines are the best-fit lines to parabolic (Marcus-type) and hyperbolic (Weller-type) functions.

free-energy analysis is preferred to the purely enthalpic examination of kinetic and thermodynamic data. In a few instances a large discrepancy between the rates measured in water and in an organic solvent was noted [69d]. These disagreements are most likely due to drastic changes in the thermodynamics of fragmentation. A large difference in solvation energy of radical ions (wherein the charge is relatively localized as in nitro or cyano groups) or small "inorganic" ionic fragments (like halide ions) may lead to large perturbations in the kinetics measured in water as compared to these in an organic medium.

In general, the activation parameters of fragmentation reactions confirm the idea that the kinetic "overheads" are low. This conclusion is based on independently measured activation parameters [74, 78, 99, 102], as well as on estimates derived from PET systems [75c, 107, 108]. Despite many possible complications, the *apparent* activation parameters measured in PET systems (by investigating the overall efficiency of fragmentation as a function of temperature) compare favorably with the thermodynamic estimates [75c, 107, 108]. A typical feature of fragmentation reactions is low or even negative entropy of activation [92, 96b, 98, 102, 106]. In all instances $\Delta S_m^{\ddagger}$'s are significantly lower than those observed in the corresponding homolytic processes. The stereoelectronic factors discussed above may contribute to that outcome, or more likely the still poorly understood solvation phenomena are responsible for the observed trends.

3.5 Outcomes of Fragmentation Reactions

The fragmentation reaction of a radical ion produces a radical and a dia-magnetic ion. The reaction constitutes a convenient and rapid method of generation of these intermediates [109]. The products derived from these intermediates and the direct observation method indicate that the electron apportionment to the fragments follows the thermodynamic prediction, i.e. the most stable ion-radical pair is formed.

The reverse reaction, the radical-ion coupling, is expected to have low activation energy if the cleavage reaction itself is endergonic (see above). The many known examples of $S_{RN}1$ reactions [110] attest to the facility of that process. Indeed, in several cases the radical–anion coupling was shown to be diffusion limited [111]. Thus, for the purpose of kinetic analysis the reversibility of cleavage cannot be dismissed a priori. In several systems where the possibility of reversibility has been explicitly tested [29, 99, 102, 106] none was found.

The absence of reversibility of fragmentation implies that other reactions of the fragments also have low activation energies. These processes include reactions of the radical and ionic fragments with both solvent and the non-cleaving radical ion component of the initial ion pair. The reactions with neutral radical ion precursors are less common. The most prevalent reactions encompass proton transfer, nucleophile-electrophile reactions, radical–radical ion coupling reactions and ET reactions between fragments and the radical ion components of the initial ion pair (Sect. 5). Dioxygen, if present, may additionally complicate the outcome [112, 113]. It is not surprising, therefore, that a variety of products are obtained in PET systems involving radical ion fragmentations [77, 78, 101–121]. The products may depend on the solvation status of the fragmenting radical ion. Thus, processes taking place in CIP may result in different products than those occurring in SSIP or free ions. In general, the processes in CIP are likely to favor reactions between the fragments and the radical ions. The lifetimes of fragments are usually too short to allow for reencounter of the reactive intermediates once they separate.

Those redox reactions between the fragments and the non-cleaving radical ion that occur in the direction opposite to the original PET warrant a special mention. Such reactions lead to "catalytic" PET systems, i.e. one of the components of ET reaction is not consumed in the process, while the other undergoes a cleavage that is equivalent to homolysis (ET from the anionic fragment to the radical cation or from the radical anion to the cationic fragment) or heterolysis (ET from (to) the radical anion (radical cation) to (from) the radical fragment). These reactions may be considered as light-assisted "catalytic" processes wherein the component not directly absorbing the photon energy undergoes a very rapid fragmentation that is equivalent to the homolysis or heterolysis which is excruciatingly slow at ambient temperatures [74]. A possible mechanistic complication which must be considered in such situations is bond reformation between the "secondary" fragments (pair or radicals or ions) leading to the apparent reversibility of the cleavage [106b]. This quantum-yield-lowering process should be distinguished from the "true" reversibility of the radical ion fragmentation.

Specific products of a given PET system are generally difficult to predict, since rate constants of the processes involved are only rarely known. Qualitative arguments are not always sufficient because the rates of many competing processes are similar and even small solvent or temperature variations may affect the outcome. The situation should improve, however, with the increasing number of kinetic and redox data that are rapidly becoming available. If predictions are still on shaky ground, the rationalization of the observed products is usually easier, and commonly serves to deduce mechanistic details of the fragmentation reaction.

4 Ion Pair Dynamics

The common practical goal of PET reactions is to increase the overall efficiency of the follow-up reactions, such as fragmentations. The overall quantum yield depends on the efficiency of using the photon energy to generate the reactive intermediates as well as on the efficiency of the cleavage step in relation to other reactions of reactive intermediates, BET being usually the most important competitor. The previous sections described the basic considerations concerning the elementary steps involved in the PET fragmentations. In practice however, PET systems are usually much more complex than the analysis of elementary steps would suggest. The complications reflect the dynamic nature of these systems. The reactive intermediates are produced in various solvation and spin states. The various ion pairs interconvert in competition with the fragmentation and back-electron-transfer reactions. Other reactions, like proton transfer, radical ion coupling, nucleophilic trapping, etc., may also interfere. A simplified scheme of a PET system includes at least five kinetically defined intermediates which may undergo fragmentation reaction (Scheme 3).

In principle, all reactions of Scheme 3 (except k_{bet}) may be reversible. A detailed kinetic analysis is usually not practical under such circumstances. In polar solvents, it is commonly assumed that solvation processes (k_{sol} and k_{sep}) are unidirectional, i.e. solvation leads to a more stable system. In nonpolar solvents contact ion pairs dominate. As discussed previously (Sect. 2.3) BET rates depend strongly on the solvation status. The fragmentation rates are usually assumed to be independent of ion pairing, although quantitative information on that respect is not available.

In photochemical systems the steady-state concentrations of reactive intermediates are very low (typically less than 10^{-9} M with conventional lamps or less than 10^{-5} M with intense laser sources). Such low concentrations essentially exclude any kinetically second-order processes between the reactive intermediates (geminate processes dominate) and minimize the probability of re-

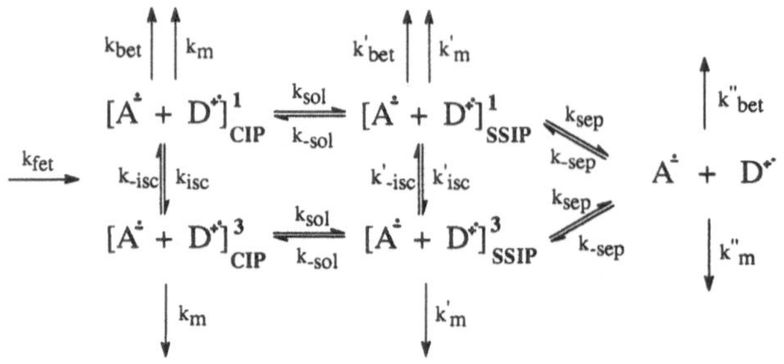

$$
\begin{array}{c}
k_{bet}\uparrow\ \ \uparrow k_m \qquad\qquad k'_{bet}\uparrow\ \ \uparrow k'_m \\[4pt]
[A^{\pm} + D^{\bullet}]^{1}_{CIP} \underset{k_{-sol}}{\overset{k_{sol}}{\rightleftharpoons}} [A^{\pm} + D^{\bullet}]^{1}_{SSIP} \underset{k_{-sep}}{\overset{k_{sep}}{\rightleftharpoons}} \quad \uparrow k''_{bet} \\
\end{array}
$$

$k_{fet}\longrightarrow$

$k_{-isc}\ \Big\updownarrow\ k_{isc} \qquad k'_{-isc}\ \Big\updownarrow\ k'_{isc} \qquad\qquad A^{\pm} + D^{\bullet}$

$[A^{\pm} + D^{\bullet}]^{3}_{CIP} \underset{k_{-sol}}{\overset{k_{sol}}{\rightleftharpoons}} [A^{\pm} + D^{\bullet}]^{3}_{SSIP} \overset{k_{sep}}{\underset{k_{-sep}}{\longrightarrow}} \qquad \Big\downarrow k''_m$

$\Big\downarrow k_m \qquad\qquad\qquad \Big\downarrow k'_m$

Scheme 3.

encounter of once separated geminal radical ions. Thus, increased solvation (transformation from CIP to SSIP to FI) increases the probability of fragmentation reaction at the expense of BET or other competing geminate reaction. The magnitudes of rates of CIP solvation (k_{sol}) and SSIP separation (k_{sep}) are mostly defined by the solvent. The structures of radical ions may contribute, however, to variations in those rates, especially for CIP solvation (k_{sol}). The SSIP separation (k_{sep}) is rather structure insensitive, providing that typical aromatic compounds are involved.

The solvent interpenetration between the reactive intermediates depends on the charge status of the pair. If one of the components is uncharged, the viscosity of the medium is the dominant factor. Typical rates of component separation in nonviscous solvents are of the order of 10^{10}–10^{9} s^{-1}. If the components have opposite charges the polarity of the solvent plays a major role. The values of k_{sol} are in the 3×10^{10}–2×10^{8} s^{-1} range, usually slightly faster in acetonitrile than in less polar solvents such as acetone, ethanol or methylene chloride [52, 53, 59]. In non-polar solvents such as benzene, chlorobenzene or hexane these rates are not measurable [53] (too slow to compete with other processes). The rates seem to be affected by the ability of the system to stabilize the charge internally within the components of the ion pair. Thus, the more "stable" ions (larger internal delocalization) are solvated slower [53, 59], although the effect may be small or even absent in some systems [52]. The reversibility of CIP solvation is low, even in CH$_2$Cl$_2$, the values of k_{sol}/k_{-sol} being in the 50–20 range [52]. Added salts are able to intercept CIP giving rise to "salt effects" [52, 53]. These effects usually manifest themselves in solvents of moderate polarity. In such situations, the reactive ions within the geminate ion pairs are exchanged by redox inactive ones.

The rates of ion separation from CIP to SSIP are apparently much less structure dependent. Again the solvent polarity plays the dominant role. The typical values of k_{sep} vary from ca. 5×10^{8} s^{-1} in acetonitrile to about 10^{5} s^{-1} in dichloromethane [50b, 122, 123]. The empirical Weller equation [123] (Eq. 8, where η is the solvent viscosity in cPs^{-1}, r is the ion separation distance within the pair and $d = \infty$) accounts well for the ion dynamics.

$$
k_{sep} = (2.3 \times 10^{9}/\eta) \cdot \exp[(-e^2/\varepsilon_s kT)(1/d - 1/r)] \tag{8}
$$

The rates associated with ion-pair dynamics are thus competitive with the rates of BET as well as with the rates of fragmentation. The quantum yield for fragmentation, therefore, usually represents a composite of all these rate constants ($\Phi_m = (k_m + k_s)/(k_m + k_s + k_{bet})$, where $k_s = k_{sep}$ or k_{sol}).

Several strategies have been developed to increase the efficiency of the follow up reaction based on perturbation of ion-pair dynamics. All of them intend to increase the fraction of more solvated ions or increase the rate of ion separation. The strategies include using polar solvents [20], adding salts [20, 114b, 126], using charged acceptors or donors to eliminate Coulombic attraction [124], increasing steric bulk [125] in the non-cleaving acceptor or donor and using secondary acceptors [112, 113] or donors [124] whose radical ions will serve as the ultimate redox agent for the cleaving component.

The spin status of the ion pair is another crucial variable affecting the overall efficiency of the process. The forward electron transfer from (or to) a dia-magnetic molecule is not affected by the spin status of the excited component. The back electron transfer, however, is forbidden within the triplet ion pairs (it would violate Pauli's exclusion principle). In situations like that the intersystem crossing will very often determine the efficiency of BET. In practice, the triplet state acceptors or donors lead to overall efficiencies that are higher than those observed with singlet state acceptors or donors [38, 78, 102, 103, 116]. An additional bonus is the fact that triplet states have longer lifetimes [2] and are efficiently ET-quenched with lower concentrations of the ground state component. Quinones and ketones are the most common triplet acceptors, while aromatic amines often serve as triplet donors.

The ion pair spin multiplicity may be a valuable tool to affect the BET rates and to probe the ion pair dynamics via magnetic field effects [36]. Even weak magnetic fields are known to influence relative probabilities of singlet and triplet reactions [34]. Chemically induced dynamic nuclear polarization (CIDNP) is a particularly informative technique [12]. Many bond scission reactions and rearrangements in cyclic radical ions have been successfully explored using this approach. Both structural data (spin densities) and approximate kinetic informations are indirectly available from such experiments [12].

5 Examples of Fragmentations by Photoinduced ET

The fragmentation of radical anions and the reverse reaction, the addition of anions to radicals, are the critical steps of $S_{RN}1$ reactions [110] which constitute perhaps the largest class of fragmentation reactions initiated by photoinduced electron transfer. These reactions are chain processes and photoinduced ET is involved *only* in the initiation step, which is usually poorly defined. The reactions may also be initiated by other means, not involving absorption of a photon. The $S_{RN}1$ reactions and related redox-activation processes have been recently extensively reviewed [72a, 110, 127] and will not be discussed here.

There are also several reviews dealing with various facets of fragmentation reactions in photogenerated radical ions [7–11] as well as more general reviews touching on that subject [1–4]. The examples discussed below were arbitrarily chosen to illustrate certain fundamental and practical aspects summarized in the previous sections.

5.1 Carbon–Carbon Bond Cleavage in Radical Cations of Diphenylethane Derivatives

The C–C bond cleavage in photogenerated radical cations of bibenzyl derivatives is probably the most studied fragmentation process [102, 103, 107, 112–115] among various PET systems (Scheme 4). The benzene, toluene or

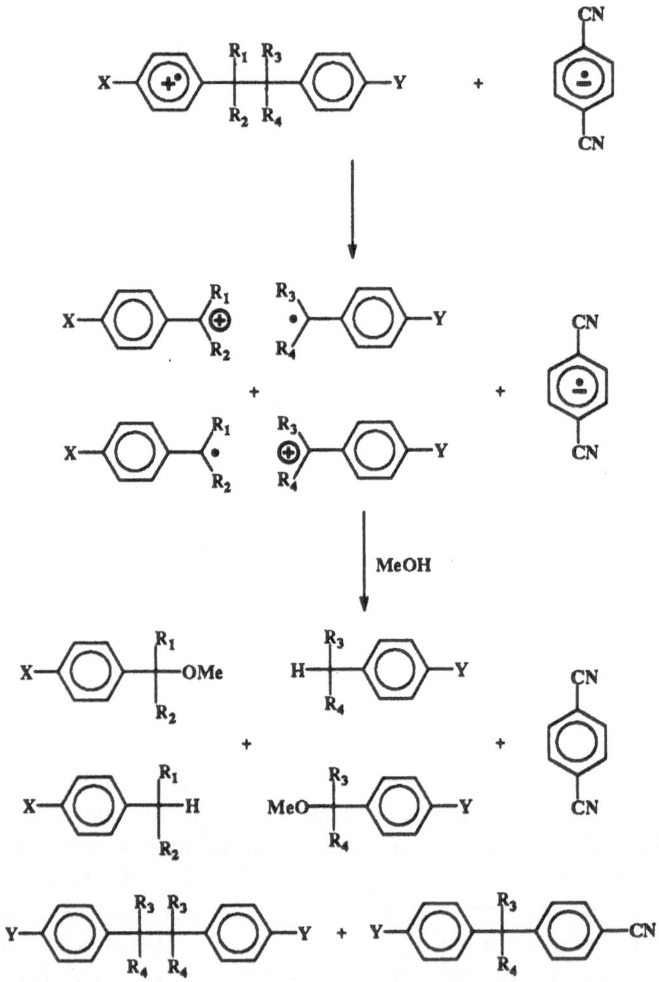

Scheme 4.

anisole moieties (X = H, Me, MeO) serve as electron donors and the excited-state cyanoaromatics such as 1,4-dicyanobenzene (shown in Scheme 4), 1,4-dicyanonaphthalene or 1,2,4,5-tetracyanobenzene serve as electron acceptors. Rotation around the scissile bond is unrestricted, and thus these systems undergo π–π cleavage, producing delocalized (benzylic) cations and radicals. The substituents at benzylic carbon (R = Ph, alkyl, OH, OR') determine the homolytic bond strength (via steric and radical-stabilizing effects) and the electron apportionment to the fragments (via cation-stabilizing effects). If one of the α-substituents is hydrogen, deprotonation reactions often compete [112, 113] leading often to protonation and further reduction of the radical anion of the acceptor.

The C–C fragmentations are thermally activated [107]. The apparent activation barriers determined from the temperature dependence of the reaction efficiency are almost identical to the estimated ΔG_m's (or ΔH_m's) [107]. The efficiency of the overall photochemical reaction becomes impractically small for processes with $\Delta H_m > 0.65$ eV. The ability of the system to delocalize the charge in the transition state facilitates the cleavage [102, 106]. Thus, electron donating substituents (Y) accelerate the reaction even in cases where the Y-substituted fragments will become radicals after the C–C bond scission [106].

In most cases only one pair of the possible four primary products (Scheme 4) is formed [107]. The electron apportionment to the fragments is determined by thermodynamic stability of the two possible ions. Only in cases where both ions have similar stability (the reduction potentials within 100 mV of each other) are all four products formed [107]. In chemically "clean" systems the ratio of cationic fragments may be used to determine the difference in their reduction potentials [107]. The "cleanness" of the system is determined by the secondary reactions of the fragments. The cations are usually trapped by MeOH, giving high yield of ethers. The reaction of diphenylmethyl or cumyl cations with MeOH (commonly present in molar concentration) is slower than the diffusion limit (10^7–10^8 s^{-1} M^{-1} [109, 128]), but apparently sufficiently rapid to prevent their reduction by the radical anion of the cyanoaromatic component. Since this redox process is exergonic by 1.5–1.8 eV for the typical bibenzyl system and should be rapid (at least within contact ion pairs), the absence of products derived from reduction of the primary cation would suggest that the cleavage takes place in SSIP or in free ions. The reduction products of the primary cationic fragment are observed in systems where the fragmentation is rapid enough [106b, 112, 113] to compete with ion solvation. In the absence of alcohol in the medium, or in cases where the benzyl fragment carries an α-OH substituent, a rapid deprotonation of the cation is often observed [112–115].

The fate of the primary radical fragment depends on its reduction potential. Thus, if the reduction of the radical by the radical anion of the cyanoaromatic component is exergonic (diphenylmethyl derivatives [107], cumene derivatives with electron-withdrawing Y substituents [106b]), the carbanion produced is rapidly protonated by the solvent (Scheme 4). If the reduction is endergonic the radical–radical anion coupling takes place [107, 106], often in the ipso position, which is followed by the expulsion of CN⁻. The characteristic radical–radical coupling products are also commonly detected [106, 107].

The reaction is often run in the presence of dioxygen which serves as a secondary acceptor, intercepting the cyanoaromatic radical anions and trapping the radical fragments [112, 113]. Alcohols, peroxides, aldehydes and ketones derived from the fragments are observed under such conditions. In the case of α-unsubstituted bibenzyls, proton loss from the radical cation followed by oxygen trapping of the radical formed and subsequent fragmentation of the radical cation of the oxygenated product may account for the cleavage products, rather than the inefficient direct fragmentation of the bibenzyl radical cation itself [112, 113].

5.2 Radical Anion Assistance in Oxidative Fragmentations of Amines

The fragmentation of C–C bonds in amine radical cations [9, 108, 116] has many similarities to cleavage reactions of bibenzyl derivatives. The unpaired electron is strongly localized on the amine nitrogen, and the bond cleavage is of the π–π type. The loss of H in the form of a proton or hydrogen atom is a common competing reaction, since the electron removal from the amine activates both the geminal N–H and the vicinal C–H bonds for heterolysis and homolysis [129]. The structures of several amines and amino alcohols undergoing facile C–C bond fragmentations are shown in Scheme 5 (the scissile bonds

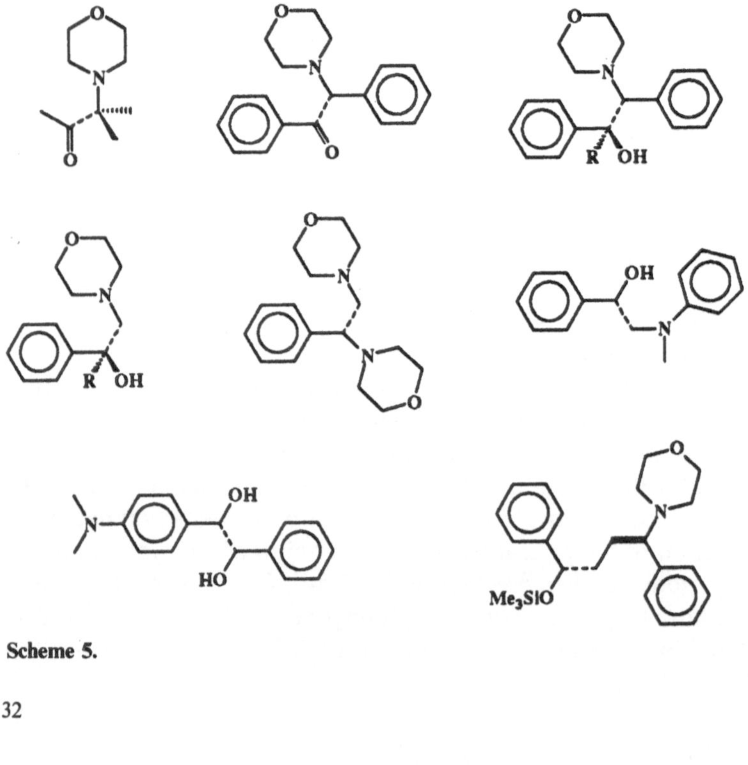

Scheme 5.

are shown with broken lines). A variety of acceptors, from inorganic ruthenium complexes via organic dyes (thioindigo) to quinones (β-lapachone) and cyano-aromatics serve as excited-state component of the PET systems. Based on thermodynamic considerations the primary fragments should consist of the amine-stabilized cation (an iminium ion if the β-C–C bond is cleaved) and the radical derived from the other fragment. Since this radical often carries an amino or hydroxy substituent (Scheme 5) it is susceptible to further oxidation followed by hydrolysis. Thus, two-electron oxidation products are often encountered. These include aldehydes, ketones, dealkylated amines etc. The chemical yields are usually high and quantum yields vary from low to moderate, often strongly depending on the solvent and identity of the acceptor.

A characteristic feature of *some* of these reactions is the dependence of their efficiency on the basicity of the radical anion [108]. The differences are especially manifested in non-polar solvents, where the CIP are expected to dominate. Some of these cleavage processes are more efficient than expected, based on the thermodynamic evaluations of the unassisted fragmentation (Sect. 3.2). Also a stereochemical preference for cleavage is observed for *erythro* isomers as compared to the *threo* isomers (Scheme 6). In benzene *erythro/threo* selectivity is high, being highest for the relatively basic radical anion of dicyanonaphthalene and lowest for the relatively nonbasic radical anion of thioindigo. The stereochemical preference disappears in acetonitrile if biphenyl is used as a co-sensitizer [108].

These observations are accounted for by radical anion (base) assisted fragmentations, where anti stereochemistry, analogous to that in elimination reactions, is favored. The intramolecular hydrogen bonding in the *threo* isomers leads to gauche arrangements of the reactive substituents, diminishing the rate of assisted fragmentation [108]. In more polar solvents and under circumstances where the free ions are produced (Ph–Ph co-sensitization), the unassisted process is observed (providing that the energetics are not prohibitive).

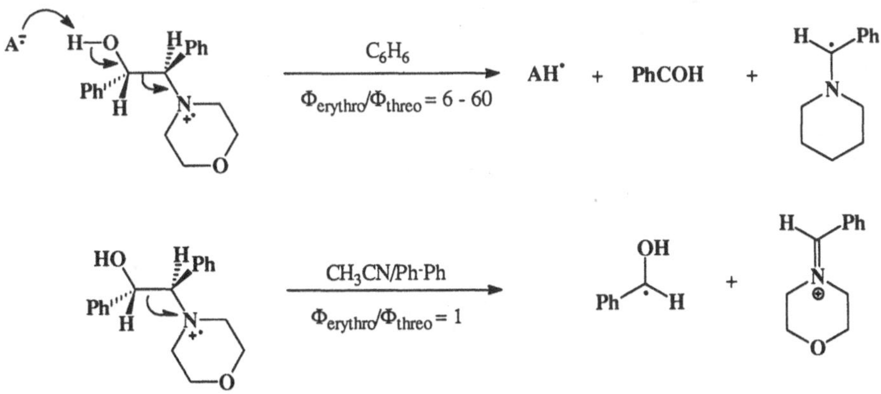

Scheme 6.

33

5.3 Radical Ions in Intramolecular ET and in Onium and Borate Salts

An interesting radical electron apportionment situation is found in photo-chemistry of some "onium" salts. The iodonium and sulfonium salts (Scheme 7) are the most widely studied, but other salts derived from halogens (chloronium and bromonium), or other elements (N, P etc) behave similarly [118–120]. These salts may act as electron acceptors, giving a neutral reduction product. The unpaired electron may be localized in one of the sigma bonds (σ^* radical anions) or in an adjacent π-system, depending on the relative energies of these orbitals. Although neutral, these species may, therefore, be considered as "radical anions" as discussed in Sect. 3.1. In some instances ET may be dissociative. The cleavage of these species may produce a radical and a neutral (most likely) or an anion and a radical cation (less likely).

The photochemistry of these salts is, however, rather complicated [7, 118–120]. Both direct and energy-transfer sensitized cleavages are observed with heterolysis and homolysis products detected, but the details are not yet completely understood. Many of these salts are able to rapidly generate Brønsted acids (by rather complicated pathways) and are used to initiate

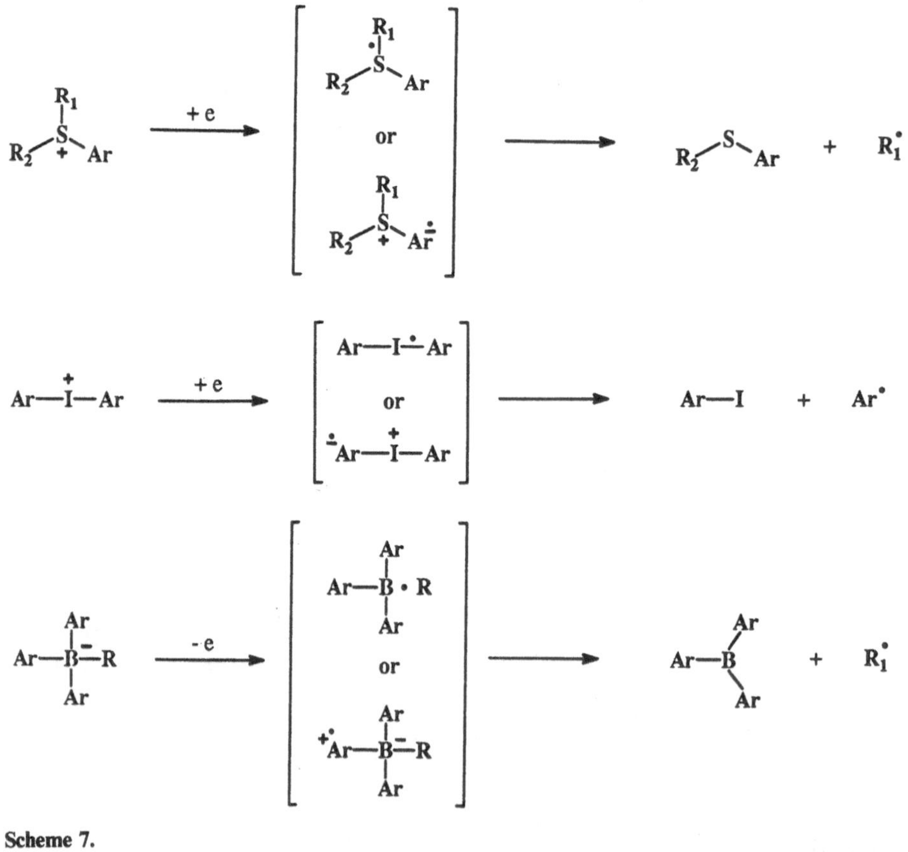

Scheme 7.

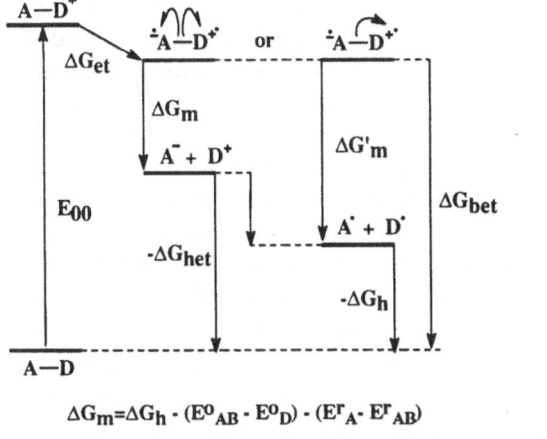

$$\Delta G_m = \Delta G_h \cdot (E^0{}_{AB} \cdot E^0{}_D) \cdot (E^r{}_A \cdot E^r{}_{AB})$$

$$\Delta G'_m = \Delta G_h + \Delta G_{bet} = \Delta G_h \cdot (E^0{}_{AB} \cdot E^r{}_{AB})$$

Scheme 8.

polymerizations, depolymerizations or other acid catalyzed reactions of importance in the imaging industry [118–120].

The borate salts (Scheme 7) constitute an analogous group of electron donors [8, 29, 121]. In this case, PET from the anions to cationic acceptors (Sect. 2.2.1) produces σ radical cations, (or alternatively, if a π-reservoir is available a π radical cation may be formed). If one of the substituents is an alkyl group the one-electron bond breaks rapidly (possibly in a dissociative process), forming a radical and a neutral. These systems also find applications in imaging.

A related, "reverse-charge" situation is found in systems where the acceptor and donor are directly connected via a scissile bond. The intramolecular electron transfer generates a zwiterionic intermediate (Scheme 8). This intermediate may be considered to be doubly activated for fragmentation [77, 102]. Thermodynamic analysis (Sect. 3.2) indicates the scissile bond to be weakened by both the oxidative and reductive contribution. Both homolytic cleavage (giving a pair of ions) and heterolytic cleavage (giving a pair of radicals) may have highly negative free-energies and high rates of fragmentation [77, 102].

The homolytic alternative may have a larger driving force than the heterolytic path if the fragments are stable ions [77]. The dynamics of such a system may be quite complex, with various deactivation pathways available [77, 102].

5.4 Tetranitromethane as a Self-Destructive Electron Acceptor

Tetranitromethane (TNM) readily forms CT complexes with aromatic systems. Irradiation of these complexes leads to rapid generation of radical cation/TNM$^{\cdot-}$ pairs [117]. The radical anion fragments rapidly ($\tau < 2$ ps) generating a triad of reactive intermediates that include the radical cation of the donor, trinitromethanide ($^-C(NO_2)_3$), and NO_2 as illustrated in Scheme 9 for methylanisole. The fragmentation is so rapid that no diffusional separation takes place. Thus, the triad is equivalent to the contact ion pair and the geminate radical pair at the same time. The dynamics of this triad has been probed in detail for various anthracene donors [52], providing direct data for the ion pair interconversions (Sect. 4). The fate of the triad is determined by the stability of the radical cation component and the solvent polarity. For relatively stable radical cations (such as those of dimethoxybenzenes) and in polar solvents (CH_3CN), radical/radical cation recombination is the dominant process. The resulting cyclohexadienyl cation may lose a proton, giving the nitration product directly, or may be trapped by the trinitromethanide, giving the adduct that can spontaneously eliminate nitroform, and lead also to the nitration product. In less polar solvents or in the case of less stable radical cations (such as anisole derivatives) the ion collapse process dominates. The cyclohexadienyl radical produced transfers a hydrogen atom to $NO_2^{\cdot}$, giving an aromatic trinitromethyl derivative.

If the radical cation contains also a scissile bond, as found in the bicumene derivatives [106d], aromatic nitration or trinitromethylation (depending on solvent polarity) may compete with bond scission. Thus, in addition to the products of aromatic substitution benzylic derivatives are observed. If the C–C bond cleavage is rapid a tetrad of reactive intermediates (two ions and two radicals) forms. The reactivity of the tetrad is dominated by the ion-collapse processes. The carbon-centered radicals are oxidized by TNM present in large excess, leading to a triad of reactive intermediates similar to the original tetrad (Scheme 10). The competition between the various processes is governed by solvent polarity as well as by the rates of C–C bond cleavage, which in turn are controlled by substitution patterns [106d].

5.5 Competition between Bond Cleavage and BET

Rapid fragmentation reactions may be used in connection with quantum yield measurements as a molecular clock to probe the electron-transfer dynamics of

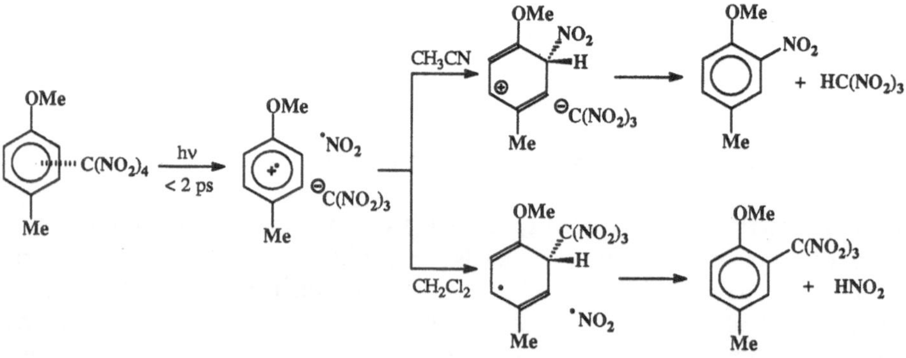

Scheme 9.

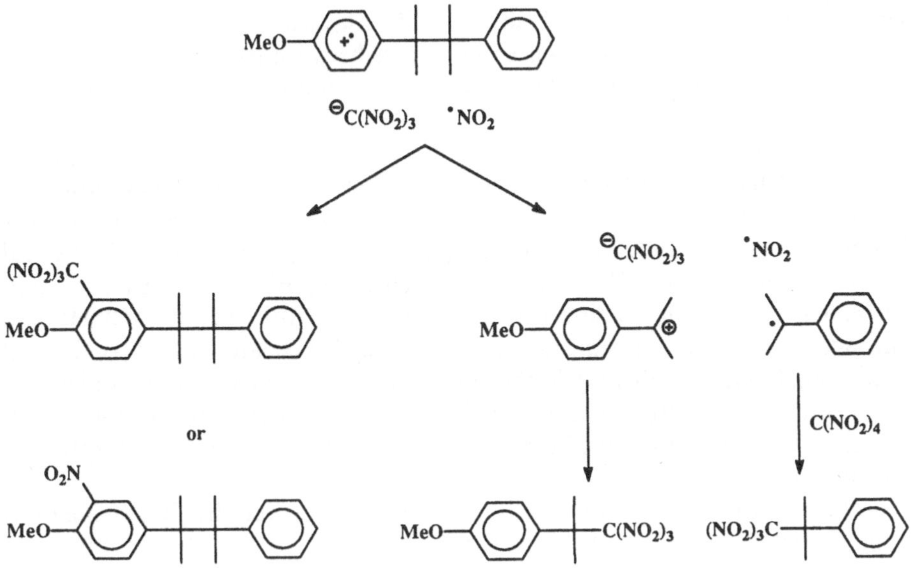

Scheme 10.

photogenerated ion pairs [102, 103, 106a, 130]. The usual assumption is that the fragmentation reaction is independent of the ion pairing and spin status of the pair, and that the efficiency of the process is determined by the competition between the fragmentation and BET ($\Phi_m = k_m/(k_m + k_{bet})$). Admittedly, the method is crude in comparison to direct observation, but in many instances it can provide reliable information in systems where complexity makes the direct method ineffective.

The C–C bond cleavage of pinacol or bicumene derivatives have been used to compare the overall efficiency of two alternative methods of generating reactive radical ions [103, 106a]. The local excitation method (Φ_{mq}), wherein the excited state of the acceptor is quenched in a diffusion-controlled process is more efficient than the irradiation of the CT bands (Φ_{mct}) of the CT complex formed by the same acceptor. This generalization applies to systems involving singlet excited states [103] as well as triplet excited states [103, 106a], the latter having the advantage of the spin-forbidden nature of BET. If the assumption of direct competition between BET and bond fragmentation holds and Φ_{mq} is extrapolated to infinite concentration of quencher (see below), the ratio of the rates of two BET processes for the two alternative methods of radical ion generation may be obtained from quantum yield measurements ($k_{bet(q)}/k_{bet(ct)}$ $= \Phi_{mct}(1 - \Phi_{mq})/\Phi_{mq}(1 - \Phi_{mct})$). In real systems the ion-pair dynamics (Sect. 4) is usually more complicated, and the data should be treated with caution. Nevertheless, the results of such analyses agree with the data obtained for BET by different methods (Sect. 2.3). Thus, the difference between BET in singlet contact ion pairs ($k_{bet(ct)}$) and singlet SSIP ($k_{bet(q)}$) is about one order of magnitude [103] and the difference in the rates between BET in singlet CIP and BET in triplet SSIP is somewhere between one and two orders of magnitude [103, 106a].

The complexity of the ion pair dynamics may be illustrated by the photo-induced ET from the excited state tetramethylbenzidine (TMB) to cyanobicum-ene (BC) derivatives (Fig. 9) [78]. At low concentrations of acceptors, only TMB[3] is quenched yielding triplet ion pairs. The fragmentation of the radical anion or ion separation (see below) competes with BET. Since BET is spin-forbidden, the overall reaction is relatively efficient. At higher concentration of the acceptor, more excited state molecules of TMB are quenched, but an increasing fraction of them is in the singlet state. This fraction gives singlet ion pairs, within which BET is more efficient. The overall efficiency of the photo-reaction, therefore, does not increase linearly with increasing concentration of the acceptor [78].

The quantum yields for disappearance of BC-a and BC-b are identical (Fig. 9), but the rates of mesolysis for the radical anions of these compounds differ by a factor of ten. Thus the quantum yields for these compounds reflect a competition between BET and ion separation. The quantum yield for disap-pearance of BC-c is higher than that for BC-a or BC-b, and is now comparable

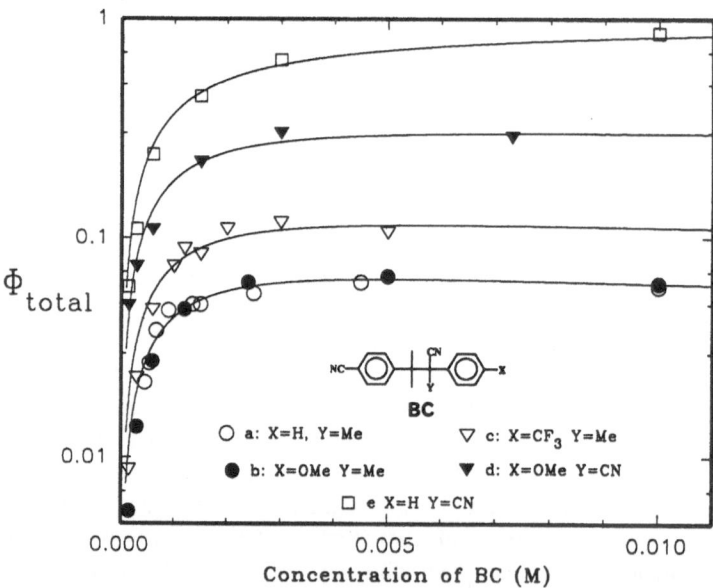

Fig. 9. The quantum yield for disappearance of BC as a function of [BC]. The lines represent the best fit to the following equation: $\Phi_{total} = \Phi_s k_{qs}[BC]/(1/\tau_s + k_{qs}[BC]) + \Phi_t k_{ic} k_{qt}[BC]/(1/\tau_s + k_{qs}[BC])(1/\tau_t + k_{qt}[BC])$, where Φ_s and Φ_t, represent absolute efficiencies of fragmentation within singlet and triplet ion pairs (($k_m + k_{sep})/(k_m + k_{sep} + k_{bet})$), k_{qs} is the rate of singlet quenching, τ_s is the singlet lifetime, k_{ic} is the rate of intersystem crossing, k_{qt} represents the rate of triplet quenching), and τ_t is the triplet lifetime

to the rate of ion separation. The rates of cleavage of BC-d and BC-e are faster than the rate of ion separation and the quantum yields reflect the competition between BET and mesolysis. In this case the difference between the rates of BET within singlet ion pairs (SSIP) and triplet ion pairs (SSIP) is only a factor of 3 or 4 [78]. In general, BET within triplet pairs is most likely controlled by intersystem crossing rates, and thus will strongly depend on the system under investigation, with such factors as presence of heavy atoms or magnetic fields playing an important role.

Acknowledgement. The research in the author's laboratory was sponsored by grants from the National Science Foundation and the Petroleum Research Fund. The contributions of my coworkers H. Abdel, S. Asel, B. Bocknack, W. Chapman, Jr., P. Funke, J. Kula, D. Malinski, J. Narvaez, T. Vallombroso and B. Watson are acknowledged with gratitude.

6 References

1. Fox MA, Chanon M (eds) (1988) Photoinduced electron transfer. parts A–D, Elsevier, Amsterdam
2. Kavarnos GJ, Turro NJ (1986) Chem. Rev. 86: 401
3. Mariano PS (ed) (1991 and 1992) Advances in electron transfer chemistry. JAI Press, vol 1–3, Greenwich, Connecticut
4. Mattay J (ed) (1990–1992) Top. Curr. Chem. vols 156, 158, 159, 163
5. Bolton JR, Mataga N, McLendon G (eds) (1991) Electron transfer in inorganic, organic and biological systems, Advances in Chemistry Series vol 228. American Chemical Society, Washington DC
6. (a) Fox MA (1990) Photochem. Photobiol. 52: 617; (b) Mattay J (1987) Angew. Chem. Int. Ed. Engl. 26: 825; (c) Davidson RS (1983) In: Gold V, Bethell D (eds) Advances in physical chemistry. Academic Press, London, vol 19, p 1; (d) Fox MA (1986) In: Volman DH, Gollnick K, Hammond GS (eds) Advances in photochemistry. Wiley, New York, vol 13, p 237; (e) Mattes SL, Farid S (1983) In: Padwa A (ed) Organic photochemistry. Marcel Dekker, New York, vol 6, p 233
7. Saeva FD (1990) Top. Curr. Chem. 156: 60
8. Schuster GB (1991) In: Mariano PS (ed) Advances in electron transfer chemistry, vol. 1. JAI Press, Greenwich, Connecticut, p 163
9. Ci X, Whitten DG (1988) In: Fox MA, Chanon M (eds) Photoinduced electron transfer, part C. Elsevier, Amsterdam, p 533
10. (a) Mariano PS (1983) Acc. Chem. Res. 16: 130; (b) Mariano PS (1988) In: Fox MA, Chanon M (eds) Photoinduced electron transfer, part C. Elsevier, Amsterdam, p 372
11. Gassman PG (1988) In: Fox MA, Chanon M (eds) Photoinduced electron transfer, part C. Elsevier, Amsterdam, p 70.
12. (a) Roth HD, Schilling MLM, Abelt CJ (1986) Tetrahedron 42: 6157; (b) Roth HD, Schilling MLM, Wamser CC (1984) J. Am. Chem. Soc. 106: 5023; (c) Roth HD, Schilling MLM, Gassman PG, Smith JL (1984) J. Am. Chem. Soc. 106: 2711; (d) Roth HD, Schilling MLM, Hutton RS (1983) J. Am. Chem. Soc. 105: 153; (e) Roth HD, Schilling MLM (1981) J. Am. Chem. Soc. 103: 7210; (f) Roth HD, Schilling MLM, Jones G (1981) J. Am. Chem. Soc. 103: 1246; (g) Roth HD, Schilling MLM (1980) J. Am. Chem. Soc. 102: 7956
13. (a) Dinnocenzo JP, Todd WP, Simpson TR, Gould IR (1990) J. Am. Chem. Soc. 112: 2463; (b) Shaik SS, Dinnocenzo JP (1990) J. Org. Chem. 55: 3434
14. See for example: (a) Roth HD (1987) Acc. Chem. Res. 20: 343; (b) Jones G II, Becker WG

(1982) Chem. Phys. Lett. 85: 271; (c) Jones G II, Becker WG (1983) J. Am. Chem. Soc. 105: 1276; (d) Peacock NJ, Schuster GB (1983) J. Am. Chem. Soc. 105: 3632; (e) Masnovi JM, Kochi JK (1985) J. Am. Chem. Soc. 107: 6781; (f) Takahashi Y, Kochi JK (1988) Chem. Ber. 121: 253; (g) Miyashi T, Kamata M, Mukai T (1987) J. Am. Chem. Soc. 109: 2755; (h) Bauld N L (1992) In: Mariano PS (ed) Advances in electron transfer Chemistry. JAI Press, vol 2. Greenwich, Connecticuti; (i) Miyashi T, Takahashi Y, Yokogawa K, Mukai T (1987) J. Chem. Soc. Chem. Commun. 175; (j) Cooksey CJ, Courtneidge JL, Davies AG, Evans JC, Gregory PS, Rowlands CC (1986) J. Chem. Soc. Commun. 1513; (k) Miyashi T, Wakamatsu K, Akiya T, Kikuchi K, Mukai T (1987) J. Am. Chem. Soc. 109: 1570; (l) Kawamura Y, Thurnauer M, Schuster GB (1986) Tetrahedron 42: 6195; (m) Nelsen SF, Kapp DL (1985) J. Org. Chem. 50: 1339; (n) Okada K, Hisamitsu K, Takahashi Y, Hanaoka T, Miyashi T, Mukai T (1984) Tetrahedron Lett. 25: 5311; (o) Dinnocenzo JP, Schimittel MJ (1987) J. Am. Chem. Soc. 109: 1561; (p) Dinnocenzo JP, Conlan DA (1988) J. Am. Chem. Soc. 110: 2324

15. See for example: (a) Kochi JK, Amatore C (1991) In: Mariano P (ed) Advances in electron transfer chemistry, vol 1. JAI Press, Greenwich, Connecticut, p 55; (b) Schlesner CJ, Amatore C, Kochi JK (1984) J. Am. Chem. Soc. 106: 3567; (c) Schlesner CJ, Kochi JK (1984) J. Org. Chem. 49: 3142; (d) Camaioni DM, Franz JJ (1984) J. Org. Chem. 49: 1607; (e) Deardurff LA, Alnajjar MS, Camaioni DM (1986) J. Org. Chem. 51: 3686; (f) Baciocchi E, Bartoli D, Rol C, Ruzziconi R, Sebastiani G (1986) J. Org. Chem. 51: 3587; (g) Baciocchi E (1990) Acta Chem. Scand. 44: 645; (h) Baciocchi E, Mattioli M, Romano R, Ruzziconi R (1991) J. Org. Chem. 56: 7154; (i) Baciocchi E, Rol C, Scamosci E, Sebastiani GV (1991) J. Org. Chem. 56: 5498

16. (a) Marcus RA (1960) Discuss. Faraday Soc. 29: 21; (b) Marcus RA (1964) Annu. Rev. Phys. Chem. 15: 155; (c) Marcus RA (1965) J. Chem. Phys. 43: 679

17. Eberson L (1987) Electron transfer reactions in organic chemistry. Springer, Berlin Heidelberg New York

18. (a) Bolton JR, Archer MD (1991) In: Bolton JR, Mataga N, McLendon G (ed) Electron transfer in inorganic, organic and biological systems, Advances in Chemistry Series vol 228. American Chemical Society, Washington DC, p 7; (b) Purcell KF, Blaive B (1988) In: Fox MA, Chanon M (eds) Photoinduced electron transfer, part A. Elsevier, Amsterdam, p 123.; (c) McLendon GL, Helms A (1991) In: Mariano PS (ed) Advances in electron transfer chemistry, JAI Press, vol 1. Greenwich, Connecticut, p 149; (d) Ulstrup JJ (1979) Charge transfer processes in condensed media. Springer, Berlin Heidelberg, New York; (e) Devault D (1984) Quantum mechanical tunnelling in biological systems, 2nd ed. Cambridge University Press, Cambridge; (f) Newton MD, Sutin N (1984) Annu. Rev. Phys. Chem. 35: 437; (g) Marcus R, Sutin N (1985) Biochim. Biophys. Acta. 811: 265; (h) Suppan P (1991) Top. Curr. Chem. 163: 95

19. Depending on the value of V, only the first few terms are of importance. In this review the first 15 terms ($w_{max} = 14$) was used for plotting

20. (a) Suppan P (1988) Chimia 10: 320; (b) Santamaria J (1988) In: Fox MA, Chanon M (eds) Photoinduced electron transfer, part B. Elsevier, Amsterdam, p 483; (c) Calef DF (1988) In: Fox MA, Chanon M (eds) Photoinduced electron transfer, part A. Elsevier, Amsterdam, p 362

21. (a) Sutin N (1991) In: Bolton JR, Mataga N, McLendon G (eds) Electron transfer in inorganic, organic and biological systems, Advances in Chemistry Series vol 228. American Chemical Society, Washington DC. p 25; (b) Wasielewski MR (1988) In: Fox MA, Chanon M (eds) Photoinduced electron transfer, part A, Elsevier, Amsterdam, p 161

22. Chanon M, Hawley MD, Fox MA (1988) In: Fox MA, Chanon M (eds) (1988) Photoinduced electron transfer, part A. Elsevier, Amsterdam, p 1.

23. (a) Alkaitis SA, Grätzel, M (1976) J. Am. Chem. Soc. 98: 3549; (b) Das PK, Muller AJ, Griffin GW, Gould IR, Tung C-H, Turro NJ (1984) Photochem. Photobiol. 39: 281; (c) Turro NJ, Tung C-H, Gould IR, Griffin GW, Smith RL, Manmade A (1984) J. Photochem. 24: 265; (d) Poizat O, Bourkba A, Buntinx G, Deffontaine A, Bridoux, M (1987) J. Phys. Chem. 87: 6379; (e) Guichard V, Bourkba A, Lautie M-F, Poizat O (1989) Spectrochim. Acta 45A: 187. (f) Fox MA (1979) Chem. Rev. 79: 253; (g) Tolber LM (1983) In: Padwa A (eds) Organic photochemistry. Marcel Dekker, New York, vol 6, p 177.

24. For a review see: Chibisov K (1981) Russ. Chem. Rev. 50: 1169

25. (a) Rehm D, Weller A (1970) Isr. J. Chem. 8: 259; (b) Weller A (1968) Pure Appl. Chem. 16: 115

26. Recently the inverted region was also observed in a second order BET reaction: McCleskey TM, Winkler JR, Gray HB (1992) J. Am. Chem. Soc. 114: 6935

27. For a bimolecular reaction $k_{et} = k_d[1 + (k_d/(K_dZ))\exp\{\lambda(1 + \Delta G_{et}/\lambda)^2/4RT\}]^{-1}$ where $K_d = k_d/k_{-d}$ (ref 17)

28. (a) Ballardini R, Varani G, Indelli MT, Sandola F, Balzani V (1978) J. Am. Chem. Soc. 100: 7219; (b) Indelli MT, Ballardini R, Scandola F (1984) J. Phys. Chem. 88: 2547
29. (a) Schuster GB (1990) Pure Appl. Chem. 62: 1565; (b) Zou C, Miers JB, Ballew RM, Dlott DD, Schuster GB (1991) J. Am. Chem. Soc. 113: 7823
30. (a) Szwarc M (1972 and 1974) Ions and ion pairs in organic reactions, vol 1 and 2. Wiley-Interscience, New York; (b) Reichardt C (1990) Solvents and solvent effects in organic chemistry, VCH, Weinheim; (c) Harris JM (1974) Prog. Phys. Org. Chem. 11: 89; (d) Mattay J. Vondenhof M (1991) Top. Curr. Chem. 159: 219; (e) Billing R, Rehorek D, Hennig H (1990) Top. Curr. Chem. 158: 151
31. Marcus Y (1985) Ion solvation. Wiley, Chichester
32. Chen J-M, Ho T-I, Mou C-Y (1990) J. Phys. Chem. 94: 2889
33. (a) Mattes SL, Farid S (1984) Science 226: 917; (b) Even in polar solvents CIP may form: Peters KS, Lee J (1992) J. Phys. Chem. 96: 8941
34. (a) Salikhov KM, Molin YN, Sagdeev RZ, Buchachenko AL (1984) Spin polarization and magnetic effects in radical reactions. Elsevier, Amsterdam; (b) Gould IR, Turro NJ, Zimmt MB (1984) Adv. Phys. Org. Chem. 20: 1
35. (a) Weller A (1982) Z. Phys. Chem. (Neue Folge) 130: 129; (b) Schulten K, Staerk H, Weller A, Werner H-J, Nickel B (1976) Z. Phys. Chem. (Neue Folge) 101: 371; (c) Werner H-J, Staerk H, Weller A (1978) J. Chem. Phys. 68: 2419; (d) Steiner UE, Ulrich T (1989) Chem. Rev. 89: 51
36. (a) Boxer SG, Goldstein RA, Franzen S (1988) In: Fox MA, Chanon M (eds) Photoinduced electron transfer, part B. Elsevier, Amsterdam, p 163; (b) Franzen S, Boxer SG (1991) In: Bolton JR, Mataga N, McLendon G (eds) Electron transfer in inorganic, organic and biological systems, Advances in Chemistry Series vol 228. American Chemical Society, Washington DC, p 149; (c) Sano H (1983) J. Chem. Phys. 78: 4423; (d) Sethi DS, Choi HT, Braun CL (1980) Chem. Phys. Lett. 74: 223; (e) Bullot J, Cordier P, Gauthier M (1980) J. Phys. Chem. 84: 3516
37. See for example: (a) Jorgensen WL, Severance DL (1990) J. Am. Chem. Soc. 112: 4768; (b) See also ref 47 and 48.
38. See for example: (a) Haselbach E, Vauthey E, Suppan P (1988) Tetrahedron 44: 7335; (b) Jones G II, Moulie N (1988) J. Phys. Chem. 92: 7174; (c) Haselbach E, Jacques P, Pilloud D, Suppan P, Vauthey E (1991) J. Phys. Chem. 95: 7115; (d) Devadoss C, Fessenden RW (1991) J. Phys. Chem. 95: 7523; (e) Angerhofer A, Toporowicz M, Bowman MK, Norris JR, Levanon H (1988) J. Phys. Chem. 92: 7164
39. (a) Gust D, Moore TA (1991) Top. Curr. Chem. 159: 103; (b) Connolly JS, Bolton JR (1988); (c) Bolton JR, Schmidt JA, Ho T-F, Liu J, Roach KJ, Weedon AC, Archer MD, Wilford JH, Gadzekpo VPY (1991) In: Bolton JR, Mataga N, McLendon G (eds) (1991) Electron transfer in inorganic, organic and biological systems, Advances in Chemistry Series vol 228. American Chemical Society, Washington DC, p 117; (d) Jordan KD, Paddon-Row MN (1992) Chem. Rev. 92: 395; (e) Zeng Y, Zimmt MB (1991) J. Am. Chem. Soc. 113: 5107; (f) Penfied KW, Miller JR, Paddon-Row MN, Costaris E, Oliver AM, Hush NS (1987) J. Am. Chem. Soc. 109: 5061; (g) Kroon J, Verhoeven JW, Paddon-Row MN, Oliver AM (1991) Angew. Chem. Int. Ed. Engl. 30: 1358; (h) Warman JM, Smit KJ, de Haas MP, Jonker SA, Paddon-Row MN, Oliver AM, Kroon J, Oevering H, Verhoeven JW (1991) J. Phys. Chem. 95: 1979
40. (a) Fox MA (1991) Top. Curr. Chem. 159: 67; (b) Nakabayashi S, Kawai T (1988) In: Fox MA, Chanon M (eds) Photoinduced electron transfer, part B. Elsevier, Amsterdam, p 599; (c) Wasielewski MR (1992) Chem. Rev. 92: 435
41. (a) Lymar CV, Parmon VN, Zamarev KI (1991) Top. Curr. Chem. 159: 1; (b) Willner I, Willner B (1991) Top. Curr. Chem. 159: 153; (c) Baral S, Fendler JH (1988) In: Fox MA, Chanon M (eds) Photoinduced electron transfer, part B. Elsevier, Amsterdam, p 541; (d) Balzani V (1992) Tetrahedron 48: 10443
42. (a) Yang X, Zaitsev A, Sauerwein B, Murphy S, Schuster GB (1992) J. Am. Chem. Soc. 114: 793; (b) Boche G (1992) Angew. Chem. Int. Ed. Engl. 31: 731
43. (a) Fox MA (1991) In: Mariano PS (ed) Advances in electron transfer chemistry, JAI Press, vol 1. Greenwich, Connecticut, p 1; (b) Pichat P, Fox MA (1988) In: Fox MA, Chanon M (eds) Photoinduced electron transfer, part D. Elsevier, Amsterdam, p 241; (c) Fox MA, Chen C-C, Younathan JNN (1984) J. Org. Chem. 49: 1969; (d) Fox MA, Younathan J, Fryxell GE (1983) J. Org. Chem. 48: 3109; (e) Fox MA, Abdel-Wahab AA (1990) Tetrahedron Lett. 32: 4533; (f) Fox MA (1987) Top. Curr. Chem. 142: 71; (g) Fox MA, Chen C-C (1981) J. Am. Chem. Soc. 103: 6757; (h) Koval CA, Howard JN (1992) Chem. Rev. 92: 411; (i) Kamat PV (1993) Chem. Rev. 93: 267

44. (a) Sankararaman S, Yoon KB, Yabe T, Kochi JK (1991) J. Am. Chem. Soc. 113: 1419 and references therein; (b) Yoon KB (1993) Chem. Rev. 93: 321
45. (a) Alfimov MV, Sazhnikov VA (1988) In: Fox MA, Chanon M (eds) Photoinduced electron transfer, part D. Elsevier, Amsterdam, p 474; (b) Carter FL (ed) (1982 and 1987) Molecular electronic devices. Marcel Dekker, New York; (c) Roncali J (1992) Chem. Rev. 92: 711 and references therein; (d) Proceedings of ICMS'88, Synth. Met. (1989), vol 27–29; (e) Law K-Y (1993) Chem. Rev. 93: 449
46. (a) Eaton DF (1990) Top. Curr. Chem. 156: 199; (b) Timpe H-J (1990) Top. Curr. Chem. 156: 167; (c) Shirota Y (1988) In: Fox MA, Chanon M (eds) Photoinduced electron transfer, part D. Elsevier, Amsterdam, p 441; (d) Shue F, Giral L, Montginoul C, Serre B (1988) In: Fox MA, Chanon M (eds) Photoinduced electron transfer, part D. Elsevier, Amsterdam, p 519; (e) Usacheva MN, Dilung II (1991) Russ. Chem. Rev. 60: 106; (f) Timpe J-H (1988) Pure Appl. Chem. 60: 1033; (g) Monroe BM, Weed GC (1993) 93: 435
47. (a) Mulliken RS, Pearson WB (1969) Molecular complexes: A lecture and reprint volume. Wiley-Interscience, New York; (b) Foster R (1969) Organic charge-transfer complexes. Academic Press, New York; (c) Nagakura S (1975) Excited states. Academic Press, New York, vol 2, p 321
48. (a) Kochi JK (1991) Pure Appl. Chem. 63: 255; (b) Kochi JK (1988) Angew. Chem. Int. Ed. Engl. 27: 1227; (c) Kochi JK (1990) Acta Chem. Scand. 44: 409; (d) Jones G II (1988) In: Fox MA, Chanon M (eds) Photoinduced electron transfer, part A. Elsevier, Amsterdam, p 245; (e) Mataga N (1991) In: Bolton JR, Mataga N, McLendon G (eds) Electron transfer in inorganic, organic and biological systems, Advances in chemistry series vol 228. American Chemical Society, Washington DC, p 91
49. (a) Hirata Y, Mataga N (1991) J. Phys. Chem. 95: 1640; (b) Mataga N, Nishikawa S, Asahi T, Okada T (1990) J. Phys. Chem. 94: 1443; (c) Miyasaka H, Ojima S, Mataga N (1989) J. Phys. Chem. 93: 3380; (d) Mataga N, Okada T, Kanda Y, Shioyama H (1985) Tetrahedron 42: 6143; (e) Ojima S, Miyasaka H, Mataga (1990) J. Phys. Chem. 94: 7534; (f) Masnovi JM, Kochi JK, Hilinski EF, Rentzepis PM (1986) J. Am. Chem. Soc. 108: 1126; (g) Hilinski EF, Milton SV, Rentzepis PM (1983) J. Am. Chem. Soc. 105: 5193; (h) Rentzepis PM, Steyert DW, Roth HD, Abelt CJ (1985) J. Phys. Chem. 89: 3955; (i) Hilinski EF, Masnovi JM, Amatore C, Kochi JK, Rentzepis PM (1983) J. Am. Chem. Soc. 105: 6167; (j) Hilinski EF, Masnovi JM, Kochi JK, Rentzepis PM (1984) J. Am. Chem. Soc. 106: 8071; (k) Masnovi JM, Kochi JK (1985) J. Am. Chem. Soc. 107: 7880; (l) Masnovi JM, Kochi JK, Hilinski EF, Rentzepis PM (1985) J. Phys. Chem. 89: 5387; (m) Masnovi JM, Kochi JK (1987) J. Phys. Chem. 91: 1878
50. (a) Gould IR, Farid S, Young RH (1992) J. Photochem. Photobiol. A: Chem. 65: 133; (b) Gould IR, Farid S (1992) J. Phys. Chem. 96: 7635.
51. (a) Takahashi Y, Sankararaman S, Kochi JK (1989) J. Am. Chem. Soc. 111: 2954; (b) Kim EK, Kochi JK (1991) J. Am. Chem. Soc. 113: 4962; (c) Kim EK, Lee KY, Kochi JK (1992) J. Am. Chem. Soc. 114: 1756; (d) Bockman TM, Karpinski ZJ, Sankararaman S, Kochi JK (1992) J. Am. Chem. Soc. 114: 1970; (e) Kim EK, Kochi JK (1989) J. Org. Chem. 54: 1692
52. (a) Masnovi JM, Kochi JK (1985) J. Am. Chem. Soc. 107: 7880; (b) Yabe T, Kochi JK (1992) J. Am. Chem. Soc. 114: 4491
53. (a) Goodman JL, Peters KS (1985) J. Am. Chem. Soc. 107: 1441; (b) Simon JD, Peters KS (1981) J. Am. Chem. Soc. 103: 6403; (c) Goodman JL, Peters KS (1985) J. Am. Chem. Soc. 107: 6549; (d) Goodman JL, Peters KS (1986) J. Am. Chem. Soc. 108: 1700; (e) Angel SA, Peters KS (1989) J. Phys. Chem. 93: 713
54. See for example: (a) Andrews LJ, Keefer RM (1988) J. Org. Chem. 53: 537; (b) Dresner J, Prochorow J, Ode W (1989) J. Phys. Chem. 93: 671; (c) Smith ML, McHale JL (1985) J. Phys. Chem. 89: 4002; (d) Gribaudo ML, Knorr FJ, McHale JL (1985) Spectrochim. Acta 41A: 419
55. (a) Calhoun GC, Schuster GB (1984) J. Am. Chem. Soc. 106: 6870; (b) Calhoun GC, Schuster GB (1986) Tetrahedron Lett. 27: 911
56. See for example: (a) Gschwind R, Haselbach E (1979) Helv. Chim. Acta 62: 941
57. (a) Beitz JV, Miller JR (1979) J. Chem. Phys. 71: 4579; (b) Calcaterra LT, Closs GL, Miller JR (1983) 105: 670; (c) Miller JR, Beitz JV, Huddleston RK (1984) J. Am. Chem. Soc. 106: 5057; (d) Miller JR, Calcaterra LT, Closs GL (1984) J. Am. Chem. Soc. 106: 3047; (e) Mclendon G, Miller JR (1985) J. Am. Chem. Soc. 107: 7811; (f) Wasielewski MR, Niemczyk MP, Svec WA, Pewitt EB (1985) J. Am. Chem. Soc. 107: 1080; (g) Irvine MB, Harrison RJ, Beddard GS, Leighton P, Sanders JKM (1986) Chem. Phys. 104: 315; (h) Harrison RJ, Pearce B, Beddard GS, Leighton P, Sanders JKM (1986) Chem. Phys. 116: 429; (i) Ohno T, Yoshimura A, Shioyama H, Mataga N (1987) J. Phys. Chem. 91: 4365; (j) Closs GL, Miller JR (1988) Science

240: 440; (k) Mataga N, Kanda Y, Asahi T, Miyaska H, Okada T, Kakitani T (1988) Chem. Phys. 127: 239; (l) Chen P, Duesing R, Tapolsky G, Meyer TJ (1989) J. Am. Chem. Soc. 111: 3993; (m) Meade TJ, Gray HB, Winkler JR (1989) J. Am. Chem. Soc. 111: 4353; (n) DeCosta DP, Pincock JA (1989) J. Am. Chem. Soc. 111: 8948

58. (a) Liang N, Miller JR, Closs GL (1990) J. Am. Chem. Soc. 112: 5353; (b) Liu JL, Bolton JR (1992) J. Phys. Chem. 96: 1718
59. O'Driscoll E, Simon JD, Peters KS (1990) J. Am. Chem. Soc. 112: 7091
60. (a) Gould IR, Ege D, Mattes SL, Farid S (1987) J. Am. Chem. Soc. 109: 3794; (b) Gould IR, Ege D, Moser JE, Farid S (1990) J. Am. Chem. Soc. 112: 4290; (c) Chung W-S, Turro NJ, Gould IR, Farid S (1991) J. Phys. Chem. 95: 7752
61. (a) Kikuchi K, Takahashi Y, Koike K, Wakamatsu K, Ikeda H, Miyashi T (1990) Z. Phys. Chem. (Neue Folge) 167: 27; (b) Kikuchi K, Takahashi Y, Hoshi M, Niwa T, Katagiri T, Miyashi T (1991) J. Phys. Chem. 95: 2378
62. Vauthey E, Suppan P, Haselbach E (1988) Helv. Chim. Acta 71: 93
63. Mataga N, Asahi T, Kanda Y, Okada T, Kakitani T (1988) Chem. Phys. 127: 249
64. In principle, each radical ion pair has its own (and different from other pairs) reorganization energy (λ_i and λ_s), coupling matrix (V), and v, thus "belonging" to a different parabola
65. Gould IR, Moser JE, Armitage B, Farid S, Goodman JL, Herman MS (1989) J. Am. Chem. Soc. 111: 1917
66. (a) Asahi T, Mataga N (1989) J. Phys. Chem. 93: 6575; (b) Asahi T, Mataga N (1991) J. Phys. Chem. 95: 1956
67. (a) Gould IR, Moody R, Farid S (1988) J. Am. Chem. Soc. 110: 7242; (b) Gould IR, Young RH, Moody RE, Farid S (1991) J. Phys. Chem. 95: 2068
68. (a) Weaver MJ (1992) Chem. Rev. 92: 463; (b) Weaver MJ, McManis GE, Jarzeba W, Barbara PF (1990) J. Phys. Chem. 94: 1715; (c) Barbara PF, Walker GC, Smith TP (1992) Science 256: 975; (d) Zaleski JM, Chang CK, Leroi GE, Cukier RI, Nocera DG (1992) J. Am. Chem. Soc. 114: 3564; (e) Weaver MJ, McManis GE III (1990) Acc. Chem. Res. 23: 294; (f) Maroncelli M, MacInnis J, Fleming GR (1989) Science 234: 1674
69. (a) Andrieux CP, Blocman C, Dumas-Bouchiat JM, M'Halla F, Savéant J-M (1980) J. Am. Chem. Soc. 102: 3806; (b) Andrieux CP, Gallardo I, Savéant J-M, Su KB (1986) J. Am. Chem. Soc. 108: 638; (c) Médebielle M, Pinson J, Savéant J-M (1991) J. Am. Chem. Soc. 113: 6872; (d) Andrieux CP, Le Gorande A, Savéant J-M (1992) J. Am. Chem. Soc. 114: 6892; (e) Savéant J-M (1992) J. Am. Chem. Soc. 114: 10595
70. (a) Cristol SJ, Bindel TH (1981) J. Am. Chem. Soc. 103: 7237; (b) Cristol SJ, Dickenson WA, Stanko MK (1983) J. Am. Chem. Soc. 105: 1218; (c) Cristol SJ, Ali MZ (1983) Tetrahedron Lett. 24: 5839; (d) Cristol SJ, Seapy DG, Aeling EO (1983) J. Am. Chem. Soc. 105: 7337; (e) Cristol SF, Bindel TH, Hoffmann D, Aeling EO (1984) J. Org. Chem. 49: 2368; (f) Cristol SJ, Aeling EO (1984) J. Org. Chem. 50: 2698; (g) Critsol SJ, Opitz RJ, Aeling EO (1985) J. Org. Chem. 50: 4834; (h) Cristol SJ, Aeling EO, Heng R (1987) J. Am. Chem. Soc. 109: 830; (i) Cristol SJ, Aeling EO, Strickler SJ, Ito RD (1987) J. Am. Chem. Soc. 109: 7101
71. Savéant J-M (1987) J. Am. Chem. Soc. 109: 6788
72. See for example: (a) Chanon M, Rajzmann M, Chanon F (1990) Tetrahedron 46: 6193; (b) Symons MCR (1981) Pure and Appl. Chem. 53: 223; (c) Shida T, Haselbach E, Bally T (1984) Acc. Chem. Res. 17: 180
73. (a) Maslak P, Guthrie RD (1986) J. Am. Chem. Soc. 108: 2628; (b) Maslak P, Guthrie RD (1986) J. Am. Chem. Soc. 108: 2637
74. Maslak P, Narvaez JN (1990) Angew. Chem. Int. Ed. Engl. 29: 283
75. (a) Wayner DDM, McPhee DJ, Griller D (1988) J. Am. Chem. Soc. 110: 132; (b) Parker VD. Handoo K, Roness F, Tilset M (1991) J. Am. Chem. Soc. 113: 7493; (c) Arnett EM, Venimadhavan S (1991) J. Am. Chem. Soc. 113: 6967; (d) Venimadhavan S, Amarnath K, Harvey NG, Cheng J-P, Arnett EM (1992) J. Am. Chem. Soc. 114: 221; (e) Okamoto A, Snow MS, Arnold DR (1986) Tetrahedron 42: 6175; (f) Parker VD (1992) Acta Chem. Scand. 46: 307
76. See for example: (a) Nelsen SF (1981) Acc. Chem. Res. 14: 131; (b) Nelsen SF, Kinlen PJ, Evans DH (1981) J. Am. Chem. Soc. 103: 7045; (c) Nelsen SF, Rumack DT, Meot-Ner (Mautner) (1988) J. Am. Chem. Soc. 110: 7945 and references therein.
77. Pienta NJ, Kessler RJ, Peters KS, O'Driscoll ED, Arnett EM, Molter KE (1991) J. Am. Chem. Soc. 113: 3773
78. (a) Maslak P, Kula J, Chateauneuf JE (1991) J. Am. Chem. Soc. 113: 2304; (b) Maslak P, Kula J (1991) Mol. Cryst. Liq. Crys. 194: 293

79. (a) Rüchardt C, Beckhaus H-D (1980) Angew. Chem. Int. Ed. Engl. 19: 429; (b) (a) Rüchardt C, Beckhaus H-D (1985) Angew. Chem. Int. Ed. Engl. 24: 529; (c) Rüchardt C, Beckhaus H-D (1986) Angew. Chem. Int. Ed. Engl. 24: 529; (c) Rüchardt C, Beckhaus H-D (1986) Top. Curr. Chem. 130: 1

80. For example see: (a) Benson SW, Cruickshank FR, Golden DM, Haugen GR, O'Neal HE, Rodgers AS, Shaw R, Walsh R (1969) Chem. Rev. 69: 279; (b) Benson SW, Garland LJ (1991) J. Phys. Chem. 95: 4915

81. See for example: (a) Clark T (1985) A handbook of computational chemistry. Wiley-Inter-science, New York; (b) Huang XL, Dannenberg JJ (1991) J. Org. Chem. 56: 6367; (c) Ades HF, Companion AL, Subbaswamy KR. (1991) J. Phys. Chem. 95: 2226

82. (a) Breslow R, (1974) Pure Appl. Chem. 40: 493; (b) Wasielewski MR, Breslow R (1976) J. Am. Chem. Soc. 98: 4222; (c) Jaun B, Schwarz J, Breslow R (1980) J. Am. Chem. Soc. 102: 5741; (d) Breslow R, Goodin R (1976) J. Am. Chem. Soc. 98: 6076; (e) Breslow R, Balasubramanian K (1969) J. Am. Chem. Soc. 91: 5182; (f) Breslow R, Chu W (1973) J. Am. Chem. Soc. 95: 411

83. (a) Bordwell FG, Bausch MJ (1986) J. Am. Chem. Soc. 108: 1979 and 1985; (b) Bordwell FG, Harelson JA Jr., Satish AV (1989) J. Org. Chem. 54: 3101 and references therein; (c) Bordwell FG, Ji G-Z (1991) J. Am. Chem. Soc. 113: 8398; (d) Bordwell FG, Ji G-Z, Zhang X (1991) J. Org. Chem. 56: 5254; (e) Bordwell FG, Harrelson JA, Zhang X (1991) J. Org. Chem. 56: 4448; (f) Bordwell FG, Zhang X, Cheng J-P (1991) J. Org. Chem. 56: 3216; (g) Bordwell FG, Cheng J-P (1991) J. Am. Chem. Soc. 113: 1736

84. (a) Arnett EM, Molter KE (1985) Acc. Chem. Res. 18: 339; (b) Arnett EM, Molter KE, Marchot EC, Donovan WH, Smith P (1987) J. Am. Chem. Soc. 109: 3788; (c) Arnett EM, Harvey HG, Amarnath K. Cheng J-P (1989) J. Am. Chem. Soc. 111: 4143; (d) Arnett EM, Amarnath K. Harvey NG, Cheng, J-P (1990) J. Am. Chem. Soc. 112: 344; (e) Arnett EM, Amarnath K, Venimadhavan S (1990) J. Org. Chem. 55: 3593; (f) Arnett EM, Amarnath K, Harvey NG, Cheng J-P (1990) Science 247: 423; (g) Arnett EM, Amarnath K, Harvey NG, Venimadhavan S (1990) J. Am. Chem. Soc. 112: 7346

85. (a) Parker VD, Tilset M, Hammerich O (1987) J. Am. Chem. Soc. 109: 7905; (b) Parker VD, Tilset M (1988) J. Am. Chem. Soc. 110: 1649; (c) Tilset M, Parker VD (1989) J. Am. Chem. Soc. 111: 6711; (d) Ryan O, Tilset M, Parker VD (1990) J. Am. Chem. Soc. 112: 2618; (e) Bausch MJ, Guadalupe-Fasano R, Gostowski R, Selmarten D, Vaughn A (1991) J. Org. Chem. 56: 5640; (f) Bausch MJ, Guadalupe-Fasano R, Peterson BM (1991) J. Am. Chem. Soc. 113: 8384; (g) Bausch MJ, Guadalupe-Fasano R, Koohang A (1991) J. Phys. Chem. 95: 3420; (h) Bausch MJ, Gostowski R, Guadalupe-Fasano R, Selmarten D, Vaughn A, Wang L-H (1991) J. Org. Chem. 56: 7191; (i) Bausch MJ, Gostowski R (1991) J. Org. Chem. 56: 6260; (i) Kern JM, Federlin P (1978) Tetrahedron 34: 661; (j) Kern JM, Sauser JD, Federlin P (1982) Tetrahedron 38: 3023; (k) Bank S, Ehrlich CL, Zubieta J (1979) J. Org. Chem. 44: 1454; (l) Bank S, Schepartz A, Giammateo P, Zubieta J (1983) J. Org. Chem. 48: 3458

86. (a) Griller D, Simôes AM, Mulder P, Sim BA, Wayner DDM (1989) J. Am. Chem. Soc. 111: 7872; (b) Sim BA, Griller D, Wayner DDM (1989) J. Am. Chem. Soc. 111: 754; (c) Wayner DDM, Dannenberg JJ, Griller D (1986) Chem. Phys. Lett. 131: 189; (d) Sim BA, Milne PH, Griller D, Wayner DDM (1990) J. Am. Chem. Soc. 112: 6635; (e) Wayner DDM, Sim BA, Dannenberg JJ (1991) J. Org. Chem. 56: 4853

87. Mann CK, Barnes KK (1970) Electrochemical reactions in nonaqueous systems. Marcel Dekker, New York

88. See for example: (a) Martens FM, Verhoeven JW, Gase RA, Pandit UK, de Boer TJ (1978) Tetrahedron 34: 443; (b) Klingler RJ, Kochi JK (1982) J. Am. Chem. Soc. 104: 4186; (c) Klinger RJ, Kochi JK (1981) J. Am. Chem. Soc. 103: 5839

89. For a computational approach to radical ion fragmentation see: (a) Clark T (1984) J. Chem. Soc. Chem. Commun. 93; (b) Villar H, Castro EA, Rossi RA (1982) Can. J. Chem. 60: 2525; (c) Bigot B, Roux D, Salem L (1981) J. Am. Chem. Soc. 103: 5271; (d) Canadell P, Karafiloglou P, Salem L (1980) J. Am. Chem. Soc. 102: 855; (e) Du X, Arnold DR, Boyd R, Shi Z (1991) Can. J. Chem. 69: 1365; (f) Takahashi O, Kikuchi O (1991) Tetrahedron Lett. 37: 4933; (g) Camaioni DM (1990) J. Am. Chem. Soc. 112: 9475; (h) Miller KE, Kozak JJ (1985) J. Phys. Chem. 89: 401

90. Dressler R, Allan M, Haselbach E (1985) Chimia 39: 385

91. The similar stereoelectronic control of C–H fragmentation in radical cations have been observed: (a) Tolbert LM, Khanna RK, Popp AE, Gelbaum L, Bottomley LA (1990) J. Am. Chem. Soc. 112: 2373; (b) Perrott AL, Arnold DR (1991) Can. J. Chem. 70: 272; (c) see also ref 15

92. Andrieux CP, Savéant JM, Zann D (1984) Noveau Chim. 8: 107

93. (a) Koppang M, Woolsey NF, Bartak DE (1984) J. Am. Chem. Soc. 106: 2799; (b) Patel KM,

Baltisberger RJ, Stenberg VI, Woolsey NF (1982) J. Org. Chem. 47: 4250; (c) Dewald RR, Conlon NJ, Song WM (1989) J. Org. Chem. 54: 261; (d) Maercker, A (1987) Angew. Chem. Int. Ed. Engl. 26: 972.

94. Beak P, Sullivan TA, J. Am. Chem. Soc. 1982, 104, 4450.

95. Kimura N, Takamuku S (1991) Bull. Chem. Soc. Jpn. 64: 2433

96. (a) Meot-Ner (Mautner) M, Neta P, Norris RK, Wilson K (1986) J. Phys. Chem. 90: 168; (b) Neta P, Behar D (1980) J. Am. Chem. Soc. 102: 4798; (c) Bays JP, Blumer ST, Baral-Tosh S, Behar D, Neta P (1983) J. Am. Chem. Soc. 105: 320; (d) Neta P, Behar D (1981) J. Am. Chem. Soc. 103: 103; (e) Behar D, Neta P (1981) J. Phys. Chem. 85: 690; (f) Norris RK, Barker SD, Neta P (1984) J. Am. Chem. Soc. 106: 3140

97. Tanner DD, Chen JJ, Chen L, Luelo VC (1991) J. Am. Chem. Soc. 113: 8074

98. (a) Wu F, Guarr TF, Guthrie RD (1992) J. Phys. Org. Chem. 5: 7; (b) Walsh TD (1987) J. Am. Chem. Soc. 109: 1511; (c) Guthrie RD, Shi B (1990) J. Am. Chem. Soc. 112: 3136

99. (a) Maslak P, Asel SL (1988) J. Am. Chem. Soc. 110: 8260; (b) Maslak P, Narvaez JN (1989) J. Chem. Soc. Chem. Commun. 138; (c) Maslak P, Kula J, Narvaez JN (1990) J. Org. Chem. 55: 2277; (e) Maslak P, Narvaez JN, Kula J, Malinski D (1990) J. Org. Chem. 55: 4550

100. A free-energy relationship describing the mesolytic fragmentation may have a general form of $k_m = A \cdot \exp(-\Delta G_m^{\ddagger}/RT)$, where the preexponential factor (A) explicitly accounts for the stereoelectronic factors (i.e. coupling between the scissile bond and the orbital bearing the unpaired electron) and $\Delta G_m^{\ddagger}$ is a quadratic (or linear) function of ΔG_m with a low intrinsic barrier (see Fig. 8).

101. Masnovi J (1989) J. Am. Chem. Soc. 111: 9081

102. Chapman, WH, Jr. (1992) Carbon–Carbon bond fragmentation in radical cations as a probe of photoinduced electron transfer. Thesis, The Pennsylvania State University, University Park

103. (a) Kochi JK, Sankararaman S, Perrier S (1989) J. Am. Chem. Soc. 111: 6448; (b) Sankararaman S, Kochi JK (1989) J. Chem. Soc. Chem. Commun. 1800

104. Dinnocenzo JP, Farid S, Goodman JL, Gould IR, Todd WP, Mattes SL (1989) J. Am. Chem. Soc. 111: 8973

105. Maslak P, Narvaez JN, Parvez M (1991) J. Org. Chem. 56: 602

106. (a) Maslak P, Chapman WH, Jr. (1989) J. Chem. Soc. Chem. Commun. 1809; (b) Maslak P, Chapman WH, Jr. (1990) Tetrahedron 46: 2715; (c) Maslak P, Kula J (1990) Tetrahedron Lett. 31: 4969; (d) Maslak P, Chapman, WH, Jr. (1990) J. Org. Chem. 55: 6334

107. Popielarz R, Arnold DR (1990) J. Am. Chem. Soc. 112: 3068

108. (a) Ci X, Whitten DG (1987) J. Am. Chem. Soc. 109: 7215; (b) Ci X, Kellett MA, Whitten DG (1991) J. Am. Chem. Soc. 113: 3893

109. (a) Steenken S, McClelland RA (1989) J. Am. Chem. Soc. 111: 4967; (b) McClelland RA, Mathivanan N, Steenken S (1990) J. Am. Chem. Soc. 112: 4857; (c) Steenken S, McClelland RA, (1990) J. Am. Chem. Soc. 112: 9648; (d) Mathivanan N, Cozens F, McClelland RA, Steenken S (1992) J. Am. Chem. Soc. 114: 2198; (e) McClelland RA, Chan C, Cozens F, Modro A, Steenken S (1991) Angew. Chem. Int. Ed. Engl. 30: 1337; (f) Mella M, Albini A (1992) J. Org. Chem. 57: 3051

110. (a) Bunnett JF (1978) Acc. Chem. Res. 11: 413; (b) Kornblum N (1975) Angew. Chem. Int. Ed. Engl. 14: 734; (c) Rossi RA, de Rossi RH (1983) Aromatic substitution by the $S_{RN}1$ mechanism. ACS Monograph 178, Washington, DC; (d) Savéant J-M (1992) New J. Chem. 16: 131

111. See for example: (a) Degrand C, Prest R (1990) J. Org. Chem. 55: 5242; (b) Savéant J-M (1990) Adv. Phys. Org. Chem. 26: 1; (c) Amatore C, Combellas C, Pinson J, Oturan MA, Robvieille S, Savéant J-M (1985) J. Am. Chem. Soc. 107: 4848; (d) Galli C, Bunnett JF (1981) J. Am. Chem. Soc. 103: 7140

112. (a) Arnold DR, Maroulis AJ (1976) J. Am. Chem. Soc. 98: 5931; (b) Lamont LJ, Arnold DR (1990) Can. J. Chem. 68: 390; (c) Arnold DR, Lamont LJ (1989) Can. J. Chem. 67: 2119; (d) Okamoto A, Arnold DR (1985) Can. J. Chem. 63: 2340

113. (a) Albini A, Mella M (1986) Tetrahedron 42: 6219; (b) Albini A, Fasani E, Mella M (1986) J. Am. Chem. Soc. 108: 4119; (c) Albini A, Sulpizio A (1988) In: Fox MA, Chanon M (eds) Photoinduced electron transfer, part C. Elsevier, Amsterdam, p 88.

114. (a) Baciocchi E, Piermattei A, Rol C, Ruzziconi R, Sebastiani GV (1989) Tetrahedron 45: 7049; (b) Akaba R, Niimura Y, Fukushima T, Kawai Y, Tajima T, Kuragami T, Negishi A, Kamata M, Sakuragi H, Tokumaru K (1992) J. Am. Chem. Soc. 114: 4460; (c) Shi M, Okamoto Y. Takamuku S (1991) J. Chem. Res. (S) 176; (d) Ishiguro K, Osaki T, Sawaki Y (1992) Chem. Lett. 743; (e) Akaba R, Kamata M, Sakuragi H, Tokumaru K (1992) Tetrahedron Lett. 33: 8105

115. (a) Reichel LW, Griffin GW, Muller AJ, Das PK, Ege SN (1984) Can. J. Chem. 62: 424; (b) Davis HF, Das PK, Reichel LW, Griffin GW (1984) J. Am. Chem. Soc. 106: 6968
116. (a) Gan H, Zhao X, Whitten DG (1991) J. Am. Chem. Soc. 113: 9409; (b) Ci X, Whitten DG (1991) J. Phys. Chem. 95: 1988; (c) Kellet MA, Whitten DG (1989) J. Am. Chem. Soc. 111: 2314; (d) Lee LYC, Xiahong C, Giannotti C, Whitten DG (1986) J. Am. Chem. Soc. 108: 175; (e) Ci X, Whitten DG (1989) J. Am. Chem. Soc. 111: 3459; (f) Bergmark WR, Whitten DG (1990) J. Am. Chem. Soc. 112: 4042; (g) Gan H, Zhao X, Whitten DG (1991) J. Am. Chem. 113: 9409; (h) Bergmark WR, Whitten DG (1991) Mol. Cryst. Liq. Cryst. 194: 239; (i) Kellett MA, Whitten DG (1991) Mol. Cryst. Liq. Cryst. 194: 275; (j) Bergmark WR, DeWan C, Whitten DG (1992) J. Am. Chem. Soc. 114: 8810; (k) Haugen CM, Bergmark WR, Whitten DG (1992) J. Am. Chem. Soc. 114: 10293
117. (a) Sankararaman S, Haney WA, Kochi JK (1987) J. Am. Chem. Soc. 109: 7824; (b) Masnovi JM, Kochi JK (1985) J. Org. Chem. 50: 5245; (c) Sankararaman S, Haney WA, Kochi JK (1987) J. Am. Chem. Soc. 109: 5235; (d) Eberson L, Radner F (1991) J. Am. Chem. Soc. 113: 5825; (f) Eberson L, Hartshorn MP, Radner F, Robinson WT (1992) J. Chem. Soc. Chem. Commun. 566; (g) Sankararaman S, Kochi JK (1991) J. Chem. Soc. Perkin Trans. 2: 1; (h) Sankararaman S, Kochi JK (1991) J. Chem. Soc. Perkin Trans. 2: 166; (i) Masnovi, JM, Kochi JK, Hilinski EF, Rentzepis PM (1986) J. Am. Chem. Soc. 108: 1126; (j) Masnovi JM, Sankararaman S, Kochi JK (1989) J. Am. Chem. Soc. 111: 2263; (k) Eberson L, Hartshorn MP (1992) J. Chem. Soc. Chem. Commun. 1563; (l) Eberson L, Hartshorn MP, Radner F (1992) J. Chem. Soc. Perkin Trans. 2: 1793; (m) Eberson L, Hartshorn MP, Radner F (1992) J. Chem. Soc. Perkin Trans. 2: 1799
118. (a) Hacker NP, Dektar JL, Leff DV, MacDonald SA, Welsh KM (1991) J. Photopolym. Sci. and Technol. 4: 445; (b) Dektar JL, Hacker NP (1987) J. Org. Chem. 53: 1833; (c) Hacker NP, Dektar JL (1987) J. Chem. Soc. Chem. Commun. 1590; (d) Dektar JL, Hacker NP (1989) J. Photochem. Photobiol. A: Chem. 46: 233; (e) Hacker NP, Dektar JL (1990) In: Hoyle CE, Kinstle JF (ed) Radiation curing of polymeric materials. ACS symposium series, vol 417. American Chemical Society, Washington, p 82; (f) Dektar JL, Hacker NP (1990) J. Org. Chem. 55: 639; (g) Dektar JL, Hacker NP (1991) J. Am. Chem. Soc. 112: 6004; (h) Hacker NP. Leff DV, Dektar JL (1990) Mol. Cryst. Liq. Cryst. 183: 505; (i) Dektar·JL, Hacker NP (1991) J. Org. Chem. 56: 1838; (j) Hacker NP, Leff DV, Dektar JL (1991) J. Org. Chem. 56: 2280; (k) Hacker NP, Welsh KM (1991) Macromolecules 24: 2137; (l) He X, Huang W-Y, Reiser A (1992) J. Org. Chem. 57: 759; (m) Buijsen PF, Hacker NP (1993) Tetrahedron Lett. 34: 1557
119. (a) Saeva FD, Breslin DT, Luss HR (1991) J. Am. Chem. Soc. 113: 5333; (b) Saeva FD, Breslin DT, Martic PA (1989) J. Am. Chem. Soc. 111: 1328; (c) Saeva FD, Breslin DT (1989) J. Org. Chem. 54: 712; (d) Saeva FD, Morgan BP, Luss HR (1985) J. Org. Chem. 50: 4360; (e) Saeva FD, Morgan BP (1984) J. Am. Chem. Soc. 106: 4121; (f) Breslin DT, Saeva FD (1988) J. Org. Chem. 53: 713; (g) Saeva FD (1987) J. Chem. Soc. Chem. Commun. 37; (h) Saeva FD (1986) Tetrahedron 42: 6123
120. (a) Schwalm R, Bug R, Dai G-S, Fritz PM, Reinhardt M, Schneider S, Schnabel W (1991) J. Chem. Soc. Perkin Trans. 2. 1803; (b) Iu KK, Kuczynski J, Fuerniss SJ, Thomas JK (1992) J. Am. Chem. Soc. 114: 4871; (c) Bi Y, Neckers DC (1992) Tetrahedron Lett. 33: 1139
121. (a) Chatterjee S, Gottschalk P, Davis PD, Schuster GB (1988) J. Am. Chem. Soc. 110: 2326; (b) Zou C, Miers JB, Ballew RM, Dlott DD, Schuster GB (1991) J. Am. Chem. Soc. 113: 7823; (c) Schuster GB, Yang X, Zou C, Sauerwein B (1992) J. Photochem. Photobiol. A: Chem. 65: 191
122. Vauthey E, Phillips D, Parker AW (1992) J. Phys. Chem. 96: 7356
123. Weller A (1982) Pure Appl. Chem. 54: 1885
124. Todd WP, Dinnocenzo JP, Farid S, Goodman JL, Gould IR (1991) J. Am. Chem. Soc. 113: 3601
125. Gassman PG, De Silva A (1991) J. Am. Chem. Soc. 113: 9870
126. Goodson B, Schuster GB (1986) Tetrahedron Lett. 27: 3123
127. (a) Julliard M, Chanon M (1983) Chem. Rev. 83: 425; (b) Chanon M, Tobe ML (1982) Angew. Chem. Int. Ed. Engl. 21: 1
128. Richard JP, Amyes TL, Vontor T (1991) J. Am. Chem. Soc. 113: 5871
129. (a) Dinnocenzo JP, Banach TE (1989) J. Am. Chem. Soc. 111: 8646; (b) Parker VD, Tilset M (1991) J. Am. Chem. Soc. 113: 8787
130. Abraham W, Glänzel A, Csongar C, Löwis M, Schnabel W, Zhu QQ (1991) J. Photochem. Photobiol. A: Chem. 60: 83

The Photochemistry of Polyhalocompounds, Dehalogenation by Photoinduced Electron Transfer, New Methods of Toxic Waste Disposal

Peter K. Freeman and Susan A. Hatlevig

Department of Chemistry, Oregon State University, Corvallis, OR 97331-4003, USA

Table of Contents

Recent work on the mechanisms of the photohydrodehalogenation of haloarenes, with an emphasis on polyhaloarenes, and the related mechanisms of phototransformations of aliphatic halocompounds are reviewed. Attention is focused on the nature of the excimer in the photochemical transformations of haloarenes without additional electron transfer reagent and on the nature of the exciplex formed in phototransformations in the presence of electron transfer reagent. Applications of surface catalyzed photochemical transformations and photohydrodehalogenation of polyhaloarenes to toxic waste disposal is discussed.

Topics in Current Chemistry, Vol. 168
© Springer-Verlag Berlin Heidelberg 1993

1 Introduction

The photodehalogenation of haloarenes has attracted the attention of scientists interested in reaction mechanisms, synthesis, and environmental processes [1]. Considerable effort has been devoted to developing our understanding of the mechanism of photodehalogenation of monohaloarenes. Since in the environment polyhaloarenes are of more importance than their monohalo parents, interest has been stimulated in applying current theory to achieving an understanding of the mechanistic features of the phototransformation of polyhaloarenes.

In the case of monochloroarenes, the initial view of the mechanistic options was that the excited singlet usually possesses sufficient energy to undergo C–Cl bond homolysis, but the triplet state formed by intersystem crossing is generally too low in energy to react in this manner without help in the form of additional thermal energy, exciplex formation, or strain relief. The C–Cl bond homolysis of the singlet generates aryl radical which leads to products (Eq. 1). Alternatively either the singlet or triplet can react with an electron rich donor molecule to produce an exciplex or radical anion/radical cation ion pair (Eq. 2). The radical anion is viewed as undergoing fission to aryl radical and chloride ion (Eq. 3).

$$ArCl \longrightarrow ArCl^1 \longrightarrow Ar^{\cdot} \longrightarrow products \tag{1}$$

$$ArCl^1 \longrightarrow ArCl^3 \xrightarrow{D} ArCl^{\cdot -} D^{\cdot +} \tag{2}$$

$$ArCl^{\cdot -} \longrightarrow Ar^{\cdot} + Cl^- \tag{3}$$

2 The Photochemistry of Polyhaloarenes

The photochemistry of pentachlorobenzene serves as a useful model for polyhaloarene photohydrodehalogenation. Direct irradiation of **1** in acetonitrile at 254 nm gives 1,2,3,5-tetrachloro- (**2**), 1,2,4,5-tetrachloro- (**3**) and 1,2,3,4-tetrachlorobenzene (**4**) as the only products at low conversions of **1**. In addition, the quantum yield of disappearance of **1** was found to increase with increasing concentrations of **1** accompanied by a steady change in the relative ratios of the regioisomers [2]. The kinetic scheme which has evolved is illustrated in Scheme 1.

In order to underline the role of the triplet state, a butyrophenone ($E_T \sim 74$ kcal/mol) [3] sensitized irradiation of **1** at 350 nm was carried out and dechlorination and formation of the tetrachlorobenzenes was observed. Since butyrophenone absorbs all the incident light at this wavelength range, the

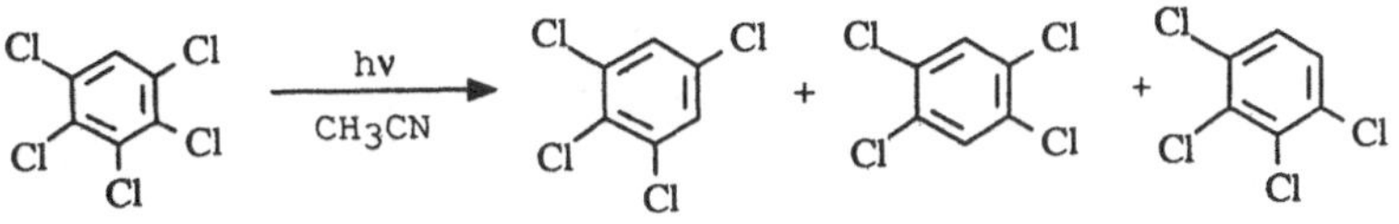

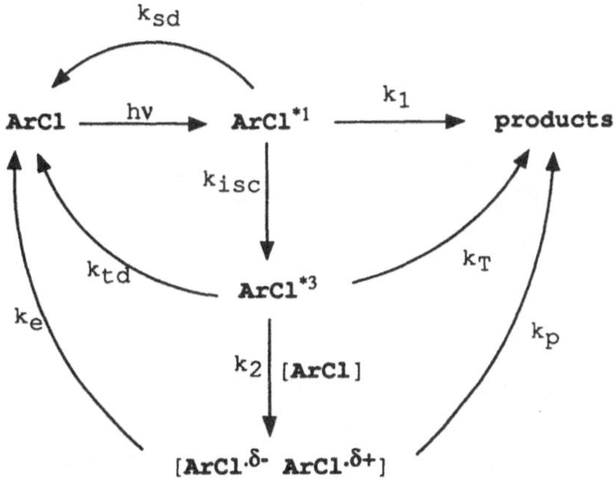

Scheme 1

reaction clearly occurs through the triplet state of **1**. In the absence of reaction from the singlet state ($k_1 = 0$), the steady-state assumption for the kinetic scheme represented in Scheme 1 gives equation 4, where $F = k_p/(k_p + k_e)$, which was originally viewed as providing a satisfactory model [1]. Actual measurement of the quantum yield of intersystem crossing of **1** in acetonitrile using the *cis-trans* isomerization of *cis*-piperylene [4], however, gave a value of 0.8 indicating that singlet reactivity cannot be ruled out entirely. Further, the reaction was quenched very efficiently by the triplet quencher fumaronitrile [5]. Figure 1 shows the Stern-Volmer quenching plot for the reaction at 300 nm. The function is linear with an intercept of 1 at low quencher concentrations, but as the concentration is increased is concave downward. Using steady-state methods, it can be demonstrated that this quenching pattern signals the presence of both singlet and triplet reactions [5]. The plateau in Fig. 1 thus represents the singlet reaction at 300 nm after the triplet reaction has been completely quenched by fumaronitrile. It thus appears that the observed quantum yield is the sum of the quantum yields of singlet and triplet processes [2].

$$\phi_{total} = \phi_{isc} \left(\frac{k_T + k_2[1]F}{k_T + k_{td} + k_2[1]} \right) \tag{4}$$

The results from quenching by fumaronitrile clearly suggest the presence of a reactive singlet state. Reaction from the singlet state could occur through direct fission of the C–Cl bond which is feasible in terms of the energy of the singlet state

49

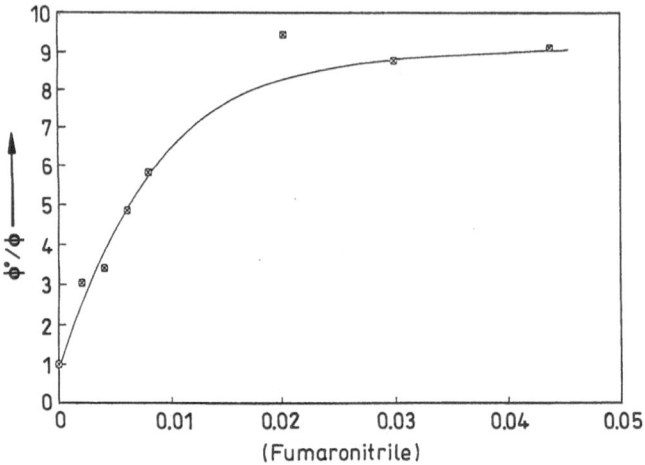

Fig. 1. Plot of inverse of relative quantum yield of photodechlorination, $\phi°/\phi$ versus concentration of fumaronitrile

($\sim$ 95 kcal/mol). In order to determine the possibility of reaction via a singlet excimer species similar to that depicted for the triplet state, the fluorescence lifetime of **1** was measured as a function of its concentration in ethanol. The lifetime ($\tau_s = 7.5$ ns) remained constant over a concentration range of 0.005–0.04 M [6], and the singlet quantum yield does not increase with increasing substrate concentration in acetonitrile, thus ruling out the intervention of a singlet excimer intermediate. The reaction from the singlet state thus occurs through the direct fission of the C–Cl bond. The contribution to the total quantum yield due to singlet reactivity (ϕ_{singlet}) may be written as in Eq. 5, while the overall quantum yield ($\phi_{\text{singlet}} + \phi_{\text{triplet processes}}$) is provided in Eq. 6.

$$\phi_{\text{Singlet}} = \frac{k_1}{k_1 + k_{\text{isc}} + k_{\text{sd}}} \tag{5}$$

$$\phi_{\text{Total}} = \frac{k_1}{k_1 + k_{\text{isc}} + k_{\text{sd}}} + \phi_{\text{isc}}\left(\frac{k_T + k_2[\mathbf{1}]F}{k_T + k_{\text{td}} + k_2[\mathbf{1}]}\right) \tag{6}$$

A plot of the total quantum yield versus the concentration of pentachlorobenzene provides a very nice linear plot (r = 0.995) (Fig. 2), suggesting that $(k_T + k_{\text{td}}) > k_2[\mathbf{1}]$. Thus, by extrapolation of [**1**] to zero, one can calculate the quantum yield independent of triplet excimer ($\phi_{\text{singlet}} + \phi_{\text{triplet}}$). Subtraction of the quantum yields for direct fission from singlet and triplet states ($\phi_{\text{singlet}} + \phi_{\text{triplet}}$) from ϕ_{Total} provides the expression for the dependence of the remainder (ϕ_{ex}) upon concentration (Eq. 7). A plot of the reciprocal ($1/\phi_{\text{ex}}$) versus the reciprocal of the concentration of the substrate is linear (r = 0.950), which is consistent with the mechanism and is illustrated in Fig. 3 [2].

$$\phi_{\text{ex}} = \phi_{\text{isc}}\frac{k_2[\mathbf{1}]F}{k_T + k_{\text{td}} + k_2[\mathbf{1}]} \tag{7}$$

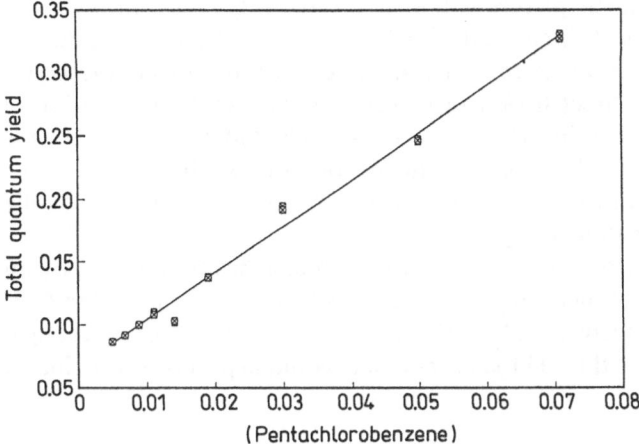

Fig. 2. Plot of the quantum yield of photodechlorination versus the concentration of pentachloro-benzene

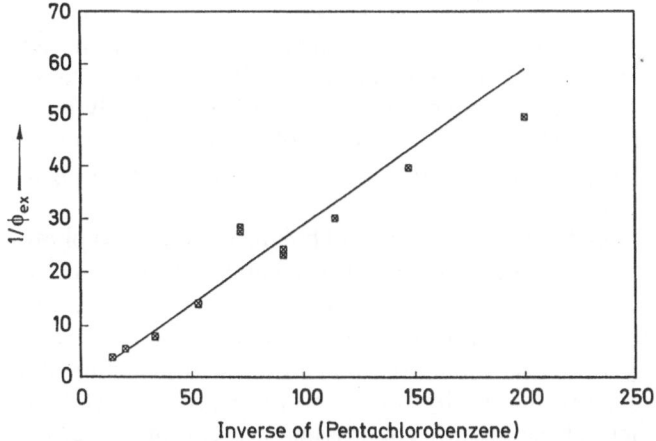

Fig. 3. Plot of the inverse of the triplet excimer quantum yield, $1/\phi_{ex}$ versus the inverse of pentachlorobenzene concentration

In reconsidering the energy requirements for conversion of triplet $(ArCl*^3)$ or excimer $(ArCl^{\delta-}ArCl^{\delta+})$ to product, a process similar to delayed luminescence [7] involving generation of higher energy singlet species via triplet-triplet annihilation processes is conceivable. Such processes, however, can be ruled out as in each case the rate of product formation would be a nonlinear function of the intensity, whereas a plot of relative product concentration versus intensity is linear (r = 0.995), which is consistent with the expression for ϕ_{Total} given in Eq. 6. Excitation of $(ArCl*^3)$ or excimer $(ArCl^{\delta-}ArCl^{\delta+})$ with a second photon can be ruled out on a similar basis.

The simplest mechanistic picture is that of a three way competition which involves direct homolysis of singlet and triplet states and C–Cl fragmentation of the radical anion-like moiety of the excimer. At 0.0050 M **1**, product is formed via direct fission of singlet, direct fission of triplet and through fragmentation of triplet excimer in the ratio 6.0 : 70.8 : 23.2, while at 0.0710 M **1**, the ratio for the same three pathways is 1.6 : 18.7 : 79.7. The major routes to product are via direct fission of triplet state and reaction through the intermediary of excimer with only a minor contribution of singlet [2].

The scope of the photohydrodehalogenation process has been tested by an investigation of the photochemistry of pentafluorobenzene (**5**). Since the C–F bond energy in fluorobenzene is 124 kcal/mol, while λ_{max} is 257 nm, suggesting a singlet energy level of less than 111 kcal/mol, one would expect direct fission to

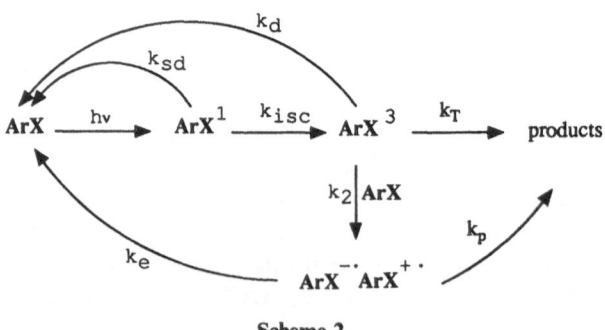

5   6

make a minimal contribution. An experiment evaluating the quantum yield dependence for hydrodefluorination at 254 nm in acetonitrile upon pentafluorobenzene concentration revealed that a plot of $1/\Phi$ vs. $1/[\text{ArF}]$ is nicely linear (correlation coefficient r = 0.999) with no evidence of a concave downward trend. This clearly supports product formation via an excimer with no competition from direct fission [8].

A basis for understanding the photochemistry of polybromobiphenyl photochemistry is provided by a consideration of the mechanistic features of the phototransformation of 4-bromobiphenyl (**6 = BpBr**) at 300 nm in various solvents [9]. 4-Bromobiphenyl is expected to undergo efficient intersystem crossing due to the heavy atom effect, and involvement of the triplet state of 4-bromobiphenyl has been shown earlier by various workers [10, 11, 12]. Since the reported quantum yield of the intersystem crossing process for biphenyl is 0.81 [13], **BpBr** was expected to have a quantum yield for intersystem crossing

Scheme 2

close to 1. Actual measurement of the quantum yield of intersystem crossing did indeed give a value of 0.98 ± 0.05. The triplet energy of 4-bromobiphenyl (66.9 kcal/mol) [14] is 12 kcal/mol lower than the phenyl C-Br bond fission energy (79.2 kcal/mol) [15]. Thus, direct formation of product from the triplet state may be inhibited. However, excimer formation through an electron transfer process from the ground state to the triplet state of **BpBr** may provide an alternative pathway to the product. A mechanistic picture reflecting these features is presented in Scheme 2 ($ArX = BpBr$, $k_T = 0$). Application of the steady state assumption provides us with Eq. 8.

$$\frac{1}{\phi_{prod}} = \frac{1}{F \cdot G} + \frac{k_d}{k_2 FG[\mathbf{BpBr}]} \tag{8}$$

where $\quad F = \dfrac{k_p}{k_p + k_e} \qquad G = \dfrac{k_{isc}}{k_{isc} + k_{sd}}$

In accord with this, the quantum yield of the reaction in any particular solvent increases with increasing concentration of **BpBr**. Also irradiation of a very dilute solution of **BpBr** (1×10^{-4} M) gave no biphenyl photoproduct. Moreover, a plot of the inverse of quantum yield versus the inverse of concentration of **BpBr** in acetonitrile was found to be linear (correlation coefficient = 0.994), consistent then with Scheme 2. The involvement of only one excited state for debromination was supported by quenching studies using cis-1,3-pentadiene [4] and fumaronitrile [16] in acetonitrile solvent. Stern-Volmer plots obtained from both quenching studies were linear with positive slopes having r = 0.992 and 0.984, respectively. This result is in harmony with the view that the triplet excimer is the sole product determining intermediate for the debromination of 4-bromobiphenyl. If reaction were to occur from the singlet state as well as via triplet excimer, the Stern-Volmer plot would be linear, with a positive slope, at low concentrations of the quencher but would flatten out as the concentration of the quencher increases.

3 Nature of the Excimer in the Phototransformations of Polyhaloarenes

Since the excimer is a key intermediate in the phototransformations of poly-haloarenes, it is of considerable importance to provide additional character-ization of bromoarene excimers. Evidence bearing on the nature of the bromobi-phenyl excimer was obtained by determining the dependence of the rate constant for the formation of excimer (k_2) upon solvent polarity. Solvents of differing polarity based on the E_T polarity scale [17] were chosen and the quantum yields were determined as a function of concentration of **BpBr** for each solvent, plots of the inverse of the quantum yield versus the inverse of the concentration of **BpBr** exhibiting a linear relationship for each solvent system. Using the least squares method, the best fitted first order equation correlating the inverse of quantum yield and the inverse of concentration for each solvent system was obtained. The above equation was then used to obtain a standard quantum yield (Φ_0) at an arbitrary reference concentration ([**BpBr**]$_0$ is equal to 0.01 M) for each system (Eq. 9). By taking the ratio of the standard quantum yield (Φ_0) to the observed quantum yield (Φ_{prod}), the slope of a plot of Φ_0/Φ_{prod} against $1/$[**BpBr**] gives k_2 as a function of k_d (Eq. 10 and 11).

$$\Phi_0 = \frac{k_2 FG\,[\mathbf{BpBr}]_0}{k_d + k_2\,[\mathbf{BpBr}]_0} \tag{9}$$

$$\frac{\Phi_0}{\Phi_{prod}} = \frac{[\mathbf{BpBr}]_0}{k_d + k_2\,[\mathbf{BpBr}]_0}\left[k_2 + \frac{k_d}{[\mathbf{BpBr}]}\right] \tag{10}$$

$$k_2 = k_d\left[\frac{1}{\text{Slope}} - \frac{1}{[\mathbf{BpBr}]_0}\right] \tag{11}$$

The value of k_d was obtained from the determination of triplet lifetimes by measuring the decay of phosphorescence and found to be insensitive to changes in solvent polarity. The k_2 values derived from Eqs. 10 and 11 were correlated with solvent parameters using the linear solvation energy relationship described by Abraham, Kamlet and Taft and co-workers [18] (Eq. 12), which relates rate constants (k) to four different solvation parameters: (1) δ_H^2 or the square of the Hildebrand solubility parameter (solvent cohesive energy density), (2) π_1^* or solvent dipolarity or polarizability, (3) α_1 or solvent hydrogen bond donor acidity (solvent electrophilic assistance), and (4) β_1 or solvent hydrogen bond acceptor basicity (solvent nucleophilic assistance).

$$\log k = (\log k)_0 + h\delta_H^2 + s\pi_1^* + a\alpha_1 + b\beta_1 \tag{12}$$

When all four solvatochromic parameters are employed (Table 1 equation A), only the terms in δ_H^2 and β have a confidence level greater than 95%, which is generally viewed as the minimum confidence level that is required to justify the inclusion of any given term in a multiple parameter linear equation [18]. Dropping the solvent dipolarity term π_1^*, since it exhibits the lowest confidence level in equation A, provides equation B. The confidence level for the participation of hydrogen bond donor term α_1 increases, but is still clearly less than 95%, so solvent electrophilic assistance is not established. Omitting both the π_1^* and α_1 terms provides equation D, which underscores the importance of the solvent cohesive energy density and solvent nucleophilic assistance. A comparison of equation D with E and G reveals that the cavity term is more important than nucleophilic solvent participation since there is a considerably greater decrease in the correlation coefficient when the cavity term is omitted. The positive term in δ_H^2 is consistent with the reorganization of solvent-solute and solvent-solvent interactions in the formation of triplet excimer. As the **BpBr**^{3} and **BpBr** components approach each other to form an excimer, for example A, a layer of

Table 1. Correlation of rate constants for excimer formation. $\log k_2 = (\log k_2)_0 + h\delta_H^2 + s\pi_1^* + a\alpha_1 + b\beta_1$

	$(\log k_2)_0$ ($\pm$)	100h ($\pm$)	s ($\pm$)	a ($\pm$)	b ($\pm$)	r	sd	Confidence level (CL%)			
								$\delta_H^2/100$	π_1^*	α_1	β_1
A.	1.95 (0.09)	0.152 (0.056)	0.097 (0.115)	0.137 (0.102)	0.216 (0.091)	0.9787	0.0417	97.5	57.5	79.0	95.9
B.	2.01 (0.06)	0.160 (0.055)		0.155 (0.098)	0.165 (0.066)	0.9793	0.0411	98.5		85.5	96.8
C.	1.87 (0.07)	0.219 (0.027)	0.128 (0.117)		0.303 (0.066)	0.9769	0.0434	100	70.1		99.9
D.	1.94 (0.03)	0.242 (0.018)			0.248 (0.043)	0.9765	0.0438	100			> 99.9
E.	2.03 (0.05)	0.268 (0.034)				0.9101	0.0843	100			
F.	2.17 (0.44)			0.438 (0.032)		0.9672	0.0516			100	
G.	2.21 (0.10)				0.392 (0.166)	0.5100	0.1749				99.9

55

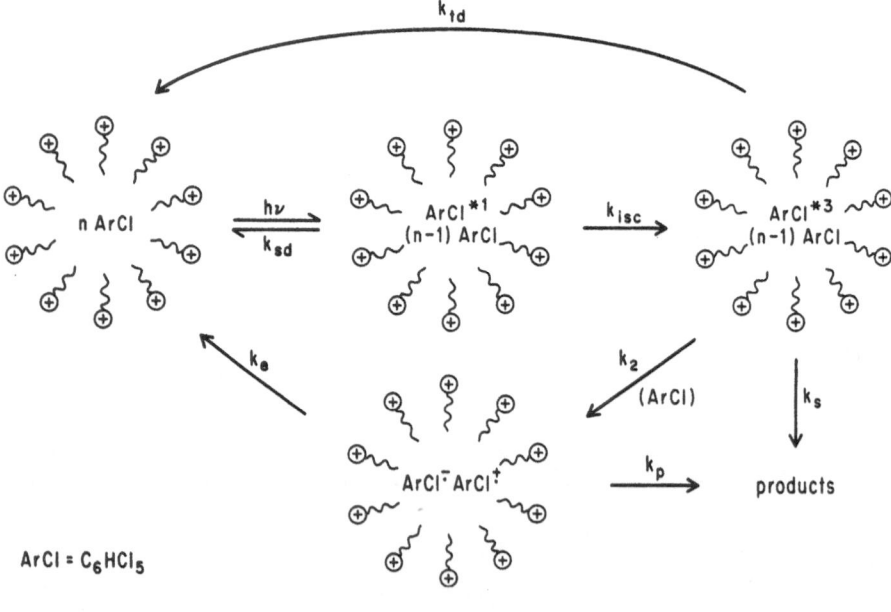

A

solvent molecules on the inside faces of the two benzene rings would be stripped away providing a larger number of solvent-solvent interactions. A small amount of charge transfer might well induce the solvent nucleophilic assistance illustrated in **A**. Overall a comparison with the solvolysis of *tert*-butyl halides reveals that the solvent effects for excimer formation are quite muted.

In a complementary study an interest in the mechanistic pathways which are operative in micellar media led to a characterization of the triplet pentachlorobenzene excimer. Pentachlorobenzene (**1**) was irradiated in 0.100 M CTAB (hexadecyltriethylammoniumbromide) aqueous solution at 254 nm. The major dechlorinated products were the tetrachlorobenzenes (**2–4**) with an accompanying trace of trichlorobenzenes. Bromotetrachlorobenzenes were observed as byproducts [19]. Building on our results in homogeneous solution, the mechanism for photohydrodehalogenation is represented in Scheme 3. Since

Scheme 3

very little direct formation of product by fission of singlet state **1** was observed in homogeneous solution, and since in the present case, in addition to the internal heavy atom effect of five chlorines, there is the external heavy atom effect of the bromide ions in the Stern layer, it was assumed that fission of singlet did not contribute to product formation.

An approach based upon the statistics of pentachlorobenzene distribution in the CTAB solution was employed [20–24]. Using Poisson statistics, the distribution of pentachlorobenzene is given by Eq. 13, where P_n is the probability of finding a micelle containing n solute molecules and S is the average number of solute molecules per micelle. The total probability of encounters between excited and ground state haloarene is given in Eq. 14.

$$P_n = \frac{S^n}{n!} e^{-S} \tag{13}$$

$$P = \sum_{n=1} (n - 1)\, P_n \tag{14}$$

The rate of excimer formation should be proportional to the total probability and thus a plot of the reciprocal of the quantum yield versus 1/P should be linear in the region where excimer formation is dominant. A plot of $1/\phi$ versus 1/P is illustrated in Fig. 4, and, as anticipated, a linear plot at high probability is revealed with the plot becoming concave downward as one moves along the x axis to values of low probability, which fits the mechanistic picture of Scheme 3.

Unexpectedly in the photolysis of pentachlorobenzene in CTAB, bromotetrachlorobenzenes (**7, 8** and **9**) were formed as byproducts. The composition of the bromotetrachlorobenzene fraction at three different concentrations of **1** is listed in Table 2. The byproducts are formed in substantial amount, increasing from 20% to 40% of the total product fraction as the concentration of pentachlorobenzene is increased.

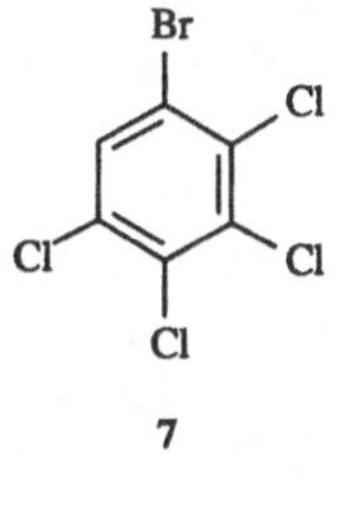

7 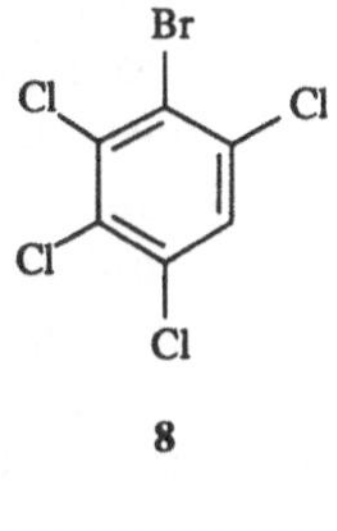   8 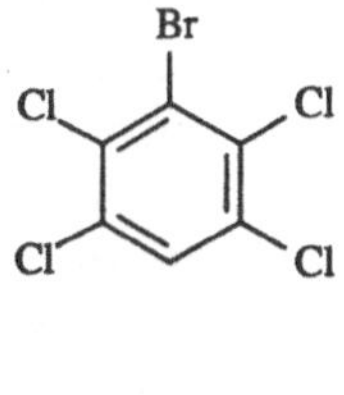   9

The bromide ion apparently has trapped the excimer. One might consider homolysis of the C–Cl bond in the triplet, followed by electron transfer from the aryl radical to chlorine to generate aryl cation which then reacts with bromide ion, using as a model the mechanism put forward to explain the phenol products from the photolysis of 4-chlorobiphenyl in water [25]. However, this mechanism seems unlikely, since the excited chloroarene is in the micellar interior and such a mechanism has not been observed in hydrocarbon solvents. In addition, this

Table 2. Distributions of byproduct bromotetrachlorobenzene

Local concentration of 1 (C_m)	Percentage ratio of products[a]		
	7	8	9
0.0230	10.0 ± 0.4	66.0 ± 2.8	22.8 ± 2.2
0.0288	9.9 ± 0.4	66.0 ± 0.3	24.2 ± 0.8
0.0721	9.1 ± 0.3	68.1 ± 1.7	22.8 ± 1.7

[a] Average of three runs with standard deviation.

Fig. 4. Plot of the inverse of the quantum yield versus the reciprocal of the total probability of encounters between excited and ground state pentachlorobenzene in CTAB solution

suggestion would not be consistent with the fact that the bromotetrachloroben-zene fraction increases with increasing amounts of reactant and thus with an increasing rate of excimer formation. Instead, a mechanism similar to that put forward by Soumillion and Wolf, who reported that photoreduction and photosubstitution occurred simultaneously in the irradiation of chlorobenzene and chloroanisoles, is favored [26]. Subsequent to excimer formation, the bromide ion reacts with the radical cationic moiety of the excimer to produce a pentadienyl radical species which loses chlorine to form product (Scheme 4).

Three additional experiments were run to test this mechanism: (a) a solution of 2.18×10^{-2} M pentachlorobenzene in CH_3CN/H_2O (8 : 2) was irradiated in

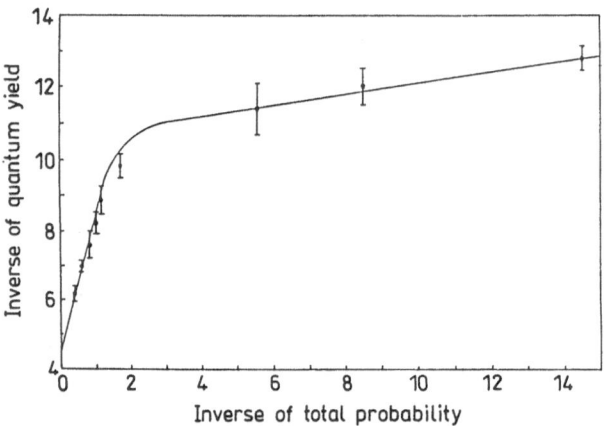

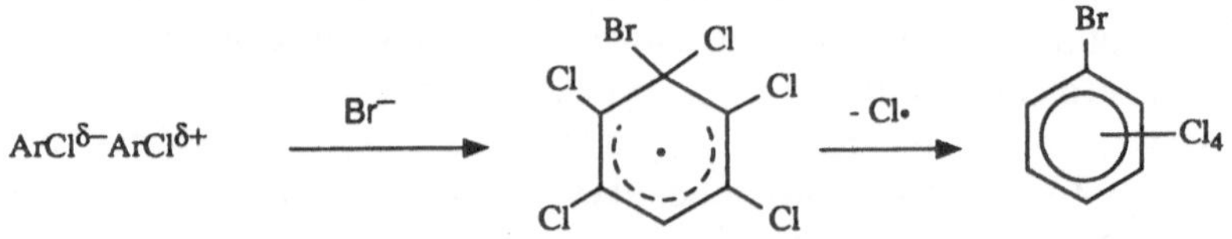

Scheme 4

the presence of excess KBr at 254 nm; a 44.5% yield of bromotetrachloroben-zenes was obtained; (b) in the irradiation of pentachlorobenzene with excess Et_3N and KBr in CH_3CN/H_2O (8:2), only tetrachlorobenzenes were produced, with no evidence of brominated products; (c) no bromotetrachlorobenzenes were formed as a consequence of the irradiation of 1 in CTAB solution in the presence of excess Et_3N. In experiment (a) the concentration of pentachloroben-zene is in the region where excimer formation is dominant [2]. The distribution of the three bromotetrachlorobenzene isomers (7:8:9 = 11.3%:66.8% :21.9%) is remarkably close to that observed in micellar media (Table 2) and supports the notion that these two different reactions result in the trapping of the same intermediate: the triplet excimer. In experiment (b), exciplex $ArCl^{-}Et_3N^{+}$ is dominant over excimer formation and in (c) radical anion $ArCl^{-}$ is formed in the micellar interior, with separation from the radical cationic partner Et_3N^{+} assured by the sphere of positively charged head groups which are interposed. In experiments (b) and (c) photochemical hydrodechlorination is achieved without the generation of excimer. It is only in experiment (a) and in the hydrodechlorination in micellar media without triethylamine that excimer and the accompanying radical cationic moiety are formed, and thus it is only in (a) that trapping with bromide occurs.

Support for this view of the trapping process can be found in the distribution of the bromotetrachlorobenzene isomers. The ability of chlorine to stabilize the pentadienyl radical can be estimated using the partial rate factors reported by Ito et al. for the phenylation of chlorobenzene ($f_o:f_m:f_p$ = 3.09:1.01:1.48) [27]. Using these factors the relative rates for reaction through radical intermediates 10, 11 and 12 in Scheme 5 would be 2.09:3.07:1 which compares well with the observed relative rates for formation of bromides 9, 8, and 7 (4.8:6.9:1). The 2-bromo product is predominant since it is only by attack at C–2 that an intermediate (11) can be generated that allows radical delocalization onto three carbons which contain chlorine. If trapping of excimer were to proceed by an S_NAr process generating intermediates such as 13, the regiochemistry should be in harmony with that predicted using partial rate factors derived from the research of Chambers et al. (*ortho:meta:para* = 12.1:4.85:1.00, calculated for ClC_6H_4F in CH_3O^-/CH_3OH) [1, 28] or Schmidt (*ortho:meta:para* = 4.20:4.29:1.00, from C_6HCl_5 in CH_3O^-, $DMSO/CH_3OH$) [29]. This mechanistic variation can be ruled out, since following either precedent, 9 is the expected major product rather than 8. Finally, an $S_{RN}1$ process can also be ruled out on a regiochemical basis, since the regiochemistry would be established by fragmentation of a free pentachlorobenzene radical anion which would clearly favor formation of 9 over 8 [30]. This is reinforced by the fact that no bromotetrachlorobenzenes are formed when free pentachlorobenzene radical anion is formed in the presence of bromide ion (experiments (b) and (c) above).

4 Photochemical Transformations of Polyhaloarenes in the Presence of Electron Transfer Reagents

In the presence of an electron rich donor molecule an alternative to direct fission or reaction via an excimer is the formation of an exciplex or radical anion/radical cation ion pair (Eq. 2). The radical anion has been viewed as the key intermediate which undergoes fission to aryl radical and halide ion (Eq. 3). With poly-haloarenes there is an additional option. A polychloroarene radical anion, for example, has two possible modes for bond fission: (a) fission to produce aryl radical and chloride ion or (b) fission to form an aryl carbanion and chlorine atom (Scheme 6). The options for fragmentation of a haloarene radical anion

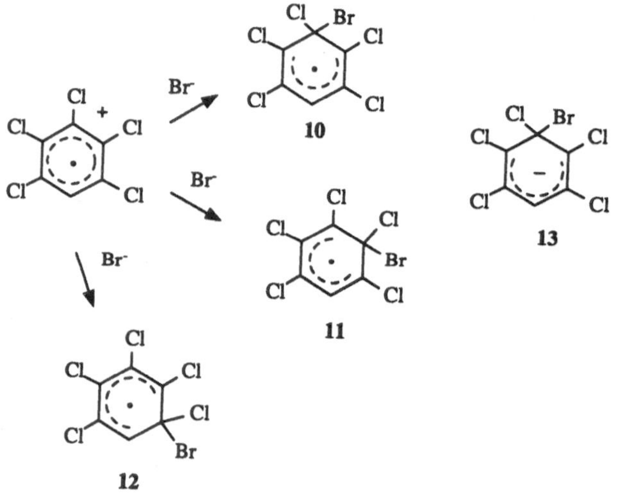

Scheme 5

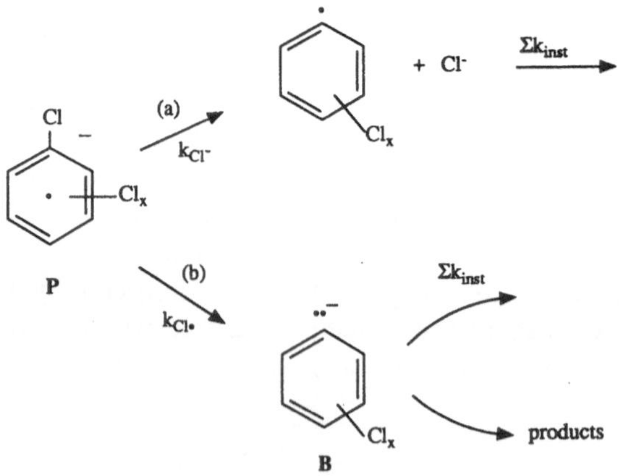

Scheme 6

may be viewed using negative chemical ionization mass spectrometry. Assuming steady-state concentrations of Cl^- and B, for the case where no further decomposition of B occurs, the ratio of the rates of Cl^- and Cl atom loss ($k_{Cl^-}/k_{Cl\cdot}$) and the dependence upon the substitution pattern ($\Sigma\sigma$) may be determined [31]. Using as standard reactions the reactions of chlorobenzene radical anion ($k_{Cl\cdot}$ and k_{Cl^-}), the Hammett relationships for loss of chloride ion (reaction constant ρ) and chlorine atom (reaction constant ρ') may be combined to give the Hammett relationship of Eq. 15, setting $C = \log(k_{Cl\cdot}/k_{Cl^-})$. A precise Hammett plot would require knowledge of the regiochemistry; however, an average value for *ortho* chlorine (0.31) [32], *meta* chlorine (0.37) [33] and *para* chlorine (0.27) [33] of 0.32 may be used. The plot of $\log(k_{Cl\cdot}/k_{Cl^-})$ vs. $\Sigma\sigma$ is simplified and exhibits a slope of 3.92 with an intercept of -5.71 (correlation coefficient = 0.998) [1].

$$\log \frac{k_{Cl\cdot}}{k_{Cl^-}} = (\rho' - \rho)\Sigma\sigma + C \tag{15}$$

We see that chloride ion loss is favored over chlorine atom loss through pentachlorobenzene with chlorine atom loss only becoming predominant at the hexachloro level of substitution. Our view of the fragmentation of these radical anions is enhanced by a consideration of the regiochemistry of photohydrodehalogenation under conditions where generation of a radical anion is the predominant pathway to product. A steady state analysis of the dependence of

Table 3. Photolysis of polychlorobenzenes[a] in acetonitrile in the presence and absence of triethylamine (TEA)[b]

	Yield (mol%)				
Prod	2	2 w TEA[c]	Prod	1	1 w TEA[c]
$1,3,5\text{-}C_6Cl_3H_3$	59.24 ± 1.49	18.77 ± 1.0	$1,2,3,5\text{-}C_6Cl_4H_2$	67.46 ± 0.85	25.32 ± 0.47
$1,2,4\text{-}C_6Cl_3H_3$	40.48 ± 1.51	75.45 ± 1.4	$1,2,4,5\text{-}C_6Cl_4H_2$	25.97 ± 0.83	66.21 ± 0.46
$1,2,3\text{-}C_6Cl_3H_3$	0.28 ± 1.49	5.78 ± 0.6	$1,2,3,4\text{-}C_6Cl_4H_2$	6.57 ± 0.02	8.47 ± 0.06

[a] 0.05 M in each case.
[b] Normalized and average of five runs with standard deviations.
[c] 1.5 M.

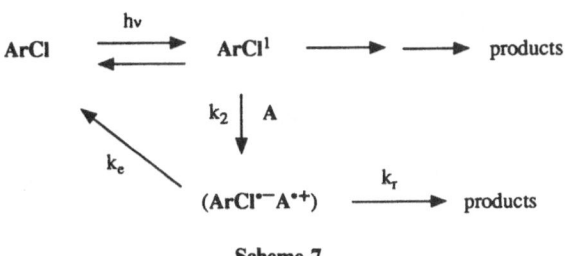

Scheme 7

the reciprocal of the quantum yield for photohydrodechlorination ($1/\Phi$) in acetonitrile at 254 nm on the reciprocal of the triethylamine concentration ($1/A$) is consistent with the mechanistic pathways outlined in Scheme 7 and allows one to determine an amine concentration where the formation of radical anion is predominant (1.5 M). Using this concentration, the regiochemistry is revealed to be sharply different in the presence and absence of electron transfer agent triethylamine (Table 3).

A rationale for the regiochemistries observed for the polychloroarene radical anions may be developed by considering the transition states for the two competing processes (Scheme 8). The loss of chloride ion in route (a) generates phenyl radical. The transition state for this process would, therefore, be expected to exhibit some radical localization at C-1. The shape of the transition state might be expected to be bent rather than planar, since heterolytic fission of the carbon-chlorine bond in a coplanar transition state would lead to an excited state (a phenyl cation with an extra electron in the π^* molecular orbital), while heterolytic fission of a bent system (such as **C**) could lead directly to a phenyl radical. Thus, the transition state for route (a) might very well possess some of the character of a delocalized anion with a bent localized radical center (**C**), while the transition state for chlorine atom loss, by a similar argument, would resemble a delocalized radical with a bent localized carbanionic center (**D**).

For cleavage processes involving the loss of chloride ion, then, one might expect to find similarities to the regiochemistry that is observed in nucleophilic aromatic substitution. Assuming for the purpose of analysis that route (a) is predominant, one can use the rate data of Chambers et al. [34] to calculate the activating effects of chlorine vs. hydrogen in nucleophilic aromatic substitution in benzene systems (*ortho* : *meta* : *para* = 12.1 : 4.85 : 1.00) and then the relative rates for nucleophilic substitution (Chart I, numbers in parentheses). Note that the

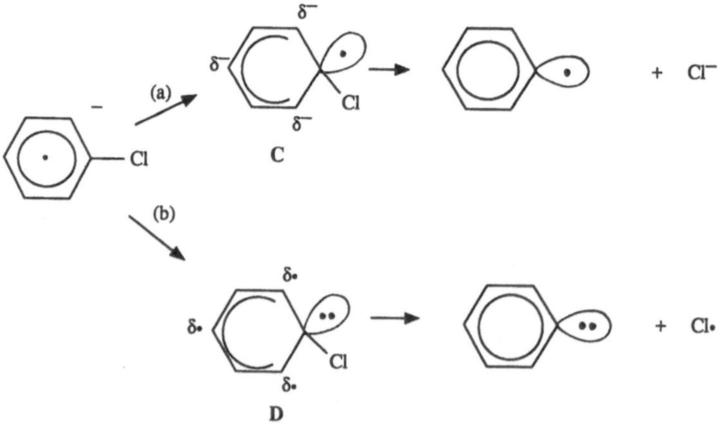

Scheme 8

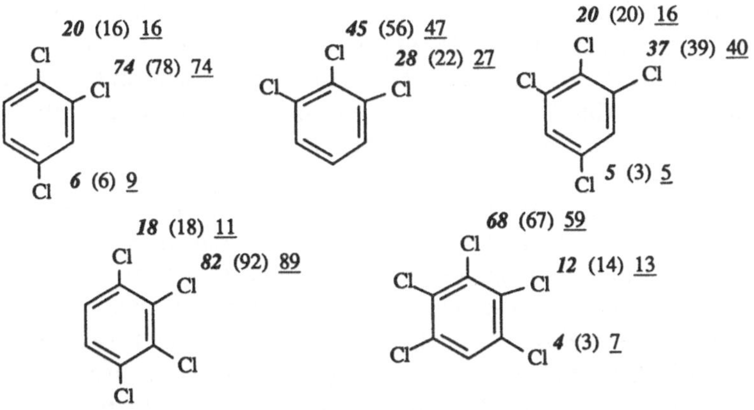

20 (16) <u>16</u>   *45* (56) <u>47</u>   *20* (20) <u>16</u>
74 (78) <u>74</u>   *28* (22) <u>27</u>   *37* (39) <u>40</u>

6 (6) <u>9</u>   *5* (3) <u>5</u>

18 (18) <u>11</u>   *68* (67) <u>59</u>
82 (92) <u>89</u>   *12* (14) <u>13</u>
4 (3) <u>7</u>

Chart I

rates for photochemical fission (bold faced numbers) and the nucleophilic aromatic substitution rates are not only in agreement on order, but overall the ratios themselves are in excellent accord. The striking agreement for pentachlorobenzene suggests that there has not been a change in the mode of fission (a → b in Scheme 8) as a consequence of the change of conditions from the chemical ionization mass spectral runs to liquid-phase conditions in acetonitrile. If there had been a change in mode of fission to route (b), the partial rate factors calculated by Ito et al. [27] for the phenylation of chlorobenzene ($f_o : f_m : f_p = 3.09 : 1.01 : 1.48$) as well as the results of phenyldefluorination of pentafluorobenzene by Allen et al. [35] suggest that the dechlorination pattern observed for pentachlorobenzene would have revealed predominant loss at C-2 (C-2 > C-3 > C-1). In addition, substituent rate factors for *ortho* : *meta* : *para* have been calculated minimizing the sum of the squared residuals for observed and calculated relative rates weighted equally for each position. The substituent rate factors obtained, *ortho* : *meta* : *para* = 8.10 : 4.5 : 1.0 provide the relative rate pattern illustrated in Chart I with underlined numbers and a very satisfactory fit.

An investigation of the dependence of the quantum yield of photohydrodefluorination upon the concentration of electron donor triethylamine is consistent with a competition between excimer formation and exciplex formation resulting from electron transfer from amine (Scheme 9). Using the optimum concentrations to favor exciplex formation, the regiochemical patterns of monodefluorination were determined for pentafluorobenzene (**14**), 1,2,3,5-tetrafluorobenzene (**15**), and 1,2,3,4-tetrafluorobenzene (**16**) at 254 nm in acetonitrile and pentane. Correcting for statistical advantages, the relative rates in acetonitrile are listed directly after the bold faced numbers, while the relative rates in pentane are in parentheses in Chart II [8]. If the fragmentation proceeds according to route (a) in Scheme 8 (Cl = F), the regiochemistry should be similar to that anticipated for nucleophilic aromatic substitution. The I_π repulsion theory, which has been used successfully, with one exception, to predict regiochemistry in S_NAr substitution in polyfluoroarenes, may be employed [36]. This approach which is based upon the assumption that the repulsion between the electron pair on fluorine and the filled pentadienyl orbital of the Wheland intermediate determines stability, leads to the predicted order of rates listed with the bold faced numbers in Chart II (1 = fastest, 3 = slowest).

While the relative rates in acetonitrile do not fit expectations, those in pentane do. Consider the two competing routes of Scheme 8 (Cl = F). In (a) a radical anion is undergoing fission to a neutral aryl radical and fluoride ion, so that in the transition state the charge should be dispersed, while in the case of pathway (b) a delocalized radical anion is undergoing fission to an aryl carbanion, which should represent a charge concentration process. A switch from a polar to a nonpolar solvent favors route (a) over route (b) and shifts the substitution pattern to that predicted by I_π theory. The substituent rate factors obtained for **14, 15** and **16** by minimizing the sum of the square of the residuals for observed and calculated relative rates, weighting each position equally are *ortho*:*meta*:*para* = 5.6:3.6:1.0 (underlined values in Chart II). The fit is satisfactory and the *o*:*m*:*p* substituent rate factor ratios are similar to those observed for photodechlorination. The orientation observed for the photodefluorination process, however, contrasts strongly with the regiochemistry expected for route (b) proceeding through pentadienyl radical transition state **D**, using as a model system radical phenyldefluorination and the partial rate factors

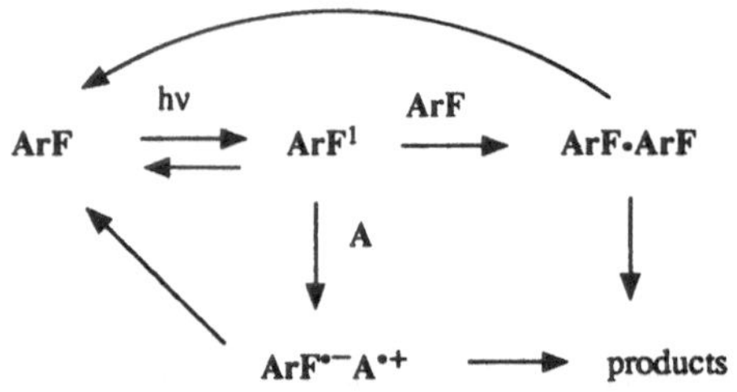

Scheme 9

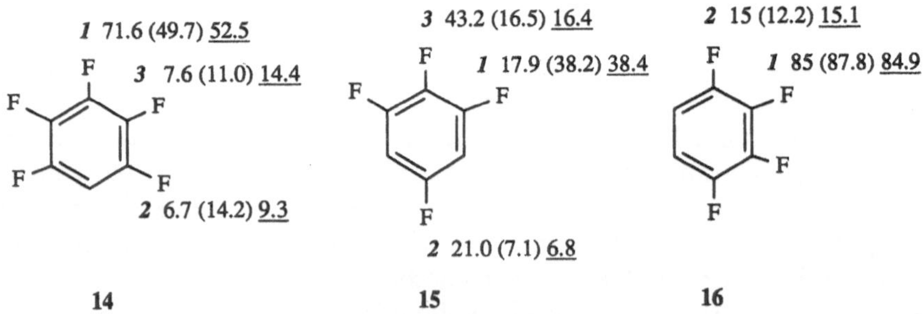

Chart II

$(o:m:p = 3.0:0.8:1.2)$ [35]. Thus, the predominant rate of attack for phenylde-fluorination is at C-2 in pentafluorobenzene and also at C-2 in 1,2,3,5-tetrafluorobenzene.

The advantage in viewing the transition state as pictured in Scheme 8 is that the close relationship to nucleophilic aromatic substitution is revealed. This description is very similar to that which invokes initial formation of a π^* radical anion, subsequent conversion to a σ^* species (enhanced by the C-halogen bond stretching process) and fragmentation [37]. A bending as well as stretching of the C-halogen bond would be expected to enhance the π^* to σ^* conversion and the subsequent fragmentation [37b,d].

The regiochemistry of the photolysis of three tetrachloronaphthalenes, 1,2,3,4-tetrachloro- (17), 1,4,6,7-tetrachloro- (18) and 1,3,5,8-tetrachloronaphthalene (19) is sharply dependent upon the presence of triethylamine. The photochemical hydrodechlorinations of tetrachlorides 17, 18, and 19 were carried out in acetonitrile at 300 nm with the extent of conversion maintained so that monodechlorination was assured. In the direct irradiation of tetrachloro 17, monodechlorination favored replacement of chlorine at C-2 over C-1, producing a ratio of 1,3,4-trichloronaphthalene (17-2H) to 1,2,3-trichloronaphthalene (17-1H) of 4.0:1.0, while photolysis of tetrachloro 18 generated an identical ratio of 4.0:1.0 for 18-1H:18-6H. Analogous photochemical hydrodechlorination of tetrachloro 19 produces 1,3,5-trichloronaphthalene (19-8H) and 1,4,6-trichloronaphthalene (19-1H) in a ratio of 2.3:1, accompanied by small amounts of two additional trichlorides (always less than 10% and presumably 19-3H and 19-5H). The photochemical hydrodehalogenation of tetrachloro 17, 18, and 19 at 300 nm in acetonitrile in the presence of triethylamine, using concentrations of tetra-chloronaphthalene and triethylamine such that all the light was absorbed by the naphthalene substrate, provided rather remarkable changes in the regiochemis-try. Amine induced photodechlorination of 17 provides an enhanced ratio of 10:1 of 17-2H:17-1H, while 18 generates a 50:1.0 ratio of 18-6H:18-1H, which represents a 200 fold switch in the regiochemistry. Parallel treatment of tetrachloro 19 provides a 25:1.0 ratio of monodechlorination products 19-1H and 19-8H, a 58 fold switch in the relative rates of hydrodechlorination [38].

Additional insight is provided by analysis of the steady state kinetics of the phototransformations of tetrachlorides 17, 18, and 19. In each case the quantum yield increased with substrate concentration [ArCl]. Plots of $1/\Phi$ versus $1/[\text{ArCl}]$ were very similar in each case. At high concentrations of ArCl the plot is linear with a positive slope, while as the concentration is decreased the plot is concave downward and approaches a linear plot with a zero slope. This is consistent with the mechanistic picture presented in Scheme 2 (**ArX=ArCl**). At high concentrations of **ArCl** excimer formation is dominant, $k_2[\text{ArCl}]F \gg k_T$ and Eq. 8 (k_d replaced by $k_d + k_T$) pertains. As the concentration of substrate is reduced excimer formation drops off and fission directly from the triplet provides a constant contribution. The view that the triplet state provides the branching point leading to product by direct fission or via excimer formation is supported by the quantum yield for intersystem crossing for **17** [4], $\Phi_{isc} = 1.00 \pm 0.04$, oxygen quenching, and linear Stern-Volmer plots of Φ_0/Φ versus concentration of triplet quencher 1,3-pentadiene.

In the photodechlorinations carried out at 300 nm in acetonitrile in the presence of triethylamine (**A**), the quantum yields are enhanced and the plots of $1/\Phi$ versus $1/[\text{A}]$ at the high end of the amine concentration range are linear, which is consistent with the mechanism presented in Scheme 10 and a bimolecular generation of exciplex which is dominant over intersystem crossing, and with steady state Eq. 16. At the low end of the amine concentration range there are constant contributions to product via the triplet state according to Scheme 2 (**ArX = ArCl**).

$$\frac{1}{\Phi} = \frac{1}{G} + \frac{k_{isc}}{k_{ex}[\text{A}]G} \quad \text{where} \quad G = \frac{k_r}{k_r + k_b} \tag{16}$$

In Scheme 10 electron transfer to singlet state is portrayed. The plots obtained are also consistent for a process which involves electron transfer to the triplet state. However, using the values of the (slope)/(intercept) for plots of $1/\Phi$ vs. $1/[\text{A}]$ for **17, 18** and **19** in the exciplex dominant region and the values of the triplet lifetimes obtained from phosphorescence measurements (112, 36, and 29 ms), values of $k_{ex} \approx 10 - 10^2 \, \text{M}^{-1}\text{s}^{-1}$ are obtained which are too low. Estimating the inherent radiative lifetime from the molar absorptivity, k_{isc} must be $10^9 - 10^{10} \, \text{s}^{-1}$, since no detectable fluorescence was observed. The slope over the intercept using Eq. 16 now yields values $10^9 - 10^{10} \, \text{M}^{-1}\text{s}^{-1}$ for k_{ex} which is more reasonable [39].

An interesting alternative to triethylamine as an electron transfer reagent is sodium borohydride. An increased photochemical reactivity for chlorobenzene,

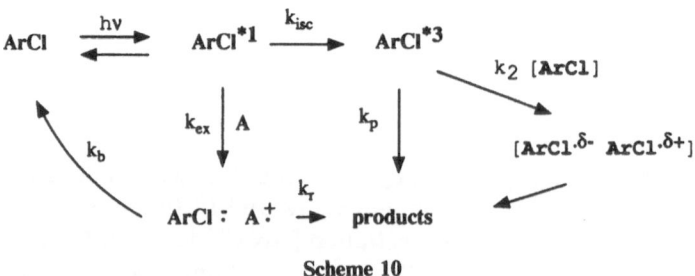

Scheme 10

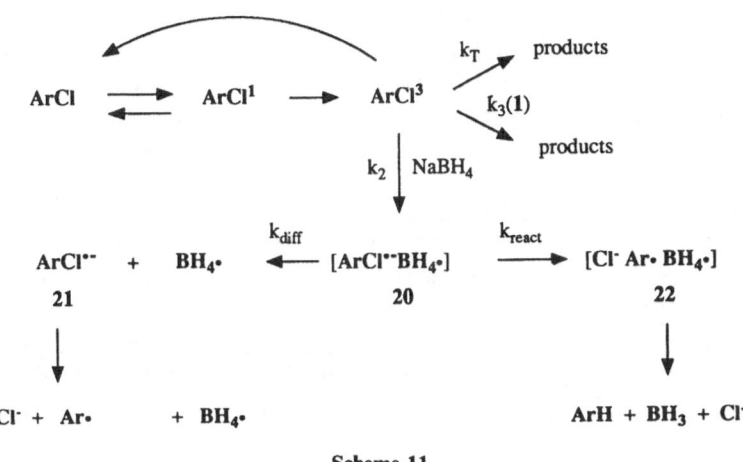

Scheme 11

chlorotoluenes and polychlorobiphenyls in the presence of sodium borohydride has been noted by Barltrop [40] and Epling [41]. Using our earlier study of the mechanism of photohydrodechlorination of pentachlorobenzene as a basis, an analysis of the dependence of the quantum yield upon the concentration of $NaBH_4$ or $NaBD_4$, and the insensitivity of the regiochemistry to increasing $NaBH_4$ concentration, the mechanistic picture which emerges is illustrated in Scheme 11 [2].

The radical anion-radical pair in the initial charge-transfer complex **20** may either diffuse away to give a free (uncomplexed) radical anion of **1** (**21**) or decompose to form an aryl radical and a radical derived from the borohydride and held within a cage as shown in **22**. The product ratios (e.g. $2:3:4 = 48.8:40.3:19.5$ at 0.20 M $NaBH_4$) rule out any substantial involvement of **21**, although the slight decrease in formation of **3** in the presence of radical trap acrylonitrile may be attributed to trapping of aryl radical from fragmentation of **21**. The deuterium incorporation results (81.6% monodeuteration), which show significant contribution of hydrogen from the borohydride in a step subsequent to bimolecular involvement of BH_4^-, strongly suggests the participation of **22**. In-cage transfer of hydrogen at this stage can explain the lack of a primary isotope effect on k_2.

5 Nature of the Exciplex

As described above, the sharp reversal of regiochemistry in the photohydro-dechlorination of pentachlorobenzene (**1**) 1,2,3,5-tetrachlorobenzene (**2**) and 1,4,6,7-tetrachloronaphthalene (**18**) in the presence and absence of triethylamine may be rationalized in terms of the fragmentation of the parent radical anion through a transition state analogous to the intermediate in S_NAr reactions. In order to provide reinforcement for this view and additional characterization of the radical anionic pathway, two new approaches were employed.

Photochemical Generation of the Aromatic Radical Anion of 1 in Micellar Media: The first method was designed to generate an uncomplexed radical anion of **1** within a micelle using triethylamine as the electron donor. Pentachlorobenzene (**1**) was dissolved in 0.200 M CTAB (hexadecyltrimethylammonium bromide) solution (bulk concentration of **1**: 0.001 M corresponding to a microscopic concentration [42] of 0.017 M). Based on Poisson statistics [43], the number of solute pentachlorobenzene molecules per micelle is expected to be predominantly ≤ 1 (Table 4). A large excess of triethylamine (0.165 M) dissolved into such a solution is expected to reside mostly in the aqueous phase and around the micellar surface [30, 43a, b]. In the irradiation of pentachlorobenzene in CTAB solution in the presence of triethylamine, electron transfer can take place from the amine residing on the surface to an excited pentachlorobenzene across the micellar interface. This would result in the formation of a radical anion of pentachlorobenzene within the micelle. Since the micellar surface is positively charged the resultant radical cation of triethylamine will be repelled away from the micellar surface to leave behind an uncomplexed radical anion of **1** within the micelle [44a] as shown in Scheme 12.

The reaction in CTAB micelles in the absence of triethylamine might be expected to depend on the occupancy number of **1** since the quantum yield of dechlorination of **1** is known to be concentration dependent [2, 45]. When the

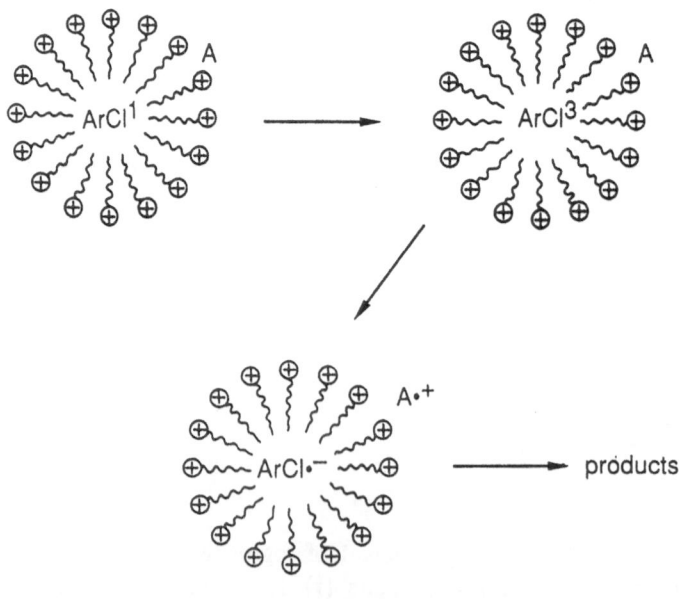

Scheme 12

Table 4. Relative ratios of products from the irradiation of **1** in various media

Medium (concentration)	Statistical probability of occupancy for occupancy numbers n = 0 – 4					Percentage ratio of products		
	0	1	2	3	4	2	3	4
CH$_3$CN (0.005 M)						48.4 ± 0.2	39.5 ± 0.2	12.1 ± 0.1
CH$_3$CN (0.071 M)						40.3 ± 0.6	40.3 ± 0.5	19.4 ± 0.8
CTAB[a] (0.017 M)[b]	0.673	0.267	0.053	0.007	0.001	49.8 ± 0.7	45.1 ± 0.3	5.10 ± 0.5
CTAB[a] (0.174 M)[b]	0.018	0.072	0.144	0.194	0.195	43.6 ± 0.6	51.0 ± 0.6	5.40 ± 0.5
CTAB[a]/TEA[c] (0.071 M)	0.673	0.267	0.053	0.007	0.001	35.0 ± 0.4	55.0 ± 0.6	10.0 ± 0.7
CH$_3$CN/TEA (0.005 M)						26.4 ± 0.3	62.5 ± 0.3	11.1 ± 0.9

[a] Concentration of CTAB in all cases were 0.20 M.
[b] Refers to microscopic concentration of **1**.
[c] 0.165 M in all cases.

concentration of **1** in the micellar solution is adjusted so that the occupancy number is either 0 or 1, the probability of excimer formation would be very remote. Conversely, when the occupancy number is at least 2, the formation of an excimer within the confines of the micelle would be expected.

However, in micellar media, in the absence of triethylamine, the dependence of the product ratio on the occupation number, although present, is not very pronounced (entries 3 and 4 in Table 4). The difference in the regiochemistries seems similar to those obtained in solution (CH_3CN) at the lowest and highest concentrations of **1** (Entries 1 and 2). In the presence of triethylamine under conditions where the occupation number is primarily less than 2, the regiochemistry of dechlorination (entry 5) is distinctly similar to the ratio obtained in solution in the presence of an electron donor (entry 6).

Nonphotochemical Generation of Radical Anions of Aromatic Halides. The second method involved using an arene radical anion as a convenient electron donor. In order to avoid side reactions, discovered when lithium naphthalenide was employed, presumably arising from coupling reactions between the donor and radical derived from acceptor radical anion, lithium *p,p'*-di-*tert*-butylbiphenylide (LiDBB) [46] was used as donor. The presence of the *tert*-butyl groups is known to prevent the side reactions encountered with naphthalene [46]. Treatment of **1** with LiDBB in THF gave the three isomers of tetrachlorobenzene as products as shown in Eq. 17.

$$+\underset{}{\bigodot}\!\!-\!\!\underset{}{\bigodot}\!\!+\ \xrightarrow[-68\,°C]{Li,\ THF}\ +\underset{}{\bigodot}\!\!-\!\!\underset{}{\bigodot}\!\!+\ Li^{+}\ \xrightarrow[-68\,°C]{\mathbf{1}}\ \mathbf{2}\ +\ \mathbf{3}\ +\ \mathbf{4} \qquad (17)$$

$$\qquad\qquad\qquad\qquad\qquad\qquad\qquad\qquad\qquad\qquad\quad 3.4\%\quad 95.7\%\quad 0.9\%$$

Remarkably, the expected major product (**3**) from the radical anion of **1** is formed almost exclusively at this temperature ($-68\,°C$). A series of reactions was carried out at various temperatures between 0 and $-68\,°C$, the results of which are shown in Table 5.

The relative rates of formation of any two isomers exhibit an Arrhenius relationship with temperature thus allowing the calculation of the expected ratio at the photochemical reaction temperature ($45\,°C$). Overall, the calculated relative ratios at this elevated temperature correlate well with those obtained photochemically in the presence of the electron donor triethylamine, thus suggesting that the products in both cases are formed through a common intermediate, a radical anion.

Furthermore, it seems reasonable to ascribe the regiochemistry observed to that of an unencumbered radical anion or solvent separated species in the radical anion reductions and the triethylamine photochemical runs. Various studies on lithium biphenylide have shown that it exists both as a contact ion pair and a solvent separated ion pair in THF, although at or below $20\,°C$ mostly as a solvent separated ion pair [47, 48]. In the case of the exciplex between the radical anion of **1** and $Et_3N^{+\cdot}$, various studies of pyrene and aliphatic or aromatic amine exciplexes [49, 50] and related excited singlet states of EDA complexes

Table 5. Relative ratio of products from the radical anion of pentachlorobenzene (1)

t °C	$1/T \times 10^{-2}$	$\log (k_3/k_2)$	$\log (k_3/k_4)$	Percentage ratio of products		
				2	**3**	**4**
0	0.366	0.988	1.87	16.7 ± 0.9	81.2 ± 2.4	2.10 ± 0.10
−25	0.403	1.11	1.63	12.8 ± 0.1	83.3 ± 0.1	3.90 ± 0.20
−40	0.429	1.35	1.94	8.00 ± 0.50	90.0 ± 1.9	2.00 ± 0.11
−55	0.459	1.50	2.16	5.90 ± 1.12	92.8 ± 0.5	1.30 ± 0.21
−68	0.488	1.76	2.31	3.34 ± 0.43	95.7 ± 0.5	0.96 ± 0.09
45	0.314	0.676[a]	1.05[a]	26.4 ± 0.30	62.5 ± 0.3	11.1 ± 0.9
45	0.314	0.683[b]	1.27[b]	25.9	62.5	6.67
45	0.314	0.173[c]	0.772[c]	50.1 ± 0.7	37.3 ± 0.6	12.6 ± 0.9

[a] Values obtained from photochemical reaction with triethylamine.
[b] Extrapolated values from Arrhenius plots.
[c] Values obtained from photochemical reaction without triethylamine.

[49, 51, 52] provide evidence that an initially formed contact ion pair would rapidly relax to a solvent separated ion pair ($k_1 \sim 10^9 \, s^{-1}$), which, in turn, would rapidly dissociate to free ions ($k_2 \sim 10^9 \, s^{-1}$) in acetonitrile. Thus, within 2 ns most of the contact ion pairs in the present case might reasonably be expected to be converted to either solvent separated ion pairs or to free ions. The conclusion that unencumbered radical anion or solvent separated radical anion/cation are product determining intermediates for the fragmentation processes observed is reinforced by the results in CTAB/Et$_3$N where the radical anion is created as an unencumbered radical anion.

Above it was noted that in the photohydrodechlorination of **17, 18, 19** without amine, regiochemistry is directed by fission of triplet and triplet excimer and with amine by fission of singlet exciplex. A characterization of the fragmentation of the radical anions related to **17, 18, 19** is achieved using a model which describes the fragmentation transition state in terms of a radical anion with a carbon-halogen bond (bent out of the plane of the arene ring) and a localized radical center at the point of fission, similar to **C** in Scheme 8 above. This would lead to transition state **E** for fragmentation at C-1 in the radical anion of **18**, for example. The stability of the transition states for the competing fragmentation pathways in the present study might be assessed by summing the charge densities at carbons containing chlorine in the NBMO (Σc_{ix}^2) using the Longuet-Higgins rule [53]. As Burdon has suggested for nucleophilic aromatic substitution, the interaction between the charge on carbon and the filled p-orbital of chlorine can be viewed as a destabilizing interaction [54]. Thus one chooses the lowest sum as representing the favored transition state and site of fragmentation. Table 6 lists the values for Σc_{ix}^2, which for each substrate, predict the major reaction pathway successfully. Thus, with a back-of-the-envelope calculation, one can rationalize the key pathways invoked in these photo-transformations.

Table 6. Assessment of relative stabilities of competing transition states

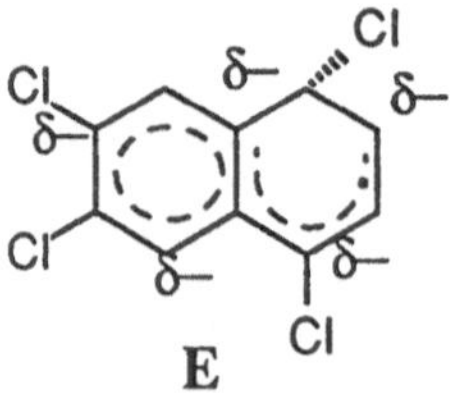

	Parent Cpd	Localized radical pos	Σc_{ix}^2
	17	1	0.7272
		2	0.6252
	18	1	0.4545
		6	0.2501
	19	1	0.0909
		3	0.1250
		5	0.5454
		8	0.3636

6 Phototransformations of Aliphatic Polyhalocompounds

Photoreactions of alkyl halides are induced by absorption of rather short ultraviolet light. The nonbonding electron on the p-orbital of the halogen is moved into the *anti*-bonding orbital of the C–X bond. The extinction coefficient of the n → σ^* transition is small due to its forbidden nature. The absorption red shifts as both the bond strength of the C–X bond and the electronegativity of the halogen decreases. Since the energy of a 254 nm photon is enough to break a C–I bond, alkyl iodides have often been used to provide fundamental information about the photochemistry of alkyl halides. Under normal photolysis, homolytic cleavage of the C–I bond occurs which results in products from several pathways which are derived by branching at the initial radical pair [55].

In polyhaloalkanes the weakest C–X bond is preferentially cleaved, and coupling is the major pathway to product. Simple polyhalomethanes such as CCl_4, CBr_4 or $BrCCl_3$ are often used synthetically to add the elements of polyhaloalkanes across double or triple bonds (Scheme 13).

$$CH_2=CH(CH_2)_5CH_3 \xrightarrow{CBr_4} CBr_3CH_2CHBr(CH_2)_5CH_3$$

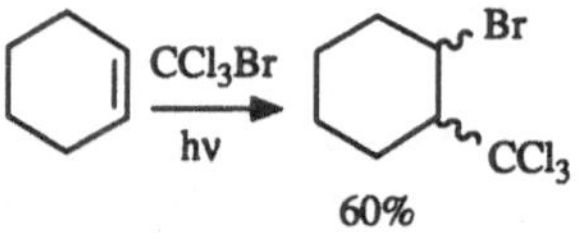

Scheme 13

Bond homolysis of alkyl halides requires less energy than heterolytic bond dissociation; however, photolytic products are formed which point to charged intermediates. Irradiation of diiodomethane in the presence of an alkene generates cyclopropanation [56, 57]. Initially, one of the C–I bonds breaks homolytically followed by electron transfer within the initially formed caged-radical pair. This highly electrophilic intermediate then reacts with the alkene in a highly stereospecific manner and with a general lack of sensitivity to steric effects. This can be seen in the reaction of the highly hindered alkene (21). The cyclopropanation of limonene (22) showed a preference for addition to the more substituted double bond. The photocyclopropanation of ca. thirty alkenes with diiodomethane also showed no formation of C–H insertion products. This rules out a carbene as an intermediate. Photoreaction of the geminal diiodo-compound (23) led to no detectable formation of 1,1-dimethylcyclopropane which would be the product of a carbene intermediate [57] (Scheme 14).

Photolysis of alkyl iodides in a variety of media can result in ions or ion-pairs which may in turn rearrange or react with the solvent. It was shown by Kropp et al. that irradiation of 1-iodo-norbornane initially gives a radical pair [57, 58]. The alkyl radical in the pair can either abstract hydrogen to form alkane or undergo electron transfer generating an ion pair, which leads to nucleophilic substitution and represents the major pathway. The strong acid character of the norbornyl cation allows it to cleave ethers and methylene chloride. Reaction was observed with essentially every solvent except for saturated hydrocarbons. Oxygen effectively competes with either the reduction or nucleophilic substitution pathway to form the corresponding hydroperoxide (Scheme 15).

Irradiation of the 2-bromo-derivatives of norbornane in methanol gave a higher ratio of radical to ionic products with an overall lower material balance than the corresponding photoreaction of the 2-iodo derivatives. In order to optimize ionic photobehavior in both the bromo- and iodo-photoreactions, hydroxide ion was added to scavenge HX, which is both a strong acid and a

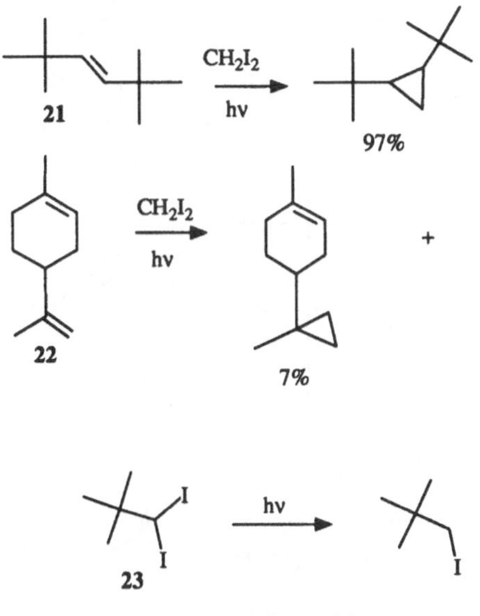

Scheme 14

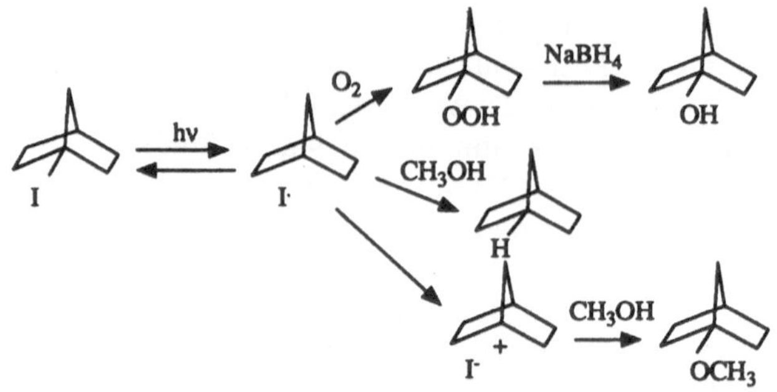

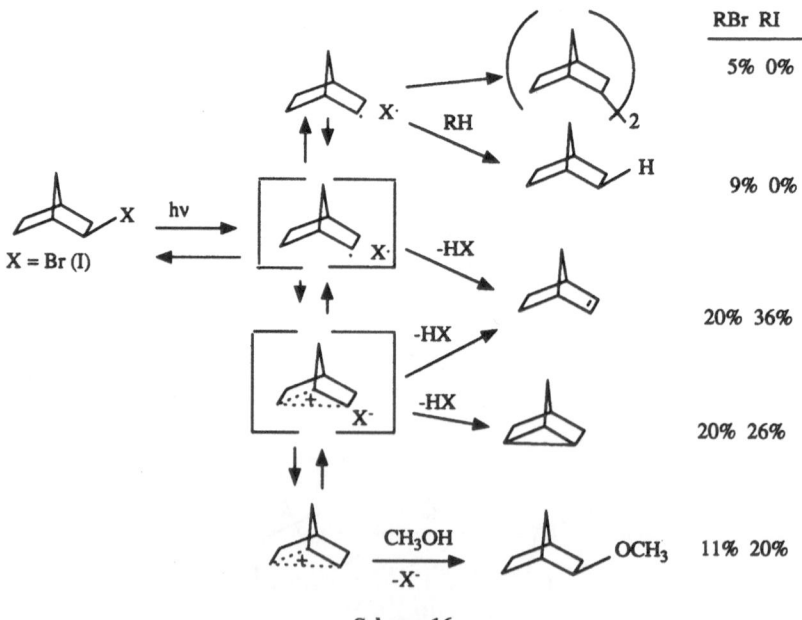

Scheme 15

source of radicals [59]. This also afforded an excellent material balance. Although more ionic product was formed, the bromides still gave higher yields of products derived from out-of-cage radical intermediates than the corresponding iodides. Initially it was thought that electron transfer might proceed more readily from the alkyl radical to iodine atom rather than to bromine atom, possibly due to iodine's greater polarizability or its lesser charge point density [57]. However, it was determined that escape from the initially formed caged-radical pair, relative to competing hydrogen atom or electron transfer, occurs more readily for bromides. This can be observed by the product yields in Scheme 16 [59].

Scheme 16

Incorporation of triethylamine into the reaction medium produced more reduction product presumably due to electron transfer from the triethylamine to the excited alkyl halide. This results in a weakly-bound amine-alkyl halide pair [57]. The alkyl halide radical anion releases X^- (Scheme 17). In a related example, it is known that solutions of aliphatic amines in CCl_4 are unstable to light quickly forming white crystalline precipitates [60]. The initial reaction is formation of a singlet radical pair via excitation of a ground state charge-transfer complex.

$$R-X \xrightarrow[h\nu]{Et_3N} R-X^{\cdot -} \cdot \overset{+}{N}Et_3 \longrightarrow R^{\cdot} + X^- \cdot \overset{+}{N}Et_3$$

Scheme 17

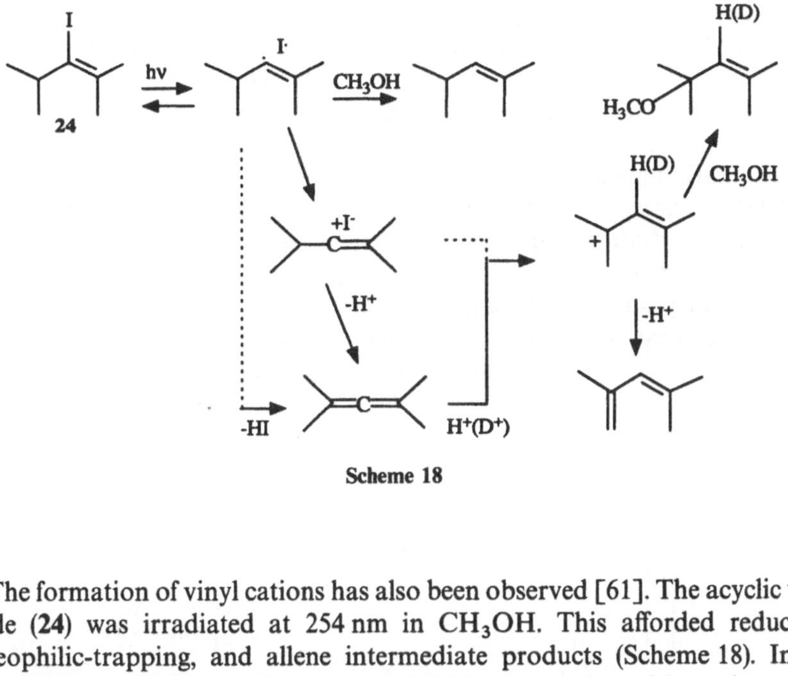

Scheme 18

The formation of vinyl cations has also been observed [61]. The acyclic vinyl iodide (**24**) was irradiated at 254 nm in CH_3OH. This afforded reduction, nucleophilic-trapping, and allene intermediate products (Scheme 18). In the presence of CH_3OD the allylic ether had 72% incorporation of deuterium at the vinylic position. On irradiation in CH_2Cl_2 or pentane, reduction and 1,3-diene product were observed.

An homologous series of cyclic vinyl iodides (**25**, n = 0 − 3) was irradiated in CH_3OH or CH_2Cl_2, and the products obtained revealed a similar mechanism picture. As the ring size decreased the allene intermediate resulting from deprotonation of the vinyl cation is too strained, and nucleophilic-trapping products predominate. Finally, the cyclopentenyl halide gave the reduction product in high yield with no detectable formation of the ionic products, presumably due to strain in the allene alternative. The exocyclic analogues

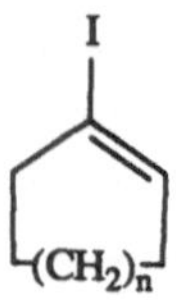

25, n = 0-3

provide skeletal rearrangements and nucleophilic substitution products which also involve carbocationic intermediates as illustrated in Scheme 19.

Exocyclic geminal dihalides underwent selective cleavage of a single carbon-halogen bond and showed no ring expansion upon irradiation (Scheme 20). The diiodide gave both reduction and nucleophilic-trapping product, while the dibromo analogue gave principally the reduction product. The dichloro analogue gave only reduction product as well as a subsequent reduction product presumably due to a secondary irradiation reaction. Overall, in the various cyclic systems the bromides gave substantially lower ratios of ionic to radical products than the iodides, while chlorides produced only radical products. In contrast, photoreactions of acyclic halides give principally unsaturated products due to a cationic elimination process [57] (Scheme 21). Since acyclic iodides primarily form elimination products, it was important to determine whether the products were due to α-elimination followed by 1,2 or 1,3-internal insertion of the resulting carbene intermediate. Using deuterium-labeling studies it was found that some amount of α-elimination was occurring [57a]. However, in none of the cases was this the major pathway to unsaturated product.

Photochemical reactions of aryl-vinyl halides also proceed via a vinylic cation [62–64]. The amounts of radical versus ionic products are affected by

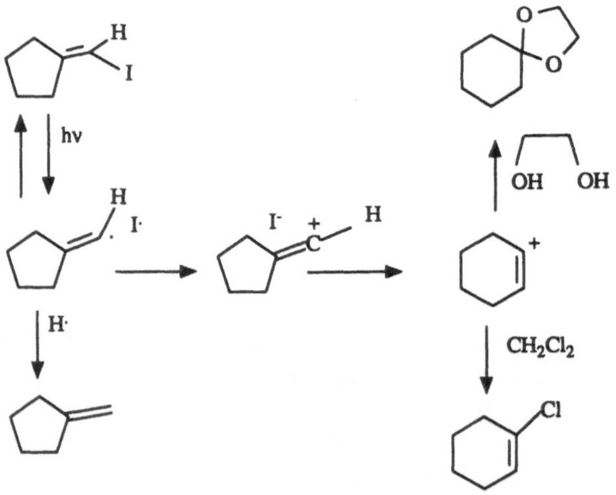

Scheme 19

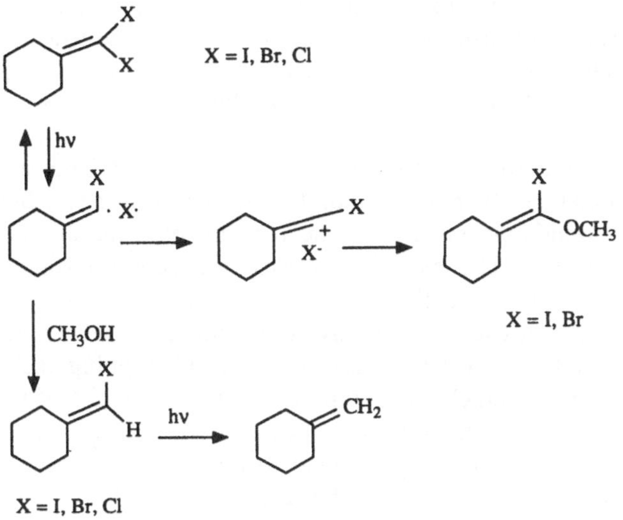

Scheme 20

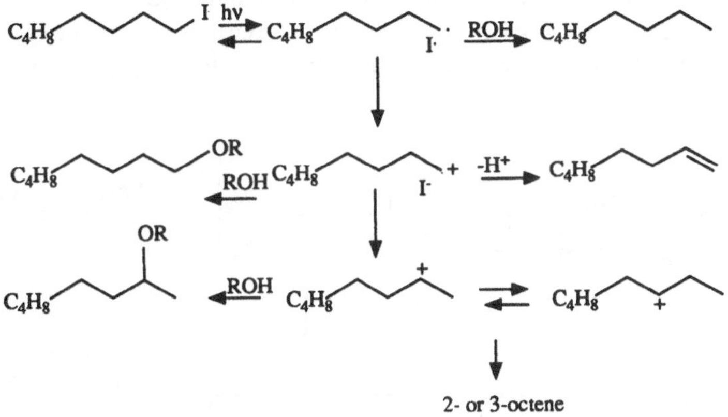

2- or 3-octene

Scheme 21

both α-, β-substituents and halogen atom nucleofuge as well as solvent and electron transfer reagents. Kitamura et al. investigated the photolysis of 1,1-diaryl-2-halopropenes (X = Cl, Br, I) in several solvents (Scheme 22) [62]. In the photolysis of the aryl-substituted vinyl halides, (26), it was found that reduction (27) and ionic products (28, 29, 30) were formed. Irradiation of the vinyl bromide in CH$_3$OD gave only 2% deuterium incorporation in the reduction product. Rearrangement products proceed via a cationic intermediate as 1,2-aryl shifts across a double bond do not occur with vinyl radicals. The overall yields are determined in the slow step of the C–X bond cleavage. The photolysis experiment gave an overall ionic to radical product ratio at constant irradiation time which decreased in the order X = Cl > Br > I. This is in contrast to thermolysis reactions which lead to heterolytic C–X bond cleavage with a rate increase in polar solvents, and the leaving group effect follows the order X = I > Br > Cl.

Changing the aryl group from *p*-anisyl to phenyl gave a nearly three-fold decrease in the ionic to radical product ratio when X = Cl, however, there was essentially no change when X = Br or I. Since Cl atom has a higher electron affinity than bromide or iodide atom the arylvinyl chloride should form a vinyl cation more readily. The greater tendency for cation formation from the chloride renders it more sensitive to α-substituent effects. Photolysis of the vinyl halides (X = Br, I) in cyclohexane gave only radical product. When the vinyl bromide was irradiated in the presence of a radical-trapping agent, copper (II) acetate, no radical product was produced, and the amount of ionic products increased. Photolysis of the vinyl halides in ethylene glycol gave a significant decrease in the product ratios of radical to ionic products due to a viscosity effect. There is a slower diffusion of the radical pair from the solvent cage increasing the importance of the competing internal electron transfer to give a higher yield of ionic products.

The photochemical reactions of some α-arylvinyl bromides (31), in acetic acid in the presence of sodium acetate and tetraethylammonium bromide (labeled with [82]Br) give nucleophilic substitution, reductive debromination, *cis*-stilbene photocyclization, and oxidation as the primary pathways to product

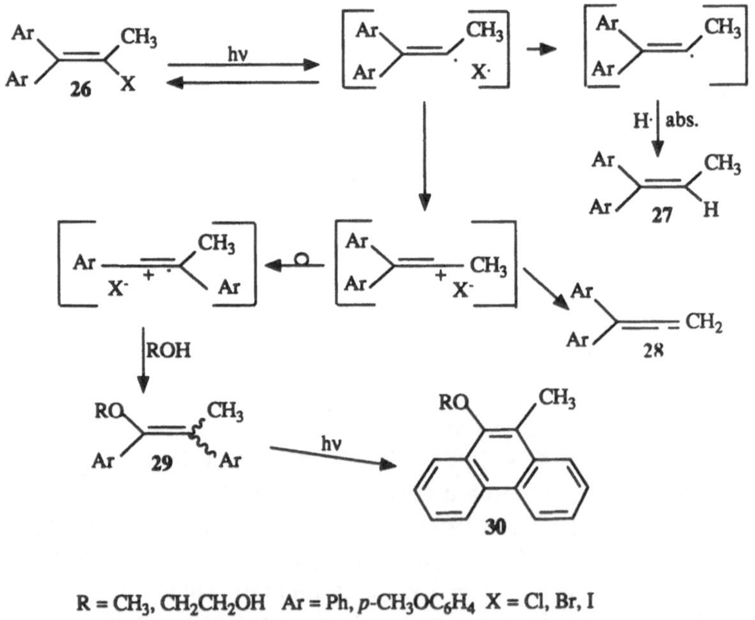

R = CH$_3$, CH$_2$CH$_2$OH Ar = Ph, *p*-CH$_3$OC$_6$H$_4$ X = Cl, Br, I

Scheme 22

[63]. The selectivity toward bromide versus acetate ions, the amounts of E/Z isomerized starting material and product, the amount of anisyl 1,2-shift, and the nature of the capturing nucleophile in the acetolysis are all in quantitative agreement with the corresponding thermolysis reactions. The experimental results for nucleophilic vinylic photosubstitution support the involvement of an intermediate linear free vinyl cation which is thermally relaxed.

Upon changing the α-substituent from p-anisyl to phenyl the decrease in amount of ionic products was not accompanied by a detectable increase in the amount of reduction products. The authors propose two alternative mechanisms: (a) reaction via an excited vinyl bromide which fragments directly to ion pair or radical pair or (b) reaction via an excited vinyl bromide which fragments directly to an ion pair or undergoes electron transfer to give vinyl bromide radical anion, which then fragments. More work is required on these alternatives to establish a clear choice.

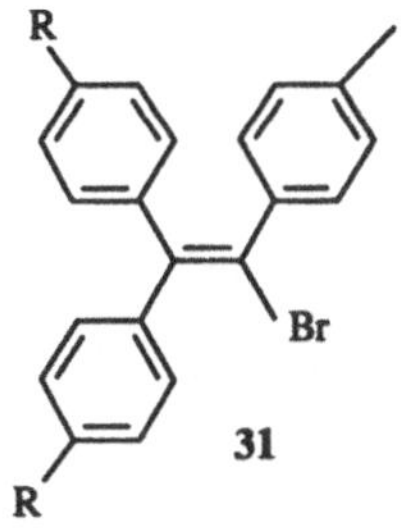

R = OCH₃ or H

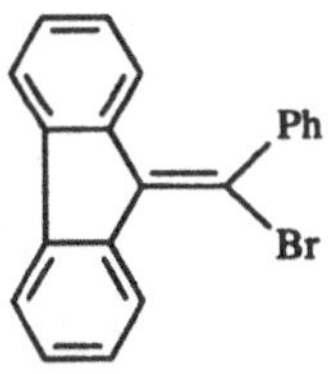

Evidence to support alternative (b) above stems from an earlier study of the photochemical behavior of vinyl bromide (**32**) in methanol with sodium methoxide, which revealed an increase in the quantum yield of formation of the reduction product with sodium methoxide compared to the irradiation with no sodium methoxide [64]. Sodium methoxide was proposed as the electron transfer reagent; however, a chain mechanism with formaldehyde radical anion as the electron transfer reagent as proposed by Bunnett seems more attractive [37g, 65].

Additional photochemical studies on α-arylvinyl bromides were undertaken in order to study their synthetic utility [66, 67]. The synthetic use of vinyl halides had been limited to conditions which required a high temperature and a protic, polar solvent. The advent of photolytically generated vinyl cations has allowed the use of arylvinyl halides to construct heterocycles in the presence of azide anion [66] and isoquinolines in the presence of cyanate ion [67] (Scheme 23 and 24).

A direct photolytic fission reaction was proposed rather than a multi-step addition-elimination reaction. The latter reaction is observed when the

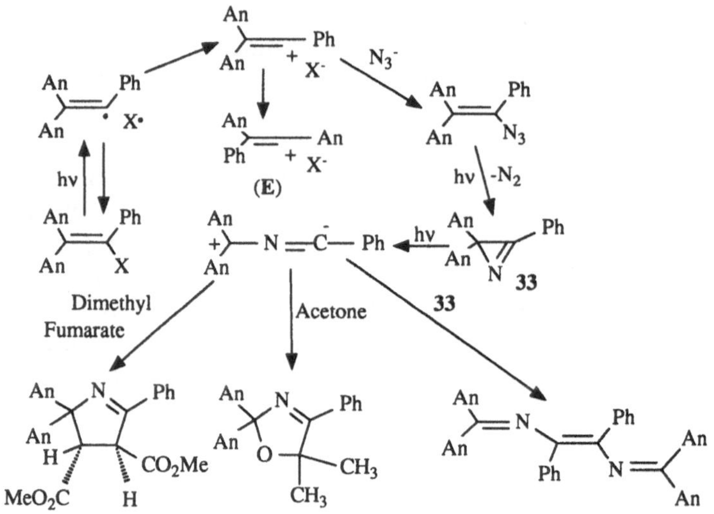

Scheme 23

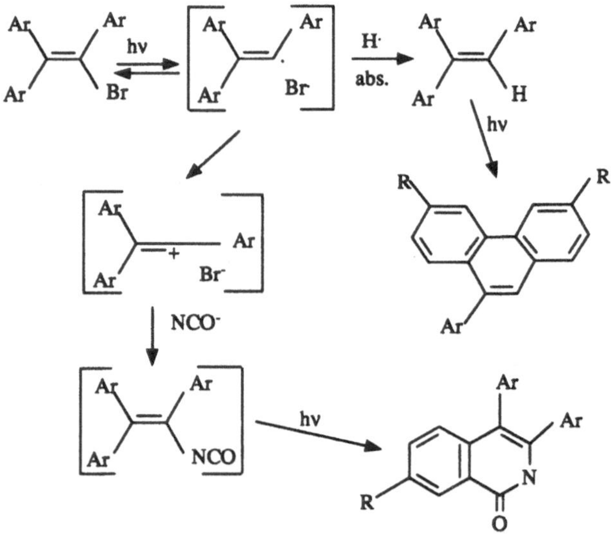

Scheme 24

β-substituent is electron-withdrawing such as CN, NO_2, and CO_2R, while fission followed by electron transfer occurs when the α-substituent is electron donating and the leaving group is a good nucleofuge. The formation of the isoquinolines was successful both in homogeneous and two-phase reaction conditions. The heterocyclic compounds were only synthesized under the latter conditions. The choice of an α-substituent on the vinyl halide had a significant effect on the yield of isoquinoline with p-$CH_3OC_6H_4$ > C_6H_5. When triphenylvinyl bromide was irradiated in the presence of cyanate ion, in a two-phase system using a phase-transfer catalyst, triphenylethene and 9-phenylphenanthrene were formed in an enhanced yield which was attributed to hydrogen abstraction by the triphenyl vinyl radical pair. This is in contrast to the findings of Lodder et al. [63].

Kitamura et al. also concluded that the ease of vinyl cation formation based on the leaving group was Cl > Br > I which corroborates the evidence found on irradiation of the 1,1-diphenyl-2-haloethenes [62], but contrasts with the work of Kropp et al. [57–59, 61]. Further evidence for a vinyl cation was uncovered in the photolysis of the α-phenyl-β,β-anisyl vinyl bromide which is illustrated in Scheme 23. A pyrroline product via intermediate **E** was formed in high yield, which reveals a 1,2-cationic anisyl shift [66].

The phototransformation of 1,1-diphenyl-2-haloethenes was reinvestigated in the presence of $NaBH_4$ and $LiAlH_4$ (Scheme 25) [68]. In the absence of hydride reagent the excited state is thought to undergo C–X bond fission to form a radical pair. Abstraction of a hydrogen atom by vinyl radical gives 1,1-diphenylethene (**34**). Another pathway to product is electron transfer followed by migration of a phenyl group and loss of a proton to give diphenylacetylene (**35**). The radical pair may also dimerize to form 1,1,4,4-tetraphenylbutadiene. Irradiation of the bromoethene in acetonitrile gave a 35% conversion with the formation of **34** and **35** in a ratio of 16:84. In the presence of an equimolar amount of $NaBH_4$ the ratio of **34** to **35** was 75:25. With 3 equivalents of $NaBH_4$ the ratio of **34** to **35** increased to 99:1. Addition of water to the reactions of bromoethene, both with and without $NaBH_4$, gave increasing amounts of diphenylacetylene. Irradiation of the bromoethene with 3 equivalents of $NaBH_4$ and a radical quencher changed the ratio of **34** to **35** from 99:1 to 76:24. Using sodium borodeuteride gave 64% monodeuterated **34**.

Irradiation of the chloroethene gave far less conversion than the bromo-ethene at 18% and a ratio of **34** to **35** of 2:98. This agrees with the findings of other photolysis reactions with aryl-vinyl halides which have been previously discussed. Irradiation of the chloroethene in CH_3CN with either 1 or 3 equivalents of $NaBH_4$, and with or without water, had very little effect on the relative ratios of **34** to **35**. Analogous photoreactions of both haloethenes in increasing amounts of $LiAlH_4$ gave increasing ratios of **34** to **35** consistent with the mechanism shown in Scheme 25.

Photosolvolysis involving heterolysis of carbon-chlorine bonds generally requires other groups, such as the aromatic rings in benzylic compounds or the carbonyl group. Cristol and Bendel have written an extensive review of these reactions [69]. Perhaps one of the more interesting systems described is that of the benzo- and dibenzobicyclo[2.2.2]octadienes which contain an aromatic chromophore and a remote halogen [70]. Under favorable conditions the excited aromatic ring transfers excitation to the reactive center which leads to a photo-Wagner-Meerwein rearrangement and/or a rearrangement with solvo-lysis (Scheme 26).

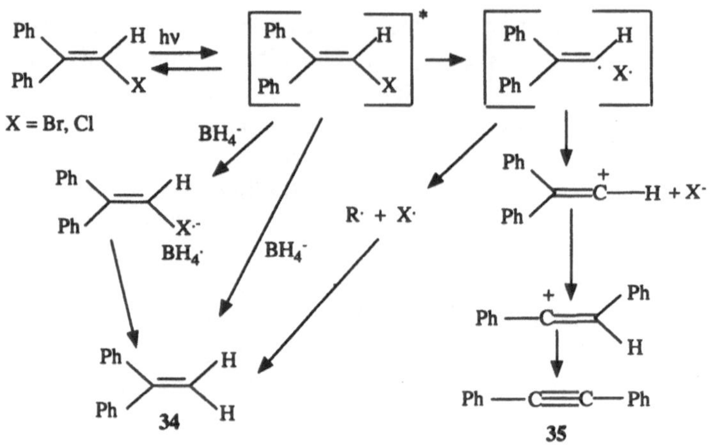

Scheme 25

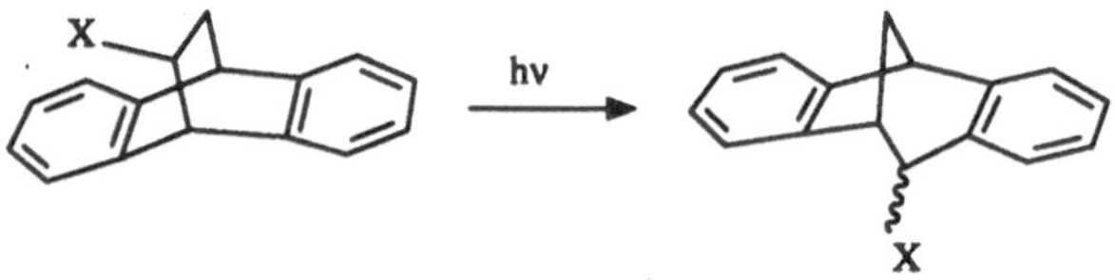

Scheme 26

In order to probe more deeply into the mechanistic details of the excitation and subsequent ring migration, several bridged-ring systems containing aromatic chromophores and nucleofugal groups were studied (X = Cl or methane sulfonate) (36–47) [71]. These studies revealed that there are stereochemical and regiochemical requirements for excitation transfer and subsequent rearrangement and solvolysis. The homobenzylic allylic epimers (36) and (37) were irradiated with 254-nm light in acetonitrile or acetic acid to give solvolysis products (36) or photo-Wagner-Meerwein products (38). Irradiation of the homobenzyl nonallylic systems showed that only the *anti*-compounds (39), (41), and (43) underwent solvolysis and photo-Wagner-Meerwein rearrangements, whereas the *syn*-compounds (40), (42), and (44) gave no such reactions. Ring migration in compounds (39), (43), (45), and (46) was shown to be *syn*-stereoselective, but not necessarily stereospecific.

Further studies on the isomeric dichlorides of (47) with different Y and Y′ substitution support the proposal that an electron transfer occurs from the photoexcited π^* orbital of the aromatic ring to the σ^* orbital of the carbon-chlorine bond [72]. When Y = H and Y′ = CN, $COCH_3$ or NO_2 no photo-solvolysis or photo-Wagner-Meerwein rearrangements were observed. Using the Weller equation it was determined that electron transfer from the presumed triplet states of the $COCH_3$ or NO_2 substituted compounds or the singlet state of the CN substituted compound would be too unfavored thermodynamically. When the reaction becomes somewhat exoergic as when Y = Y′ = Cl only loss of *anti*-Cl is observed. More highly exoergic electron transfers (Y = Y′ = benzo or OCH$_3$) lead to loss of both *syn*- and *anti*-chlorine. It was also found that when Y = Y′ = OCH$_3$, triplet sensitization led to photoreaction. Triplet state involvement was verified by complete quenching by 0.1 M piperylene. Examination of the product ratios point to two pathways of decay of the initial radical anion-radical cation state [73, 74] (Scheme 27). In the first pathway loss of chloride ion prior to migration produces a biradical cation. This allows migration of either the electron deficient ring or the "normal" ring. A second pathway proceeds by a suprafacial (*syn*) migration concerted with loss of the chloride ion.

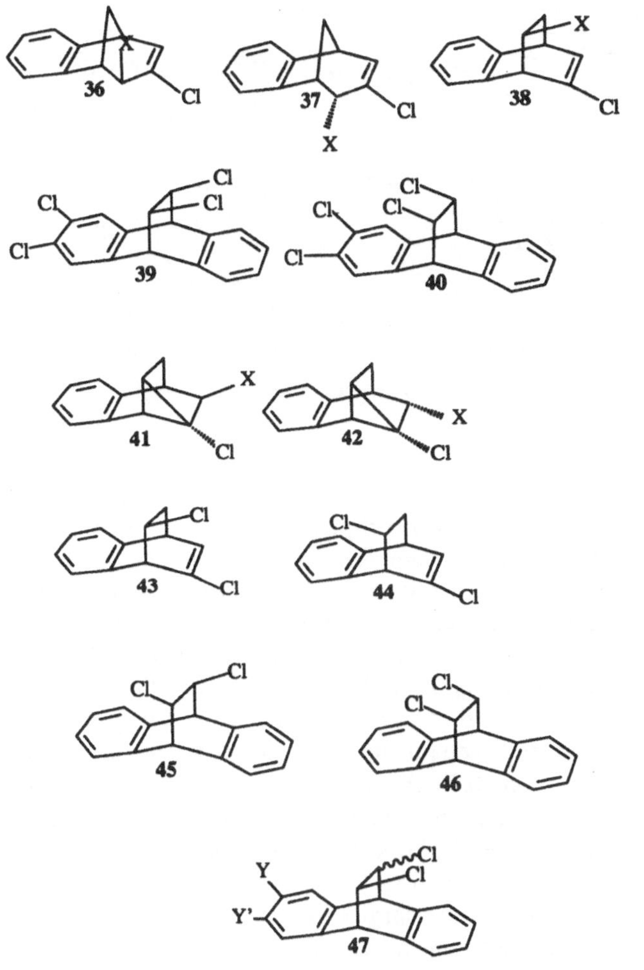

A "*meta*" effect has been proposed by Zimmerman and Sandel in which there is excess electron density at this position (as well as a smaller amount in the *ortho*

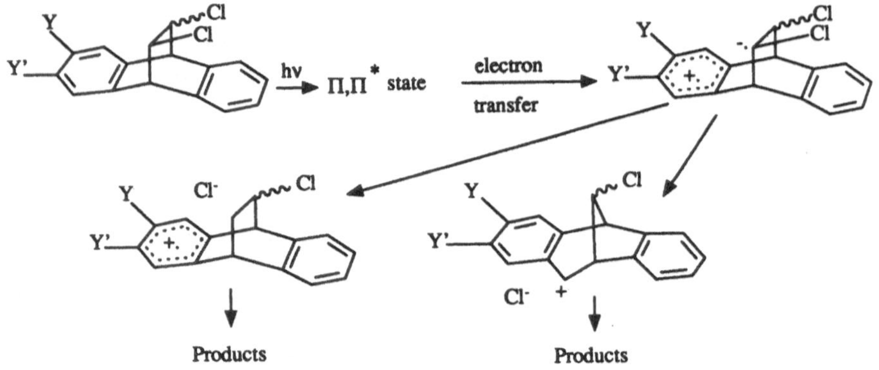

Scheme 27

position) in the first excited singlet state of anisole [75]. By extending this effect to a "*homometa*" situation it was proposed that orbital overlap is also crucial to the product outcome [76]. There was a clear preference for the loss of chlorine *anti-* and *homometa-* upon irradiation of **47** when Y = H and Y′ = OCH$_3$.

Photophysical studies were conducted on **45**, **46**, and **47** and the corresponding non-halogenated analogues [77]. The fluorescence lifetime was determined to be 27 ns for dibenzobicyclo[2.2.2]octadiene, and addition of two chlorines (**46**) reduced the lifetime to ca. 0.15 ns in cyclohexane. In the more polar acetonitrile the lifetime was reduced even further to ca. 0.005 ns. The quantum yield for ionic products changed from 0.03 in cyclohexane to ca. 0.21 in acetonitrile. The veratrolobenzobicyclo[2.2.2]octadiene compounds (**47** Y = Y′ = OCH$_3$) followed a similar trend with the largest quantum yield of ionic products occurring when the substituents were *anti-cis* to the veratrole ring. Both fluorescence and reactivity quantum yields show large solvent effects which supports involvement of a charge-transfer mechanism. The fluorescence emission spectra of these systems are broad and structureless with a red-shift of the fluorescence maxima upon chlorine substitution, especially for the veratrolobenzo compounds. This may be rationalized by the assumption that the S$_1$ states of the substituted compounds have some degree of charge transfer.

Cristol and his colleagues have contributed greatly to the understanding of the photochemical transformations of compounds which undergo light-induced solvolysis and photo-Wagner-Meerwein reactions. There are several other studies which are important additions to this body of work which are not covered in this review, and we refer the reader to these references [78–83].

Morrison et al. have also studied the photolytic behavior of chloro-substituted benzobicyclo-compounds and have proposed (π* + σ*) LUMO mixing in the singlet excited state which initiates C–Cl bond cleavage prior to a rapid electron transfer [84, 85]. Quantum efficiencies of C–Cl cleavage in β-substituted (**48**) and (**49**) are greater for *exo-* than for *endo*-cleavage [86, 87]. The products are characterized by extensive *syn*-migration, and the formation of alkenes and insertion products. There was a reversal in the *exo/endo* quantum efficiency ratio in the photolysis of the γ-substituted benzobicyclo-compounds *exo*-(50) and *endo*-(51) [88]. The total amount of hydrocarbons formed from both 50 and 51 is greater than that for the 48 and 49 series, and there is a very minor free-radical component. Overall, there is extensive rearrangement from photolysis of these systems, and this has been rationalized as a result of the

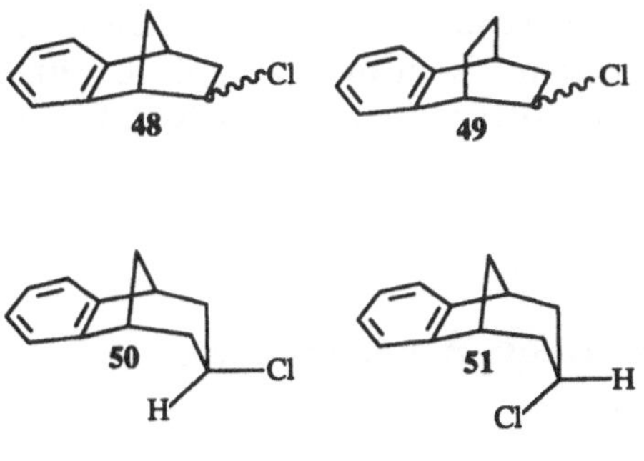

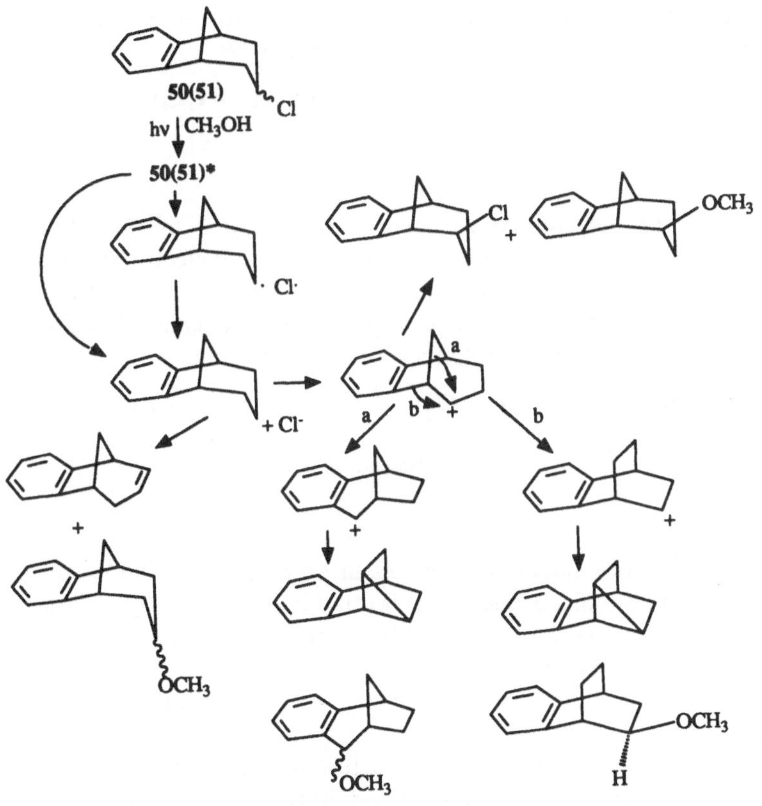

Scheme 28

sequential formation of several cationic intermediates. The origin of the photolytically generated cations is again ascribed to either an initial heterolytic cleavage or a homolytic cleavage followed by rapid electron transfer (Scheme 28).

7 Applications to Toxic Waste Disposal

The photoassisted (≈ 300–500 nm) heterogeneous catalytic degradation of halogenated organic compounds is one of the more successful methods for decontaminating polluted water. This method involves the continuous illumination of a photoexcitable solid catalyst, such as TiO_2, ZnO, or Fe_2O_3, which can bring about reactions of molecules adsorbed on its surface [89–94]. The photoexcitation of a TiO_2 particle, for example, promotes an electron from the valence to conduction band. In an aqueous solution and in the absence of oxygen, adsorbed water or hydroxyl ions are oxidized to hydroxyl radicals (•OH) by the valence band holes, h_{vb}^+. The hydroxyl radicals were shown to be mostly surface-adsorbed species which go on to initiate the radical reactions of the halogenated organic substrates [90, 91] (Scheme 29).

$$\cdot OH + C_2H_nCl_{6-n} \longrightarrow H_2O + (C_2H_{n-1}Cl_{6-n})\cdot$$

Scheme 29

Oxygen was determined to serve two functions in the photodecay of haloethanes [90]. It scavenges any conductance band electrons which prolongs the life of the valence band holes, however, its most important function is to generate peroxyl radicals en route to organic acids (Scheme 30). Although $O_2^{-}\cdot$ is formed, it does not react with halocarbons [90, 91].

$$(C_2H_{n-1}Cl_{6-n})\cdot + O_2 \longrightarrow (C_2H_{n-1}Cl_{6-n})-O-O\cdot$$

Scheme 30

Degradation of chlorinated ethanes in an illuminated TiO_2 solution gives mostly C_2 acids, but also some C_1 acids (mainly CO_2 due to C–C bond rupture). Primary and secondary peroxyl radicals follow the Russell mechanism with subsequent hydrolysis to give organic acids (Scheme 31). Tertiary peroxyl radicals may react via two pathways (Scheme 32). A bimolecular radical-radical reaction gives molecular oxygen and two oxyl radicals which are reduced to the corresponding alcohol. Alternatively, the tertiary oxyl radicals may undergo fast β-cleavage of either the C–C or C–Cl bond. Both of the oxygenated products can hydrolyze to their corresponding acids [90].

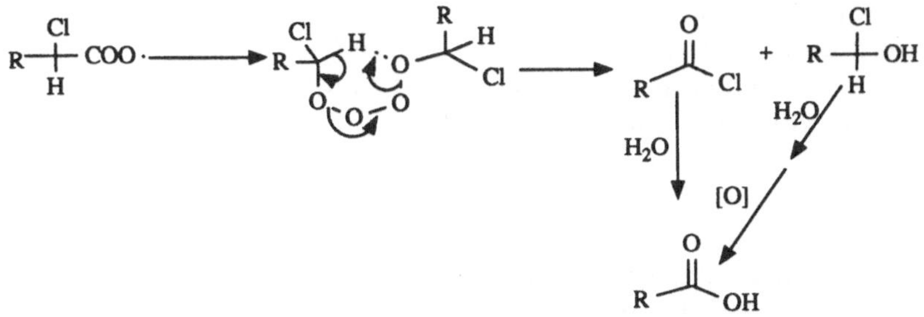

Scheme 31

$$2\ R\text{–}CCl_2O\text{–}O\cdot \longrightarrow 2\ R\text{–}CCl_2O^{\cdot} + O_2$$

$$R\text{–}CCl_2O^{\cdot} + RX \longrightarrow R\text{–}CCl_2OH + RX(-H)^{\cdot}$$

$$R\text{–}CCl_2O^{\cdot} \longrightarrow R^{\cdot} + CCl_2O \longrightarrow Cl^{\cdot} + R\text{–}CClO$$

Scheme 32

Few reactions of electrically neutral organic compounds by photoelectrochemical means have been reported, and reduction has been observed only when oxygen has been specifically removed [91]. Reduction of a cationic organic substrate by conduction band electrons can occur, as in the case of methylviologens, for example [91, 92]. Chlorinated organic acids which have no oxidizable hydrogen, such as trichloroacetic acid were found to degrade, albeit in small yields [90]. This was attributed to oxidation via the valence holes which is known as a photo-Kolbe process (Scheme 33).

Purging a liquid containing haloalkanes or haloalkenes, which have been photochemically reacted in the presence of TiO_2, with nitrogen gas and subsequently passing the purge stream

$$CCl_3COO^- + h_{vb}^+ \longrightarrow [CCl_3COO^{\cdot}] \longrightarrow {}^{\cdot}CCl_3 + CO_2$$

Scheme 33

through a saturated barium hydroxide solution produces a barium carbonate precipitate [93]. This demonstrates complete oxidation of carbon to CO_2. The complete mineralization of trichloroethylene and chloroform, for example, can be seen in Scheme 34.

$$CCl_2\text{=}CHCl + H_2O + 3/2\ O_2 \xrightarrow[\text{2) Ba(OH)}_2]{\text{1) TiO}_2,\ \text{near UV}} 2\ CO_2 + 3\ HCl$$

$$CHCl_3 + H_2O + 1/2\ O_2 \xrightarrow[\text{2) Ba(OH)}_2]{\text{1) TiO}_2,\ \text{near UV}} CO_2 + 3\ HCl$$

Scheme 34

A study of the photodecay of 12 organochlorine compounds in water was performed using TiO_2, Pt-loaded TiO_2, and TiO_2 with added H_2O_2 [94]. For the most part degradation rates were found to follow first-order kinetics and were enhanced by Pt-loading and the addition of H_2O_2. The addition of 10^{-3} $mol\,l^{-1}$ of H_2O_2 in a $5 \times 10^{-4}\,mol\,l^{-1}$ solution of dichloroethylene shortened the half-life 25.5 times.

A simple kinetic model for homogeneous reactions, $-dc/dt = kc^n$, was applied to the photoxidation of 23 organic compounds in an aqueous solution containing TiO_2 [89]. The concentration of the organic substrate as a function of irradiation time was recorded, and the relative efficiencies of photooxidation were determined. A plot of the percent photooxidation versus irradiation time gave a slope of $n = 0$ (zero-order kinetics) at high substrate concentrations (> 3 mM) and a slope of $n = 1$ (pseudo-first-order kinetics) at low concentrations (< 1.4 mM). At high substrate concentrations all catalytic sites of the TiO_2 surface are occupied, therefore an increase in substrate concentration does not alter the efficiency of the photooxidation. However, at low concentrations the efficiency is proportional to the substrate concentration and shows apparent first-order kinetics.

By comparing relative efficiencies of photooxidation some general trends become apparent [90, 93]. The photolysis of chlorinated alkanes seemed less efficient than that of chlorinated alkenes and aromatic compounds. The higher substituted alkanes were less able to be photooxidized, and CCl_4 was not degraded. The relative reactivities of halomethanes were shown to be $CHBr_3 > CHCl_3 > CH_2Cl_2 > CCl_4$. Bromination or fluorination did lower the photooxidation efficiency, and this was thought to be due to weaker or slower adsorption of these compounds on TiO_2. A decrease in pH also decreased the rate of photooxidation since protons will react with the TiO_2 surface.

Although photocatalytic treatment holds promise for degradation of halogenated compounds, problems remain to be solved. For example, the photocatalytic treatment of chlorobenzenes in a suspension of TiO_2 produced a range of complex products including biphenyls upon long illumination times [93]. The mechanisms of photodecay or photorearrangement presented in this review reveal dependence upon the structure and substitution of the halogenated compound, the solvent, the presence of an electron transfer reagent, and the presence of oxygen. The photohydrodehalogenation reaction may very well provide an important basis for the development of an efficient method of toxic waste disposal. There are, however, two hurdles to overcome. In the case of pentachlorobenzene, for example, the triplet state undergoes fission directly to product and is the gateway to excimer formation. Thus, in an aqueous system exposed to the air, oxygen would be expected to quench the triplet state and retard the reaction. Secondly, hydrogen abstraction to complete the reduction process could be derailed, since the O–H bond in water (119 kcal/mol) is too strong to donate hydrogen to an aryl radical, and, consequently, even more toxic dimers may be produced.

However, these two problems might be solved quite nicely by carrying out the photohydrodehalogenation in micellar media. In a reasonably dilute micellar environment the odds that a substrate molecule and an oxygen molecule would occupy the same micelle are low. The phosphorescence of halonaphthalenes can be observed in aerated, aqueous CTAB solution at room temperature, for example [95]. Hydrogen abstraction should no longer be a concern, since aryl radical abstraction from a CTAB alkyl tail provides an exothermic route to product (− 15 kcal/mol) [96]. This approach has been tested using the photolysis of pentachlorobenzene in a CTAB aqueous micellar solution exposed to air. A sun lamp was used in the irradiation of pentachlorobenzene in aqueous 0.20 M CTAB solution. Pentachlorobenzene disappears after 24 hours, and tetrachlorobenzene is produced rapidly and is then subsequently converted to trichlorobenzene completely within 72 hours. Trichlorobenzene is dechlorinated after 48 hours and is expected to be converted to dichlorobenzene completely. Dichlorobenzene did not yield monochlorobenzene at all since dichlorobenzene does not absorb in the sun lamp spectral range. Work is continuing in our laboratory to gain further knowledge of the mechanisms of the photodegradation of environmental pollutants. This knowledge is important in order to design efficient and safe methods of toxic waste disposal.

Acknowledgments: The research described here from our laboratory was made possible through an enjoyable collaboration with the coworkers cited. Their contributions and the support of the NIEHS is gratefully acknowledged.

8 References

1. Freeman PK, Srinivasa R, Campbell J-A, Deinzer ML (1986) J Am Chem Soc 108: 5531
2. Freeman PK, Ramnath N (1988) J Org Chem 53: 148
3. a. Pitts JN, Burley DR, Mani JC, Broadbent AD (1968) J Am Chem Soc 90: 5900, 5902;
   b. Kochevar IE, Wagner PJ (1972) J Am Chem Soc 94: 3859
4. Lamola AA, Hammond GS (1965) J Chem Phys 43: 2129
5. Turro NJ (1978) Modern molecular photochemistry. Benjamin Cummings, Menlo Park, CA, p 254
6. Ethyl alcohol has a greater E_T value than acetonitrile (51.9 vs. 46.0), and the rate constant for excimer formation of 4-bromobiphenyl is 60% greater in ethyl alcohol than in acetonitrile, J-S. Jang (unpublished work)
7. Birks JB (1970) Photophysics of aromatic molecules, Wiley: New York, p 372
8. Freeman PK, Srinivasa R (1987) J Org Chem 52: 252
9. Freeman PK, Jang J-S, Ramnath N (1991) J Org Chem 56: 6072
10. Bunce NJ, Safe S, Ruzo LO (1975) J Chem Soc Perkin Trans 1 1607
11. Pedersen CL, Lohse C (1975) Acta Chemi Scand B 33: 649
12. Ruzo LO, Zabik MJ (1975) Bull Environ Contam Toxicol 13: 181
13. Sandros K (1969) Acta Chemi Scand 23: 2815
14. O'Donnell CM, Harbaugh KF, Fisher RP, Winefordner JD (1973) Analyt Chem 45: 609

15. Egger KE, Cocks AT (1973) Helv Chim Acta 56: 1516
16. a. Wagner PJ (1971) J Acc Chem Res 4: 168; b. Wagner PJ, Kochevar I (1968) J Am Chem Soc 90: 2232; c. Wettack FS, Renkes GD, Renkly MG, Turro NJ, Dalton JE (1970) J Am Chem Soc 92: 1318
17. a. Dimroth K, Reichardt C, Seipman T, Bohlman F (1963) Justus Liebig Ann Chem 661: 1; b. Reichardt C (1971) ibid 752: 64; c. Reichardt C (1965) Angew Chem Int Ed Engl 4: 29; d. Totter WJ (1977) Bull Environ Contam Toxicol 18: 726
18. Abraham MH, Doherty RU, Kamlet MJ, Harris JM, Taft RW (1987) ibid 1097: 913
19. Freeman PK, Lee Y-S (1992) J Org Chem 57: 2846
20. Yekta A, Aikawa M, Turro NJ (1979) Chem Phys Lett 63: 542
21. Infelta PP (1979) Chem Phys Lett 61: 88
22. Selinger BK, Watkins AR (1978) Chem Phys Lett 56: 99
23. Dorrance RC, Hunter TF (1974) J Chem Soc Faraday Trans 70: 1572
24. Infelta PP, Grätzel M (1979) J Chem Phys 70: 179
25. Moore T, Pagni RM (1987) J Org Chem 52: 770
26. Soumillion JP, Wolf BD (1981) J Chem Soc Chem Commun 436
27. Ito R, Migita T, Morikawa N, Simamura O (1965) Tetrahedron 21: 955
28. a. Chambers RD, Close D, Williams DLH (1980) J Chem Soc Perkin Trans 2: 778; b. Chambers RD, Waterhouse JS, Williams DLH (1977) J Chem Soc Perkin Trans 2: 585
29. Schmidt RD (1991) M.S. Thesis, Oregon State University
30. Freeman PK, Ramnath N (1991) J Org Chem 56: 3646
31. a. Bursey MM, McLafferty FW (1967) J Am Chem Soc 89: 1; b. Bursey MM, McLafferty FW (1966) J Am Chem Soc 88: 4484; c. Bursey MM, McLafferty FW (1966) J Am Chem Soc 88: 529
32. Jones DAD, Smith GG (1964) J Org Chem 29: 3531
33. Wells PR (1968) In: Linear free energy relationships, Academic Press, New York, Chapter 2
34. a. Chambers RD, Close D, Williams DLH (1980) J Chem Soc Perkin Trans 2: 778; b. Chambers RD, Waterhouse JS, Williams DLH (1977) J Chem Soc Perkin Trans 2: 585
35. Allen KJ, Bolton R, Williams GH (1983) J Chem Soc Perkin Trans 2: 691
36. a. Burdon J, Gill HS, Parsons IW, Tatlow JC (1980) J Chem Soc Perkin Trans 1: 1726; b. Burdon J, Parsons IW, Gill HS (1979) J Chem Soc Perkin Trans 1: 1351; c. Burdon J, Parsons IW (1977) J Am Chem Soc 99: 7445; d. Burdon J (1965) J Tetrahedron 21: 3373; e. Burdon J, Childs AC, Parsons IW, Tatlow JC (1982) J Chem Soc Chem Commun 534
37. a. Bethell D, Compton RG, Wellington RG (1992) J Chem Soc Perkin Trans 2: 147; b. Symons MCR, Bowman WR (1988) J Chem Soc Perkin Trans 2: 583; c. Amatore C, Oturan MA, Pinson J, Savéant J, Thiébault A (1985) J Am Chem Soc 107: 3451; d. Dressler R, Allan M, Haselbach E (1985) Chimia 39: 385; e. Bays JP, Blumer ST, Baral-Tosh S, Behar D, Neta P (1983) J Am Chem Soc 105: 320; f. Riederer H, Huttermann J, Symons MCR (1978) J Chem Soc Chem Commun 313; g. Bunnett J (1992) J Acc Chem Res 25: 2; h. Bowman WR, Taylor P (1990) J Chem Soc Perkin Trans 1: 919; i. Symons MCR, Bowman WR (1990) J Chem Soc Perkin Trans 2: 975; j. Moreno M, Gallardo I, Bertrán J (1989) J Chem Soc Perkin Trans 2: 2017; k. Bunce NJ, Gallagher JC (1982) J Org Chem 47: 1955
38. Freeman PK, Clapp GE, Stevenson BK (1991) Tetrahedron Lett 5705
39. Bunce NJ, Pilon P, Ruzo LO, Sturch DJ (1976) J Org Chem 41: 3023
40. Barltrop JA, Bradbury D (1973) J Am Chem Soc 95: 5085
41. a. Epling GA, Florio EJ (1986) J Chem Soc Chem Commun 185; b. Epling GA, Florio E (1986) Tetrahedron Lett 27: 675
42. a. Pownall HJ, Smith LC (1973) J Am Chem Soc 95: 3136; b. Shinoda K, Soda T (1963) J Phys Chem 67: 2072
43. a. Katusin-Razem B, Wong M, Thomas JK (1978) J Am Chem Soc 100: 1679; b. Ericksson JC, Gillberg G (1966) Acta Chemi Scand 20: 2019
44. a. Dainty C, Bruce DW, Cole-Hamilton DJ, Camilleri P (1984) J Chem Soc Chem Commun 1324; b. Soumillion PJ, Wolf BD (1981) J Chem Soc Chem Commun 436
45. Freeman PK, Ramnath N, Richardson AD (1991) J Org Chem 56: 3643
46. a. Freeman PK, Hutchinson LL (1983) J Org Chem 48: 4705; b. Freeman PK, Hutchinson LL (1980) ibid 48: 4705 and references therein
47. Szwarc M (1974) In Ions and ion pairs in organic solutions; Szwarc M (ed), John Wiley New York Vol 2, Chapter 1
48. a. Nichols D, Sutphen C, Szwarc M (1968) J Phys Chem 72: 1021; b. Canters GW de Boer E (1973) Mol Phys 26: 1185

49. Mataga N, Okada T, Kanda Y, Shioyama H (1986) Tetrahedron 42: 6143
50. Weller AZ (1982) Phys Chem Neue Folge 130: 129
51. Gould IR, Moody R, Farid S (1988) J Am Chem Soc 110: 7242
52. Goodman JL, Peters KS (1986) J Phys Chem 90: 5506
53. Dewar MJS, Dougherty RC (1975) The PMO theory of organic chemistry: Plenum: New York pp 78
54. Burdon J, Gill HS, Parsons IW, Tatlow JC (1980) J Chem Soc Perkin Trans 1: 1726
55. Saplay KM, Damodaran NP (1983) J Scient and Indust Res 42: 602
56. Kropp PJ, Pienta NJ, Sawyer JA, Polniaszek RP (1981) Tetrahedron 37: 3229
57. For reviews: a. Kropp PJ (1984) J Acc Chem Res 17: 131; b. Lodder G (1983) In: Patai S, Rappoport Z (eds) The chemistry of halides, pseudohalides and azides. Wiley, Chichester, p 1605
58. Kropp PJ, Poindexter GS, Pienta NJ, Hamilton DC (1976) J Am Chem Soc 98: 8135
59. Kropp PJ, Adkins RL (1991) J Am Chem Soc 113: 2709
60. Stevenson DP, Coppinger GM (1962) J Am Chem Soc 84: 149
61. Kropp PJ, McNeely SA, Davis RD (1983) J Am Chem Soc 105: 6907
62. Kitamura T, Shinjiro K, Taniguchi H (1982) J Org Chem 47: 2323
63. van Ginkel FIM, Cornelisse J, Lodder G (1991) J Am Chem Soc 113: 4261
64. Verbeek JM, Cornelisse J, Lodder G (1986) Tetrahedron 42: 5679
65. Bunnett JF, Wamser CF (1967) J Am Chem Soc 89: 6712
66. Kitamura T, Kobayashi S, Taniguchi H (1984) J Org Chem 49: 4755
67. Kitamura T, Kobayashi S, Taniguchi H (1990) J Org Chem 55: 1801
68. Zupančič N, Šket B (1991) Tetrahedron 47: 9071
69. Cristol SJ, Bindell TH (1983) in Organic photochemistry Padwa A (ed), 6, Marcell Dekker New York, p 327
70. Cristol SJ, Dickenson WA, Stanko MK (1983) J Am Chem Soc 105: 1218
71. Cristol SJ, Seapy DG, Aeling EO (1983) J Am Chem Soc 105: 7337
72. Cristol SJ, Bindel TH, Hoffman D, Aeling EO (1984) J Org Chem 49: 2368
73. Cristol SJ, Aeling EO (1985) J Org Chem 50: 2698
74. Cristol, SJ, Opitz RJ (1985) J Org Chem 50: 4558
75. Zimmerman HE, Sandel VR (1963) J Am Chem Soc 85: 915
76. Cristol SJ, Aeling EO, Heng R (1987) J Am Chem Soc 109: 830
77. Cristol SJ, Aeling EO, Strickler SJ, Ito RD (1987) J Am Chem Soc 109: 7101
78. Cristol SJ, Ali MZ (1985) J Org Chem 50: 2502
79. Cristol SJ, Opitz RJ, Aeling EO (1985) J Org Chem 50: 4834
80. Cristol SJ, Braun D, Schloemer GC, Vanden Plas BJ (1986) Can J Chem 64: 1081
81. Cristol SJ, Vanden Plas BJ (1989) J Org Chem 54: 1209
82. Cristol SJ, Mahfuza BA, Sankar IV (1989) J Am Chem Soc 111: 8207
83. Cristol SJ, Vanden Plas BJ (1991) J Phys Org Chem 4: 541
84. Morrison H, Nash JJ, (1990) J Org Chem 55: 1141
85. Morrison H, Singh TV, de Cardenas L (1986) J Am Chem Soc 108: 3862
86. Morrison H, Miller A, Bigot B (1983) J Am Chem Soc 105: 2398
87. Morrison H, Muthuramu K, Pandey G, Severance D, Bigot B (1986) J Org Chem 51: 3358
88. Morrison H, Muthuramu K, Severance D (1986) J Org Chem 51: 4681
89. Sabin F, Türk T, Vogler A (1992) J Photochem Photobiol A: Chem 63: 99
90. Mao Y, Schöenich C, Asmus K-D (1991) J Phys Chem 95: 10080
91. Fox, MA (1987) In Topics in current chemistry, 142; Steckhan E, (ed), Springer-Verlag: New York, p 99
92. Ollis DF, Pelizzetti E, Serpone N (1989) In: Sepone N, Pelizzetti E (eds), Photocatalysis fundamentals and applications. John Wiley, New York, p 603
93. Ollis DF (1985) Environ Sci Technol 19: 480
94. Hisanaga T, Harada K, Tanaka K (1990) J Photochem Photobiol A: Chem 54: 113
95. Turro NJ, Liu K-C, Chow M-F, Lee P (1978) Photochem Photobiol 27: 523
96. Egger KW, Cocks AT (1973) Helv Chim Acta 56: 1516

Photoinduced Electron Transfer Employing Organic Anions

Jean-Philippe Soumillion

Laboratory of Physical Organic Chemistry and Photochemistry, Catholic University of Louvain, Place Louis Pasteur, 1, B-1348 Louvain-la Neuve, Belgium.

Table of Contents

Topics in Current Chemistry, Vol. 168
© Springer-Verlag Berlin Heidelberg 1993

Jean-Philippe Soumillion

Reductive properties of excited state organic anions are discussed, starting from the preliminary presentation of other related properties: spectral characteristics, ion pair associations and photoejection ability.

Electron transfer reactions may be correlated by the Marcus treatment and some peculiarities of the use of this model, in the case of oxyanions, are pinpointed.

Photoreactions employing an electron transfer are discussed. Among these are recent examples of photochemical $S_{RN}1$ reactions, photoalkylations of carbanions and photoreductions initiated by oxyanions and radical anions. Anions used in their ground state as electron donating quenchers are also considered. Intra ion pair electron transfers as well as the use of anion-like precursors in charge transfer complexes or charge transfer excited states are presented.

1 Introduction

The reductive properties of anions have been recognized for rather a long time and reviewed among other properties, in the case of aromatic radical anions or dianions by N.L. Holy [1, 2]. Photochemistry of anions was first drawn to our attention by M.A. Fox in an important review covering the subject in 1979 [3]. Their reactivity in the field of electron transfer photochemistry and their photoreductive properties were underlined in this paper and their involvement in photoinduced electron transfer has been the object of other reviews [4, 5].

Several advantages of working with anions in photoelectron transfer may be expected. Often bathochromic when compared to their neutral precursors, they allow selective excitation in the presence of numerous coreactants. Moreover, the presence of the negative charge is responsible for an increased reducing power and after occurrence of electron transfer in the presence of a neutral acceptor, a radical-radical anion pair, free of electrostatic attraction is formed. This means that these species will diffuse more freely into the solution. An increased probability of reaction (k_r) is thus expected in the case of charge shifts than in the case of charge separations where the formation of isolable photoproducts is often severely limited by the back electron transfer restoring the starting ground states (k_b).

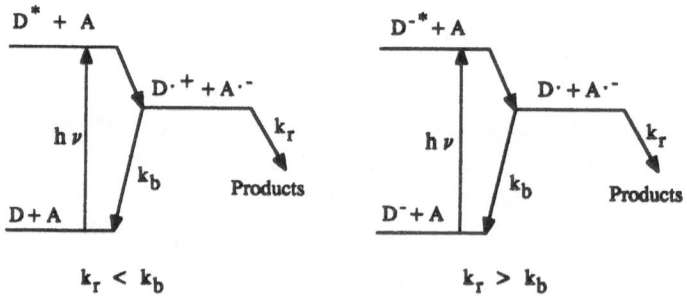

In the first part of this paper some peculiarities of electron transfer employing anions will be pinpointed as for instance ion pairing phenomena and electron photoejection. In the following section, the correlation between electron transfer rate constants and the reaction free energy, in cases where anions are reacting partners, will be considered.

Two ways are open for the photoinitiation of electron transfer reactions employing anions: excitation of the anion itself or the use of anionic quenchers. The next two parts of this review will deal with reactions initiated by these two ways.

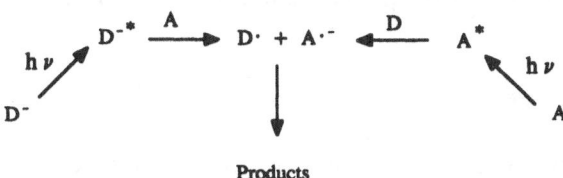

Products

In the last section, particular reacting systems will be envisaged: photoinduced intra-ion-pair electron transfer, photoinduced electron transfer occurring through excitation of charge transfer complexes including anions or electron transfers originating from the negative end of charge transfers excited states.

2 Anion Electron Transfer Peculiarities

2.1 Absorption Spectra

The absorption spectrum of an anion is often bathochromic when compared to the corresponding neutral molecule. This corresponds to a higher energy non-bonding HOMO and often to increased delocalization in the deprotonated species. When excited anions are used as an electron source, this allows a more selective excitation and may make it possible to use of visible light. The price for this selectivity is of course paid by some decrease in the energy content of the excited species. The decreased oxidation potential due to the anionic character will thus be partly compensated by the decrease of the available excitation energy. On the other hand, this makes energy transfer more difficult to activate.

Two types of transitions may be expected in organic anions. $\pi \rightarrow \pi^*$ Transitions are well recognized in carbanions [6], or in phenolates and enolates with high extinction coefficients [7, 8, 9, 10]. In carboxylates, the n, π^* and the π, π^* states coexist and this may lead to wavelength dependent photochemistry [11]. In certain cases, the shift to longer wavelength observed for the anion is accompanied by a drastic change in the excited states ordering. This occurs with a striking effect on reactivity as for instance when 4-hydroxybenzophenone is ionized: the absorption maximum moves from 275 nm in cyclohexane to 360 nm in alkaline aqueous solution. In cyclohexane, the hydrogen abstraction quantum yield is 0.9 while it drops to 0.02 in isopropanol where the photochemistry is in fact that of the anion since the molecule is rapidly deprotonated in the excited state. The hydrogen abstracting n, π^* state is still the lowest in the hydroxylated form of the molecule while a charge transfer state is located below the n, π^* state in the anion [12, 13]. These observations show that when anions are used instead of the neutral molecule, reactivity differences are to be expected not only based on the change of the electronic redistribution of the charge in the excited state but also on the possible change in the nature of the excited state.

2.2 Ion Pairing Effects

The reactivity of an anion as an electron donor will of course be related to its state in the solution. Ion properties are strongly solvent dependent and may differ greatly according to the strength of their interaction with the surrounding molecules. This is also true in the case of excited state anions as shown by their solvent dependent photophysics. In weakly polar solvents interionic interactions may dominate resulting in ion pair formations and a strong influence of the counter-cation on the anion properties will result. In polar solvents, ions are solvated and their mutual influence is considerably diminished.

The ion pair concept, introduced by Bjerrum [14], was critically reviewed by Szwarc [15] and definitions were given based on the mutual geometry of ions and solvent. The existence of 'loose' and 'tight' ion pairs was suggested by Winstein [16] and Sadek [17] and it is now common to speak about free ions (FI) as well as of solvent-separated ion pairs (SSIP) or contact ion pairs (CIP), having in mind the oversimplified picture:

CIP   SSIP   FIP

More subtle distinctions were made by Y. Marcus [18] according to the number of solvent molecules (two for solvent-separated and one for solvent-shared ion pairs) between the two ions. The tight and loose ion pair terminology is also used indicating that more than two kinds of ion pair may be considered: a CIP may have a different average interionic distance, depending on the nature of its environment and temperature. Similarly, the SSIP of a particular salt may vary according to the size and geometry of the solvating or complexing surrounding molecules [19].

In this context, effects of the counter-ion and of the solvent needs to be taken into account, if the photophysical and photochemical properties of anions are examined. Ion pairing effects are important when studying electron transfer photochemistry of the anions in the following aspects. The absorption wavelength will change and this determines a part of the energy available for the electron transfer reaction. The emission wavelength is of course also modified and this must be taken into account when fluorescence data needs explanation. The lifetimes are different and this may be a limiting factor of the photoreactivity of anions. The association level of ions will also influence their oxidation potentials: if an anionic charge is stabilized by an opposing cation in a tight ion pair, the oxidability of the anion will be reduced.

Absorption characteristics of ion pairs were reviewed by Smid [19] and Hogen Esch [20]. The absorption wavelength of an anion may be sensitive to the size of the associated alkali metal [21]. This occurs mainly in weakly polar solvents in which CIP are viable states. The spectrum of an anion belonging to a SSIP ion pair is much less influenced by the cation radius. Generally, when changing the cation from a soft and large caesium cation to a hard and small lithium, a hypsochromic shift is observed [7, 19]. This is attributed to a higher association (and stabilization) of the ion pair in the ground state when compared to the excited state: this difference increases when the cation radius decreases.

The ion pair status is not predictable only on the basis of the cation used. For highly delocalizable carbanions like 9-fluorenyl in THF, the CIP fraction was found to increase rapidly in the order Li < Na < K < Cs [22] and the caesium salt was found to be the most bathochromic in the series. The degree of association of carbanion ion pairs in non polar solvents is highly dependent upon the extent of charge delocalization [23]. With highly delocalizable anions SSIP will be formed with Li cation. A completely different sequence of ion pairs was shown in the case of naphtholate anions: the CIP fraction increases in the order K < Na < Li [24]. These anions are thus to be considered as more localized anions in order to agree with the preceding observations.

The solvent effects on the absorption spectra of ion pairs were studied by many authors and the direction of the observed shift depends on the change (increase or decrease) of dipole moment upon the electronic transition [25]. Generally a bathochromic shift is observed with an increase of solvent polarity. When going from a polar solvent to a less polar one, the association in the ground state increases more strongly than in the excited state: this may be understood if the ion pair switches progressively from SSIP to CIP status. Observations of this type were often made, together with cation effects, as for instance in the case of alkali phenolates and enolates [7], fluorenyl and other carbanion salts [22] or even for aromatic radical anions [26, 27].

The emission properties of anions as for instance carbanions of indene [28], xanthene or thioxanthene [29], fluorene [30, 31], diphenyl propene or pentadiene [32] were also discussed. It was shown that in certain cases, important changes may occur between ground and excited states, increasing the percentage of SSIP when the relaxed excited state is reached [33]. Two different types of behaviour were underlined. With highly delocalized carbanions like idenyl or fluorenyl salts, a red shift is observed, in absorption and a blue shift in emission, when the cation radius is increased [28, 31]. With less delocalizable nitrogen-centered anions like carbazolyl salts [34] or alkali naphtholate [24, 35], absorption and emission were found to shift together to the red. The observations are summarized in Fig. 1. In the case of naphtholates and nitrogen-centered anions, the cations are probably retained near the oxygen or nitrogen lone pairs even in the excited state. The opposite tendencies between these anions and more delocalizable carbanions is probably a consequence of the relative hardness and softness of the involved species [36].

In certain cases the absorption and emission frequencies may be distinguished for CIP and SSIP so that the concentration ratios may be measured [22, 31]. However, the use of time resolved measurements are necessary when superimposed spectra of the two ion pairs are found. This type of measurement was made in the case of β-naphtholate anion [24, 35] and the main results are

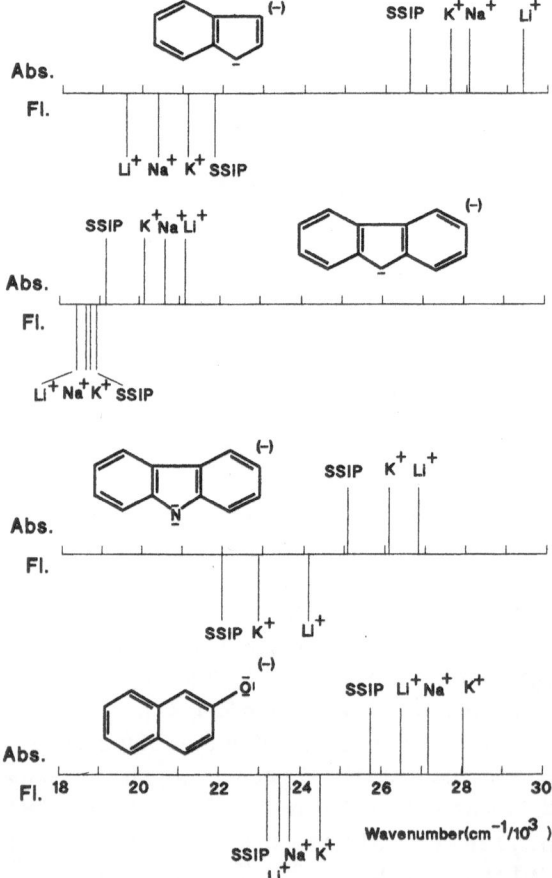

Fig. 1. Absorption and emission wavelengths for anions in ethereal solvents [24, 34]. The SSIP value for naphtholate is the value observed in the presence of a crown ether

presented in Table 1. A monoexponential decay was observed in protic solvents as well as in aprotic polar solvents. The lifetimes are however different because free anions are found in solvents like DMF or DMSO while hydrogen bonded anions are present in methanol. Blue shifts are observed when a polar solvent is replaced by a protic one and this effect parallels the observation made in less polar solvents when the cation radius decreases. This probably means that the stabilization of the anion ground state by an hydrogen bond parallels the stabilization by the electrostatic field of a cation.

A biexponential decay was observed in weakly polar solvents like dioxane, THF or dimethoxyethane: CIP and SSIP ion pairs coexist in these solvents with a majority of CIP in the case of lithium naphtholate while the SSIP dominates in the case of the potassium salt. In the presence of a crown ether and sodium cation, a monoexponential decay was observed and this allowed the attribution of the longest decay time to the SSIP.

Superimposed spectra (absorption and emission in the case of naphtholate anion) leads to wavelength effects. The choice of the excitation wavelength will affect the emission spectrum maximum and the fluorescence intensity. This is related to the percentages of CIP and SSIP excited at each wavelength, the last one being slightly bathochromic when compared with the first one. This type of observations has been confirmed in a rather similar situation where excited state ion pairs formed between naphthol and triethylamine are found to coexist in the CIP and SSIP status [37]. Analysis of emission data for ions in the excited state needs to be done with great care since the emission intensity for a given set of excitation and emission wavelengths is shown to depend on four factors: the relative concentration of the ion pairs in the ground state, the molar extinction coefficients of the two ion pairs at the excitation wavelength, the relative values of the fluorescence quantum yields of CIP and SSIP and the relative fluorescence intensities at the selected emission wavelength.

Table 1. Decay times (in nanoseconds) of β-naphtholate anions as a function of solvent and cation [24]

Solvent	Cation	τ (ns)	A[a]
Water	Na	9.2	
Methanol	Na	9.6	
THF	Li	6.4	
THF	Na	7.0	0.43
		13.8	0.24
THF	K	6.5	0.18
		14.4	0.39
THF	Na (CE)[b]	19.2	
DMF	Na	17.0	
DMSO	Na	17.8	

[a] Preexponential terms in cases of double exponential decays as obtained with an excitation wavelength of 330 nm and an emission measured at 430 nm.

[b] In the presence of dibenzo-18-crown-6.

Considering the appropriate kinetic scheme, a kinetic analysis of the excited state dynamics may be performed and values for the exchange rate constants between CIP and SSIP were estimated in the range of $1 \times 10^7 \, s^{-1}$ for alkali naphtholate in the excited state and in solvents like THF [24]. These are rather low values: higher rate constants (in the range of $10^9 \, s^{-1}$) were found for the solvation (change from a CIP to a SSIP) of radical ion pairs formed between alkylbenzenes and tetracyanoanthracene [38], alkylbenzenes and tetracyano-benzenes [39], polyaromatics and tetracyanoethylene [40] or trans-stilbene and fumaronitrile [41]. Similar values are expected for SSIP to CIP "desolvation" on the basis of observations made on benzophenone-diethylaniline systems [42]. These values are however related to the "solvations or desolvations" of highly delocalized radical ion pairs in a very polar solvent (acetonitrile) and we do not think that they are directly comparable to the preceding ones related to rather hard naphtholate anions in weakly polar solvent.

Another type of ion "pairing" effect likely to influence the electron transfer reactivity of dianions is the so called triplet association (triplet being here taken in the sense of triple association between two cations and one dianion). When Coulombic interactions in such multiple ions are considered, the electrostatic stabilization is enhanced when the dianion charges are close together, an unexpected observation if the electrostatic repulsions within the dianion are considered [43]. Multiple association of this type probably affects the reductive reactivity of the anionic species.

The previously discussed characters will influence the electron transfer rates implying anions. One of the simplest examples was given by the rate constant difference observed in reactions in which pyrene (Py) reacts with an electron or an electron-cation pair [44, 45]. The same type of difference was measured in the exchange between the radical anion of biphenyl (B) and pyrene (Py) [46]. The reduced reactivity is the consequence of the cation proximity in the ion pair.

$$Py + e^- \longrightarrow Py^{\cdot-} \qquad\qquad k = 7 \times 10^{10} \, M^{-1} s^{-1}$$

$$Py + e^-, Na^+ \longrightarrow Py^{\cdot-}, Na^+ \qquad\qquad k = 1 \times 10^{10} \, M^{-1} s^{-1}$$

$$Py + B^{\cdot-} \longrightarrow Py^{\cdot-} + B \qquad\qquad k = 5 \times 10^{10} \, M^{-1} s^{-1}$$

$$Py + B^{\cdot-}, Na^+ \longrightarrow Py^{\cdot-}, Na^+ + B \qquad k = 0.5 \times 10^{10} \, M^{-1} s^{-1}$$

When fluorescence quenchings due to photoinduced electron transfer from excited β-naphtholate are measured in THF, the cation effect is also clearly seen (see Table 2) [24]. The excited species are mixtures of CIP and SSIP whose percentages are not exactly known so that individual quenching rate constants are not available. However, the Stern Volmer constants clearly show that if an electron transfer reaction is to be runned in THF, the best choice, on kinetic grounds, is the potassium salt. When comparing the values for the sodium naphtholate the wavelength effects are also evidenced. In entries 2–4, the Stern Volmer value increases with the emission wavelength showing that, in the mixture of CIP and SSIP excited at 335 nm, the SSIP which emits more

Table 2. Quenching by naphthaene of the β-naphtholate fluorescence in THF [24]

Entry	Cation	λ_{ex} (nm)	λ_{em} (nm)	k_{SV} (M^{-1})
1	Li	358	407	3
2	Na	335	400	6
3	Na	335	420	11.2
4	Na	335	480	16.8
5	Na	368	420	15
6	Na	380	420	21.4
7	K	376	430	75

efficiently in the bathochromic part of the fluorescence spectrum is also more efficiently quenched than the CIP. When entries 3, 5 and 6 are compared, the Stern Volmer value is found to rise with the excitation wavelength showing that, at the more bathochromic wavelength, an increased fraction of SSIP is excited and quenched.

2.3 Electron Photoejection

Through the presence of a negative charge, an anion is expected to be a better electron donor than the corresponding neutral molecules since due to electron-electron repulsions the valence electrons are shielded from the nucleus. Photo-excitation often places the anion in a state of excess energy relative to the one of free radical plus electron: electron photoejection occurs from the excited state of some organic anions. This property has indeed often been observed in the case of negatively charged molecules. Implications of these observations in photo-induced electron transfer is not necessarily direct but they obviously demonstrate the ease of electron loss from photoexcited anions.

Aromatic radical anions were shown to photoeject electrons as in the case of pyrene-sodium in THF, giving electron-sodium cation pairs (presumably SSIP) as products [47]. The initial reactant may be restored through back electron transfer or pyrene dianion may be formed through a further electron transfer to the starting radical anion. Electron transfer scavenging of this electron-Na$^+$ pair by neutral aromatics was also demonstrated [44, 48, 49].

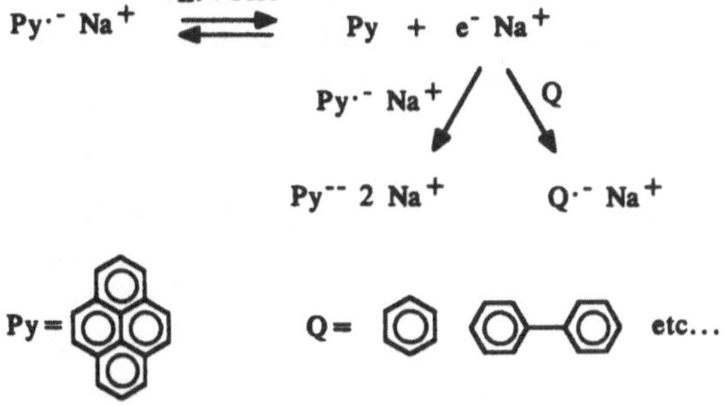

Photochemically induced electron ejection from anions has been used in the preparation of radical anions in the cavity of an ESR spectrometer. Subtle structural differences between the radical anions prepared by photoirradiation and by other ways were outlined [50, 51]. Two modes of radical anion photoinitiated formation were used. An electron is photoejected from the excited fluorene anion and adds to a neutral fluorene molecule in a subsequent step (A) or a dianion obtained from the potassium mirror reduction of tetrabenzocyctloocatetraene is directly photooxidized to the corresponding radical anion (B). This last method was also used to generate the otherwise inaccessible radical anion of pentalene, starting from the corresponding dianion irradiated in an ESR probe [52].

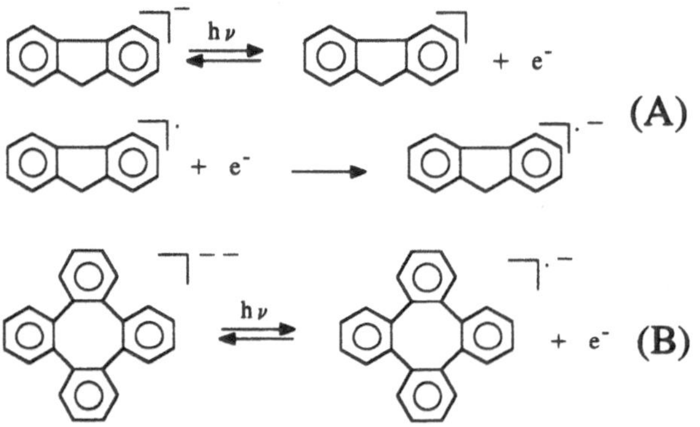

In the case of the cyclooctatetraene dianion (COT^{--}), the efficient electron photoejection under UV light [53] is followed by a disproportionation reaction between two radical anions, restoring a dianion and a neutral substrate [54]. Attempts were made to use these dianions irradiated in liquid ammonia at the surface of a semiconductor TiO_2 electrode in order to generate photocurrents: an anionic photogalvanic cell was found to work with stability during five days of irradiation. An analogous system was tested with cyclooctatetraene dianions covalently attached to antimony-doped SnO_2 semiconductor electrode. This system was found to be less stable than the preceding one [55, 56].

Closed shell anions were also examined and fluorene carbanions substituted in the 9 positions shown to photoeject electron under UV light in EPA glasses at 77 K: this process was demonstrated as biphotonic and strongly dependent on the ion pair status: slight modifications in the EPA glasses composition modifies the reaction efficiency [57, 58].

The 1,1,4,4-tetraphenylbutane disodium salt was found to eject electrons scavenged by added 1,1-diphenylethylene [59, 60]. In the case of *trans*-stilbene dianion, the photoejection of an electron was accompanied by isomerization in *cis*-stilbene radical anion while biphenyl was used as electron scavenger [61]. *cis*-Stilbene itself was used as scavenger of electron photoejected from aromatic radical anions like sodium perylenide [46]. These scavenging experiments are photoinduced electron transfer from excited dianions to neutral acceptors mediated through the ejected electron.

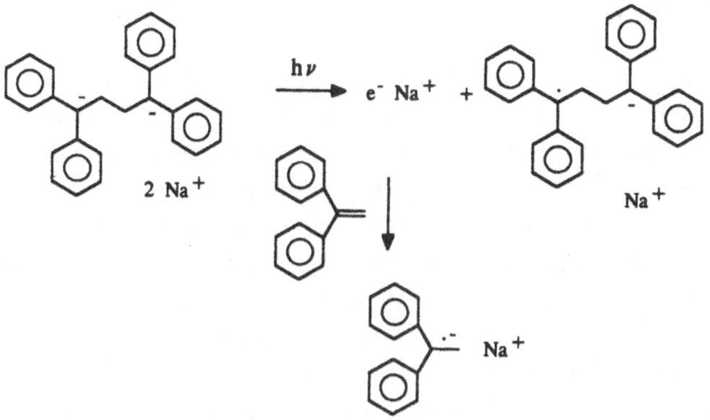

As may be expected oxyanions of molecules known as antioxidants were also found to produce solvated electrons under irradiation. Phenolate anion in water leads to aqueous electron photoejection under 254 nm irradiation [62] and to benzosemiquinone through phenoxy radicals as shown by CIDNP arguments [63]. In the case of β-naphtholate [64] the electron is photoejected from the unrelaxed singlet state [65, 66] with a low quantum yield (see Table 3) [67].

Solvated electrons do not inevitably interfere in photoinduced electron transfer. Their observations are often made under laser irradiations in order to detect these transients efficiently. Under these conditions processes may occur in a multistep and biphotonic way [68], the triplet state being one of the possible intermedites [69]. The two photon process of electron ejection may dominate under pulsed laser conditions of high excitation energy while a monophotonic process prevails under continuous laser intensity conditions. These differences may explain the quantum yields observed for instance for the electron photo-ejection from excited phenolate in water under different irradiation conditions (0.23 [70], 0.17 [71], 0.37 [72]). When using conventional light sources, a relatively low yield of solvated electron is to be expected [69, 72].

Implications of ejected electrons in conventional photochemistry may be expected with some anions showing a CTTS (charge transfer to solvent) transition (as for ferricyanide anion) and giving fast and monophotonic ejected

Table 3. Photophysics of the β-naphtholate anion: life-times and quantum yields [67].

τ (ns)	Φ_{e-s}	Φ_{S1}	Φ_F
9	0.08	0.61	0.16
9	0.04	0.69	0.18
12	0	1.0	0.36

electron [73]. In a phenoxide photolysis, a phenoxy radical is formed giving a sequence of other reactions implying sodium borohydride: this reaction is effectively quenched in the presence of a solvated electron scavenger [74] and phenolate has been shown to photoeject electrons quickly and by a one photon process [73].

Carboxylates as the phenylacetate anion also eject electrons in methanol [75] giving benzyl anion after recombination between solvated electron and benzyl radical [76]. In phenyl substituted carboxylate anions (from benzoate to phenylbutyrate) in water the quantum yield of photoejected electron was found between 0.002 and 0.03: these values increase with increasing excitation energy and with the number of CH_2 separating the phenyl and carboxylate group [77]. In the case of phenylalanine and tryptophan in water, the mechanism seems to differ according to the conditions: biphotonic and from a triplet state in neutral solution or monophotonic in basic medium [78, 79, 80]. In certain cases, the quantum yield for electron ejection is found to increase with pH [79]. The anion of bromouracil also gives hydrated electron [81].

A disodium phenyl phosphate was also reported to photoeject electrons from the relaxed and unrelaxed S_1 state. This may occur with different efficiencies, according to the pH, from the neutral, anionic or bis anionic species and a wavelength effect on the quantum yield was observed [82]. Hydrated electron has also been detected in the photolysis of ascorbic acid [83].

An important peculiarity of these electron photoejections is the dependency of their quantum yield on the nature of the solvent. Large differences may be found between two organic solvents even when the very similar THF and dioxane are compared [84]: very efficient electron ejection from tetraphenylethylene dianions in THF and no electron detected in dioxane. The ejection quantum yield rapidly drops when the solvent is no longer water as shown for phenol (H_2O:0.021 MeOH:0.0018 i-PrOH:0.00045) [85]. This reduces the impact of electron ejection on the photoelectron transfers occurring in other solvents than water.

3 Kinetics and Thermodynamics of Electron Transfer

In order to extract rate constants for electron transfer from observed fluorescence quenching measurements the following mechanistic scheme is commonly used. In this scheme, presented here in the case of an excited anionic donor, k_d and k_{-d} are the rate constants for the diffusive formation and dissociation of an encounter complex within which the electron transfer takes place (k_{el}).

$$D^{-*} + Q \underset{k_{-d}}{\overset{k_d}{\rightleftharpoons}} [D^{-*} \ldots Q] \overset{k_{el}}{\longrightarrow} D^{\cdot} + Q^{\cdot -} \qquad (1)$$

The steady state approximation applied to Eq. (1) gives the following expression for the observed fluorescence quenching rate constant:

$$k_q = \frac{k_d}{1 + k_d/(k_{el}K_d)} \qquad (2)$$

where $K_d = k_d/k_{-d}$ corresponds to the diffusional equilibrium constant.

Conventional Hammett free energy relationships have sometimes been applied to electron transfer rate constants concerning anions. This was made with the measurements of the quenching rate constants of aromatic triplet carbonyls [86] or tris (2,2'-bipyridine) ruthenium II [87] by substituted phenolate anions in aqueous acetonitrile and in water respectively. With the carbonyl excited states, quenchings by phenols were compared to quenchings by phenolates: with phenols, a significantly negative Hammett slope ($\rho^+ = -1.4$) was found showing the electron accepting character of the carbonyls while with the more donating phenolates, the quenchings were found to be controlled by the diffusion and thus insensitive to the substituent effects. For the quenching of $Ru(bpy)_3^{2+*3}$ by phenolates, the electron transfer rates were smaller than the diffusion and a negative Hammett slope was again found ($\rho^+ = -3.0$).

The Eyring equation for k_{el} may be introduced in Eq. (2) leading to:

$$k_q = \frac{k_d}{1 + (k_d/K_d v)\exp(\Delta G_{el}^{\neq}/RT)} \qquad (3)$$

with $v =$ frequency factor of the unimolecular rate constant (k_{el}) in the case of a barrierless transfer and $\Delta G_{el}^{\neq} =$ free energy of activation of the electron transfer step.

The standard free energy difference of a photoinduced electron transfer reaction between a donor and an acceptor may be estimated by the well known Eq. (4) [88, 89]:

$$\Delta G^\circ = 23.06 (E_{ox} - E_{red} - E_{00}) + (Z_1 - Z_2 - 1)\frac{e^2}{\varepsilon r_{AD}} \qquad (4)$$

where E_{ox} and E_{red} are the redox potentials of the donor and acceptor, E_{00} is the excitation energy of the excited reactant, Z_1 and Z_2 are the charge numbers of donor and acceptor respectively. e is the charge of an electron, ε is the dielectric constant of the solvent and r_{AD} is the reaction distance.

Rate constants for electron transfer may be related to the free energy ΔG° of the reaction through the classical Marcus equation Eq. (5), where $\Delta G_0^{\neq}$ is the intrinsic activation barrier of the reaction process [90, 91].

$$\Delta G_{el}^{\neq} = \Delta G_0^{\neq}(1 + \Delta G^\circ/4\Delta G_0^{\neq})^2 \qquad (5)$$

This relation implies the existence of a so called Marcus inverted region in which the electron transfer rate constant no longer increases but decreases with the exothermicity of the process. Rehm and Weller have concluded from examination of numerous cases that the inverted region does not exist and that another expression than the Marcus relation has to be used [88, 92]. This relation (6), based empirically on their results does not implicate an inverted region and is as follows:

$$\Delta G_{el}^{\neq} = [(\Delta G^{\circ}/2)^2 + (\Delta G_0^{\neq})^2]^{1/2} + \Delta G^{\circ}/2 \tag{6}$$

Two factors were recognized by Marcus as contributing to the reorganization term $\Delta G_0^{\neq}$: an inner contribution $(\Delta G_0^{\neq})_v$ corresponding to differences in vibrational frequencies between the reduced and oxidized forms of the reagents and an outer contribution $(\Delta G_0^{\neq})_s$ related to the solvent reorganization. Always, according to the Marcus treatment, this last contribution can be estimated by the Born treatment, leading to Eq. (7).

$$\Delta G_0^{\neq} = e^2/4(1/2r_D + 1/2r_A - 1/r_{AD})(1/n^2 - 1/\varepsilon) \tag{7}$$

Some examples of correlations of electron transfer rate constants photoinduced from excited organic anions have been presented. In benzene solvent, intra ion pair electron transfer rates between borate anions and excited cyanine cation have been measured and correlated with the reaction free energy [93]. In a non polar medium such as benzene, the solvent reorganization energy is easily calculated to be nearly zero so that the 0.1 eV obtained from this correlation for $\Delta G_0^{\neq}$ was attributed entirely to the inner contribution $(\Delta G_0^{\neq})_v$.

The radical anions of anthraquinone and 9,10-dicyanoanthracene are fluorescent in the absence of oxygen. Their quenching rate constants by various aromatic acceptors has been measured in DMF and correlated with the free energy of the reaction [94, 95]. A Rehm-Weller type curve has been found but although the inverted region was not evidenced, Weller equation was found less suited for the treatment of the data. The quenchings were treated according to the Marcus model and a very large $\Delta G_0^{\neq}$ was found to fit the results. This was attributed to a folded conformation of the anthraquinone excited radical anion, a geometry quite different from the planar ground state and justifying an important (inner) bond reorganization energy.

In a series of measurements devoted to the β-naphtholate fluorescence quenchings, the results were also treated by the Marcus model and although the curves obtained were of the Rehm-Weller shape, the $\Delta G_0^{\neq}$ value proposed by Weller did not fit the results [96]. Again this value was found to be larger (5.8 Kcal) and the Marcus model more appropriate to understand the results (Fig. 2).

It was of course of interest to review in this context the use of the Rehm-Weller empirical equation, Eq. (6), versus the Marcus model of Eq. (5). The method used to correlate the results for the β-naphtholate quenchings was thus applied to the important Weller series of fluorescence quenching measurements made in acetonitrile [88] and this shows that 60 values (on a total number of 65) fit quite well with a mean intrinsic activation barrier of 5.5 Kcal·M^{-1} [96] (see Fig. 3). Moreover, this value is not far from the expected value for the outer contribution $(\Delta G_0^{\neq})_s$ as calculated by Eq. (7) and it has often been underlined that the solvent reorganization term is effectively the dominant contribution in

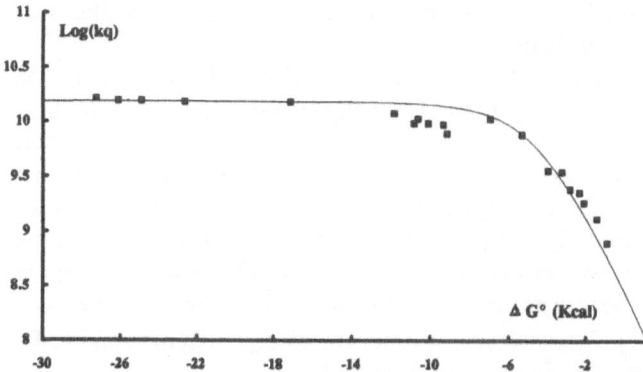

Fig. 2. Fluorescence quenching rate constants of excited β-naphtholate anion correlated with the electron transfer free energy [96]

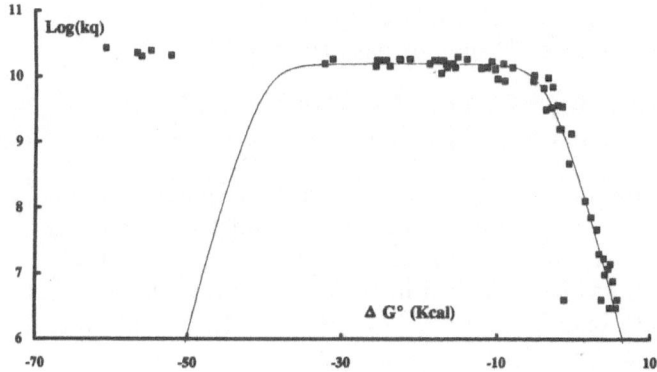

Fig. 3. Rehm Weller results (fluorescence quenchings in acetonitrile) [88] correlated by the Marcus model with $\Delta G_0^{\neq}$ of 5.5 Kcal. M^{-1} [96]

these examples involving aromatic molecules. The 5 "unfitted" quenching rate constants, located in the extreme exothermic region of the graph concerned quenchings by tetracyanoethylene, a quencher quite different from the other aromatic compounds: there are probably reasons to believe that in this case an extra reorganization energy (probably an inner contribution) is required in order to understand the results.

If correct, this may be instructive in the discussions about the inverted Marcus region. In these freely diffusing systems, the exothermicity needed to demonstrate the inverted area is much more difficult to reach than previously expected. When this region is approached, some new efficient acceptors or donors are needed. Generally, this will introduce new uncertainties as far as the quenching mechanism or the $\Delta G_0^{\neq}$ calculation is concerned: in the Weller series, the TCNE illustrates this difficulty.

When dealing with organic anions, a problem often encountered is to find correct values for the oxidation potential. Electrochemical methods frequently lead to irreversible values generally related to the high reactivity of the radical species formed after oxidation. One way to circumvent this problem is to use a series of electron transfer quenching measurements involving the anion and to adjust the oxidation potential in order to fit the results to a Marcus correlation. This has been done using the Weller equation to calculate the oxidation potentials of a series of borate anions [93] and the Marcus model in the case of naphtholate anions [96]. A reasonable estimation of E_{ox}, with discrepancies of 0.2 V as a maximum can be obtained by this method.

Another problem, often important when dealing with Marcus correlations may arise when the potentials are known in one solvent while the kinetic experiments are made in another one. Calculation of ΔG requires inclusion of the effect of solvent change on the electrochemical potentials. The usual procedure considers the difference in the solvation energies of the ions according to the Born equation. For instance, when using acetonitrile measurements to evaluate values in benzene, Eq. (8) may be used [97, 98].

$$(E_{ox} - E_{red})_{PhH} = (E_{ox} - E_{red})_{MeCN} - e^2/2(1/r^+ + 1/r^-)$$
$$\times (1/\varepsilon_{PhH} - 1/\varepsilon_{MeCN}) \tag{8}$$

This is an unimportant contribution when solvents of similar polarities are concerned (as DMF and acetonitrile) but very significant corrections are calculated if one changes from the polar acetonitrile to benzene (from 1 to 1.5 V). This type of correction has been applied in the case of intra ion-pair electron transfer between borate anions and excited cyanine in order to make a correct estimation of the electron transfer free energy [93].

When using the Marcus treatment in the case of anions, great care must be taken in order to include in the method the possible specific interactions between anions and solvent. This was for instance evidenced in the study of naphtholate anions in a protic solvent (methanol) where it was necessary to include the effect of the hydrogen bonding between solvent and anion in the Marcus treatment of the results [96]. The hydrogen bonding status of the ground state of this anion is not the same than in the excited state: excitation corresponds to a redistribution of the charge whose density decreases on the oxygen and increases in the aromatic part of the molecule. The effect of this change was evidenced in the correlations of two different sets of fluorescence quenchings: naphtholate anions excited states were quenched by various aromatic neutrals and on the other hand an excited aromatic acceptor (perylene) was quenched by ground state naphtholate anions.

$$NO^{-*} + ArH \longrightarrow NO^{\cdot} + ArH^{\cdot -} \qquad \text{1st set of quenching constants}$$

$$Pe^* + NO^- \longrightarrow NO^{\cdot} + Pe^{\cdot -} \qquad \text{2nd set of quenching constants}$$

NO^- = naphtholate anion
ArH = various aromatic quenchers
Pe = perylene

The oxidation potentials of the naphtholate anions, as extracted from the two different sets of measurements did not agree. Agreement was only obtained when the breaking of an hydrogen bond was taken into account in the Marcus treatment. When naphtholate anion is used in the ground state (perylene quenching) the hydrogen bond is broken during the electron transfer step itself and in this case an extra reorganization energy linked to this H-bond rupture must be introduced in the calculations. When the naphtholate anions are used in the excited state, the H-bond is broken by the excitation and this has consequences on the way of evaluating the E_{00} term of the electron transfer free energy. In the cases of neutral molecules with small Stokes shifts, the available excitation energy is calculated by the best possible estimation of the 0–0 transition. This is normally found as the mean value between excitation and emission energy. In protic solvent, an anion like naphtholate has lost the stabilization energy of the broken H-bond when reaching its Franck Condon excited state. When going back to the ground state, the H-bond is reconstructed during its relaxation: a dissymmetry is expected between excited and ground state relaxations. The energy really available for the electron transfer, starting from the relaxed excited anion, lies closer to the absorbed energy than to the one emitted by fluorescence. A correct estimation of $\Delta G°$ is only possible if this is accounted for.

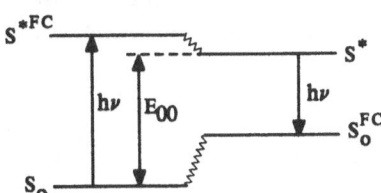

In this section devoted to the free energy correlations in anionic electron transfer photochemistry it is worth mentioning that many non-photochemical reactions involving anionic nucleophiles like Grignard reagents, alkoxides, thiolate etc. . . . have been proposed to proceed via an initial electron transfer and subsequent radicaloid steps instead of following classical polar mechanisms. The use of Marcus type correlations in order to estimate the range of rate constants for these reactions has been proposed and discussed by L. Eberson [89, 99]. Even if some of the necessary thermodynamic data are often rather uncertain, it is possible to estimate a correct order of magnitude for the electron transfer rates and to accept or reject on this basis the proposal of an electron transfer mechanism.

4 Reactions Employing Anion Excited States

4.1 $S_{RN}1$ Reactions

The best known reaction implying anions and photoinduced electron transfer is the $S_{RN}1$ reaction. In this synthetically useful process a radical chain reaction (steps (9) to (11)) is running: ArX is an electron accepting substrate, Nu is the nucleophile and ArNu is the substituted product:

$$ArX^{\cdot -} \longrightarrow Ar^{\cdot} + X^{-} \tag{9}$$

$$Ar^{\cdot} + Nu^{-} \longrightarrow ArNu^{\cdot -} \tag{10}$$

$$\frac{ArNu^{\cdot -} + ArX \longrightarrow ArNu + ArX^{\cdot -}}{Nu^{-} + ArX \longrightarrow ArNu + X^{-}} \tag{11}$$

This is the generally accepted scheme but, according to a recent reexamination of various experimental results a reconsideration of the mechanism is possible. In certain cases at least, a nucleophilic substitution on the intermediate radical anion has also to be considered [100] and with another chain mechanism made of steps (11) and (12), the aryl radical intermediate no longer needs to be involved:

$$Nu^{-} + ArX^{\cdot -} \longrightarrow X^{-} + ArNu^{\cdot -} \tag{12}$$

In contrast to this proposal, A.L.J. Beckwith has recently established the photostimulated $S_{RN}1$ mechanism using diphenylphosphide or phenylthiolate anions and an organic bromide as an accepting substrate. A radical clock was used whose rearrangement was the demonstration of the radical intermediate [101].

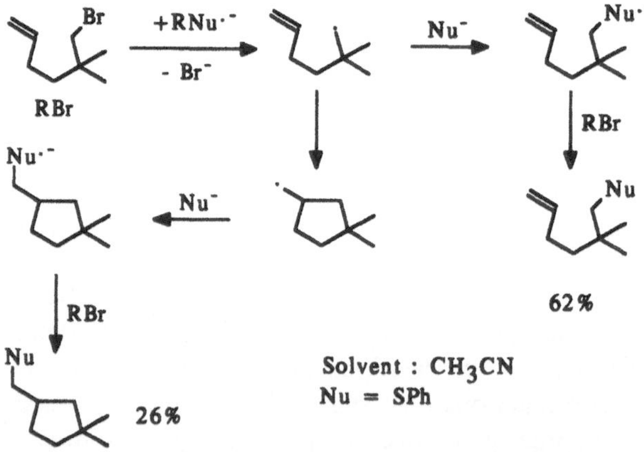

Solvent : CH_3CN
Nu = SPh

Large quantum yields often obscure the role of the initiation step and the involvement of an excited anion or excited acceptor is not often clear. An excited anion is however highly probable since its bathochromic absorption makes it the best candidate for absorption in the commonly encountered reaction conditions. In many $S_{RN}1$ examples, wavelengths > 350 nm are used with pyrex vessels containing an anion and haloaromatic acceptors [102] lacking significant absorptions in this part of the spectrum. The activation of photochemical $S_{RN}1$ reactions may often be written according to Eqs (13) and (14).

$$Nu^- \xrightarrow{h\nu} Nu^-* \qquad (13)$$

$$Nu^-* + ArX \longrightarrow Nu^\cdot + ArX^{\cdot -} \qquad (14)$$

1-Iodoadamatane was found unreactive with the nitromethane anion under light stimulation, but the $S_{RN}1$ substitution product (1-adamantylnitromethane) was, however, obtained when the reaction was activated in the presence of acetone enolate [103]. This underlines quite well the initiation role of the anion. The enolate is able to transfer an electron to the iodoadamantane acceptor under photostimulation while the nitromethane anion is a too weaker electron donor in order to initiate the photoreaction. A clear example of the involvement of the anion excited state in the activation of $S_{RN}1$ process was recently given: diphenyl phosphide anion absorbs strongly in the visible region (λ = 475 nm) and his $S_{RN}1$ reaction with organic sulfides was found to promote carbon-sulfur cleavage in the ordinary laboratory light [104]. Another proof of the involvement of the anion was given by the fluorescence quenching of the 1,3 diphenylindenyl anion by bromobenzene as acceptor. In this case, the photoarylation quantum yield of the reaction was found to follow the classical kinetic relation with the bromobenzene concentration confirming that the activation is effectively the photoinduced electron transfer from the anion to the acceptor [105].

The situation may however be more complicated if charge transfer complexes (CTC) between the substrate and the anion are formed. This was suggested [104] in the case of the $S_{RN}1$ reaction between ethyl-, phenyl- and diethyl phosphite anion in DMF where the UV spectrum of the mixture shows an enhanced absorption attributed to a CTC, the last one being the possible mediator for the photochemical activation of the reaction.

The possibility of a thermally activated electron transfer from an anion to an acceptor is not always excluded. As explained by Bordwell [106], this will of course depend on the oxidation potential of the anion or on its related basicity and this author has shown that the electron transfer reactivity of carbanions rapidly decreases with a decrease of basicity. It is thus possible that among the dark $S_{RN}1$ reactions some of them are activated by an initial ground state electron transfer from the anion to the accepting substrate.

The $S_{RN}1$ subject has already been reviewed [4, 107, 108, 109, 110, 111] and the large variety of the used anions (carbanions, nitronates, thiolates, enolates, anions of group V etc. . .) with several electron accepting substrates (chloronitro- or dinitro- compounds, dimethylsulfoxide, organic halides, etc. . .) was

Table 4. Recent examples of photo $S_{RN}1$ implying activation by photoinduced electron transfer from excited anion

Anion	Electron accepting substrate	Solvent	Products	Ref:
		NH_3	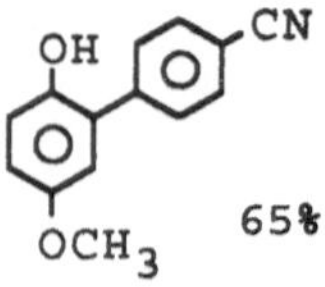 65%	[112]
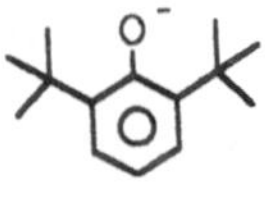	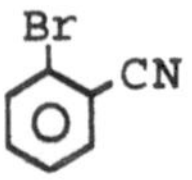	NH_3	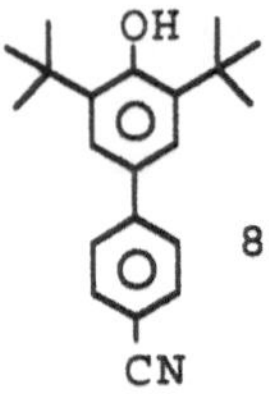 88%	[112]
	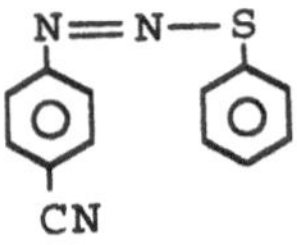	DMSO	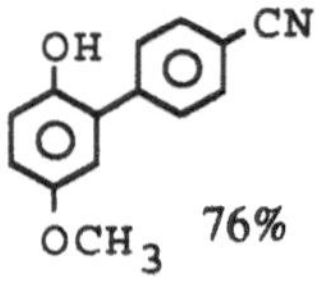 76%	[113]
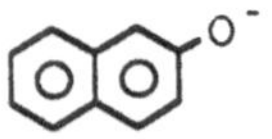	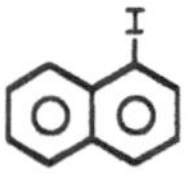	NH_3	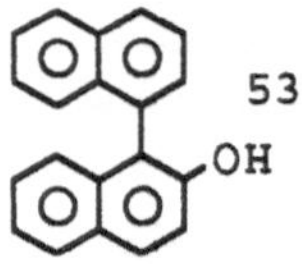 53%	[114, 115, 116]
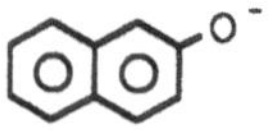	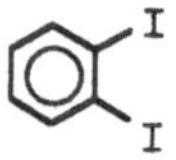	NH_3	45% 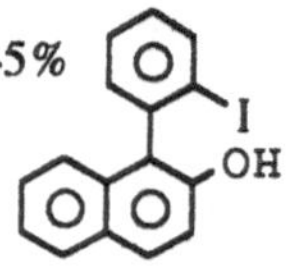 24% 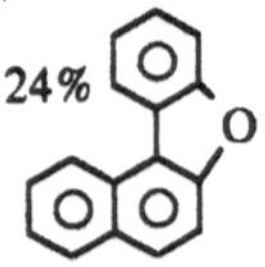	[117]
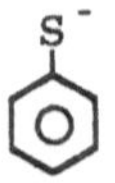	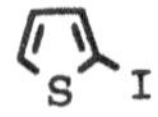	CH_3CN	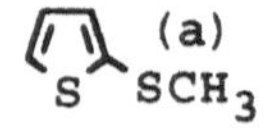 (a) 10% 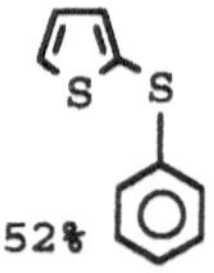 52%	[118]
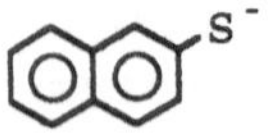	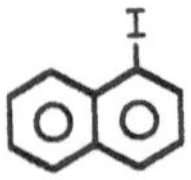	NH_3	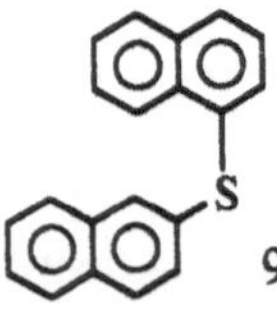 95%	[116]
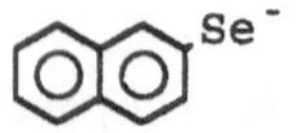		NH_3	85% 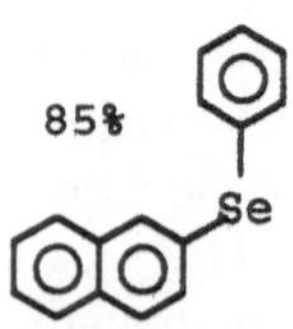	[116]

Table 4. (Continued)

Anion	Electron accepting substrate	Solvent	Products	Ref:
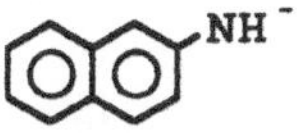	 I	NH$_3$	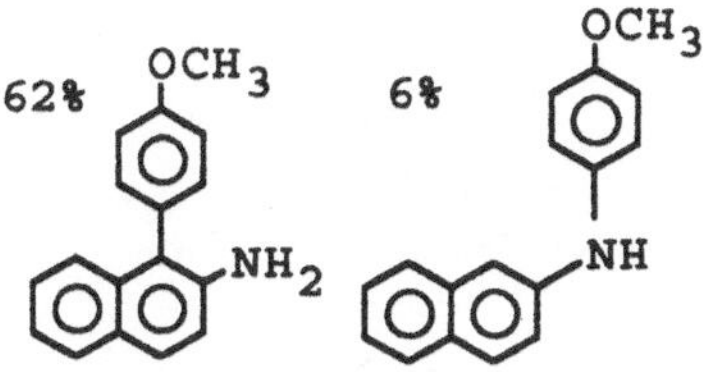 62% OCH$_3$ ... NH$_2$ 6% OCH$_3$ ... NH	[116]
(C$_6$H$_5$)$_2$P$^-$	I (bicyclic)	NH$_3$	87% (C$_6$H$_5$)$_2$P	[119, 120, 121]
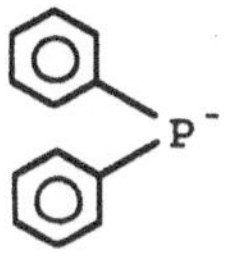 (C$_6$H$_5$)$_2$P$^-$	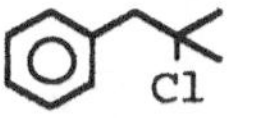 Cl	NH$_3$	 36% P(C$_6$H$_5$)$_2$	[122]
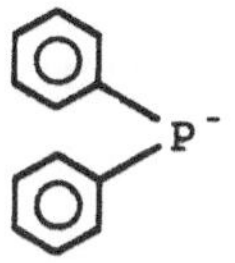 (C$_6$H$_5$)$_2$P$^-$	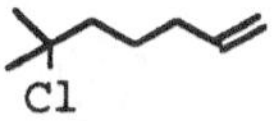 Cl (alkene chain)	NH$_3$	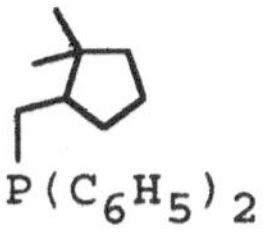 P(C$_6$H$_5$)$_2$	[122]
(C$_6$H$_5$)$_2$P$^-$	SO$_2$CH$_2$CH$_3$	DMSO	PCH$_2$CH$_3$ 71%	[104]
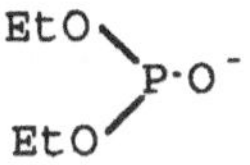 EtO$\,$P-O$^-$ / EtO	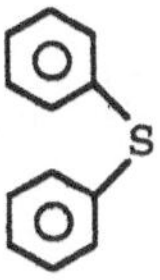 S	DMSO	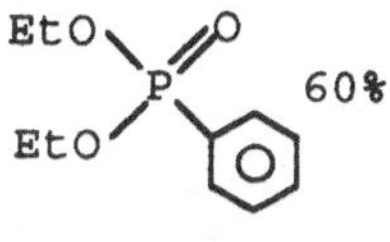 EtO P=O / EtO 60%	[104]
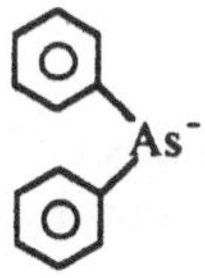 As$^-$	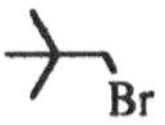 Br	NH$_3$	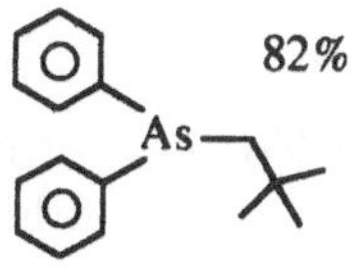 82% As	[123]
 O$^-$	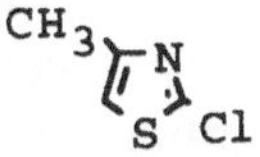 CH$_3$ thiazole Cl	NH$_3$	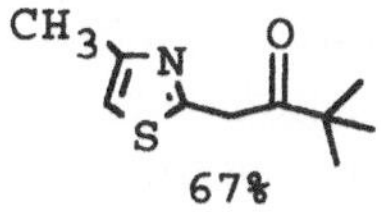 CH$_3$ thiazole C=O 67%	[124]

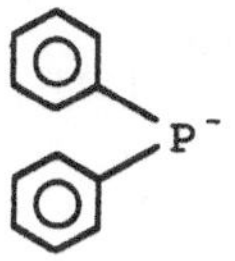

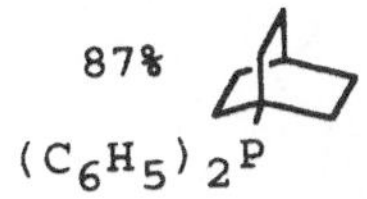

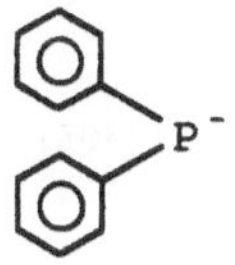

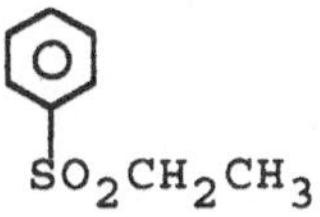

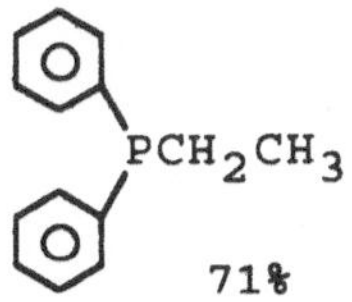

113

Table 4. (Continued)

Anion	Electron accepting substrate	Solvent	Products	Ref:
 (pinacolate anion structure)	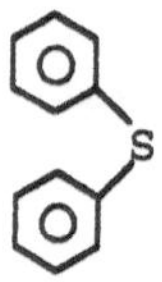 (phenyl phenyl sulfide structure)	NH_3	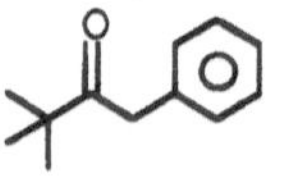 (product structure) 98%	[104]
 (pinacolate anion structure)	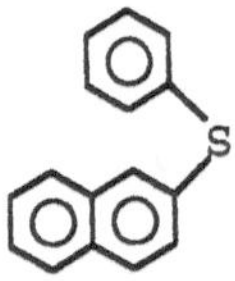 (phenyl naphthyl sulfide structure)	DMSO	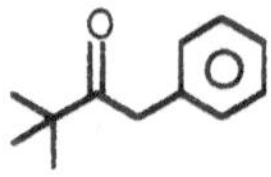 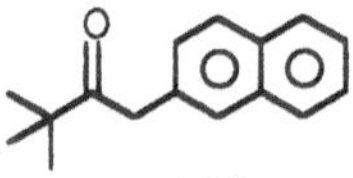 (product structure) 64% (product structure) 18%	[104]
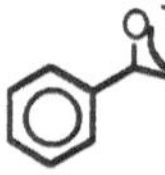 (benzoyl enolate structure)	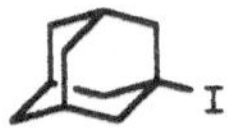 (adamantyl iodide structure)	DMSO	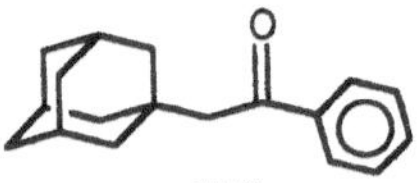 (product structure) 65%	[103]
 (cyclohexyl nitrile carbanion, CN)	 (4-bromoanisole, Br / OCH_3)	NH_3	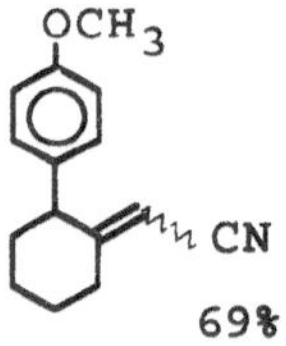 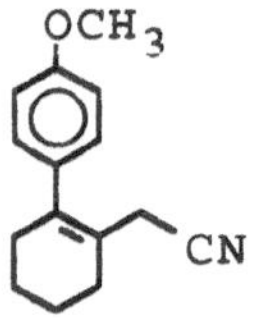 (product structures with OCH_3 and CN) 69%	[125]
$^-CH_2NO_2$	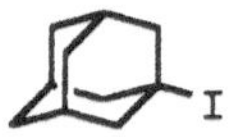 (adamantyl iodide structure)	DMSO	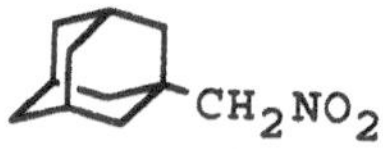 (product structure) CH_2NO_2 87% (b)	[103]
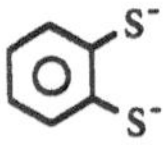 (benzene dithiolate, S^- S^-)	 (dichloro structure, N Cl / N Cl)	NH_3	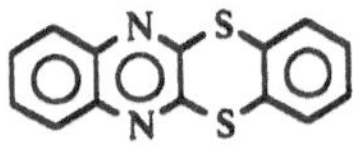 (product structure) 100%	[126]

(a) Reaction was quenched with methyl iodide.
(b) Reaction performed in the presence of acetone enolate.

sufficiently underlined. To discuss the scope of this reaction is not the subject of this review coming after several others. Some more recent reactions pertaining to anion initiation are however summarized in the form of Table 4.

In the case of the halothiazole presented in Table 4, if the pinacolate anion is replaced by an α-cyano-carbanion, the expected product is still formed, but in this case the dark reaction works better than the photostimulated one, suggesting that products are photodegraded [124]. The authors interpretation was that in this case a classical thermal aromatic nucleophilic substitution (S_NAr) mechanism with an addition-elimination scheme governs the reaction.

The regiochemistry of the coupling products between aryl radicals and ambident nucleophiles was at the center of the recent work of Pierini et al. [116] (see Table 4). Changing the heteroatom of the nucleophile into a softer one by going down in the periodic table (from naphtholate to naphthylthiolate for instance) leads to an increase in heteroatom substitution: C substitution increases when the nucleophile heteroatom is more electronegative or if the aryl moiety of the nucleophile is a more delocalized one.

Unsymmetrical hydroxybiaryls were conveniently prepared by $S_{RN}1$ reactions in which the thiolate group of phenylazosulfides is the leaving group [113]. Here again the C-arylation regiochemistry was verified with the aryloxide nucleophile. In this reaction, a photoinduced electron transfer from the phenolate to the diazosulfide is the probable initiating step. However, care must be taken in the mechanistic interpretation: it was indeed demonstrated that diazosulfides are reactive, under $S_{RN}1$ photochemical conditions with tetrabutyl ammonium cyanide [127, 128], although electron transfer is impossible between the poor electron donating cyanide and the diazo substrate. The starting reaction in this case is the photolysis of the diazosulfide itself.

Photostimulated $S_{RN}1$ reactions of bridgehead halides [119, 120] and of tertiary chlorides [122] has been examined and considered as a possible route to nucleophilic substitution of these normally unreactive substrates. In the case of the reaction of 1-iodoadamantane with ketone enolates, the photosubstitution product yield was improved in the presence of 18-crown-6 [103]. This is probably related to the ion pairing effect already discussed in a preceding section.

When vinylogous anions of α-cyano compounds like the carbanion of cyclohexenylideneacetonitrile are made to react with haloaromatics under $S_{RN}1$ photostimulated conditions, arylation on the aliphatic ring, at the C_γ position from the cyano group is only observed (see Table 4) [125]. This shows that the coupling of the anion with the aryl radical only gives the most stable radical anionic adduct in which the nitrile group and the double bond are conjugated. With the odd electron located in a π^* orbital, in this case, no fragmentation of the radical anion was observed.

In $S_{RN}1$ reactions implying aryl radicals and acetonitrile anions, the phenylacetonitrile radical anionic intermediate undergoes bond rupture showing that in this case a σ^* radical anion is probably formed [110].

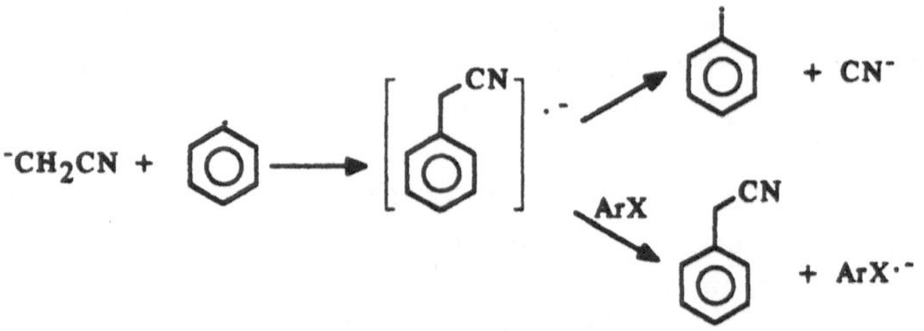

$^-CH_2CN$ + [diagram] $\longrightarrow$ [radical anion intermediate] $\longrightarrow$ [product] + CN^-

ArX

[product with CN] + ArX·$^-$

In a study of $S_{RN}1$ photoreactions of halothiophenes with phenylthiolates anions, phenylthiothiophene radical anions are formed and the yield of the expected phenylthiothiophene product is limited by the bond rupture in this intermediate: the thiophenylthiolate anion is formed and detected as thiomethyl ether after quenching of the reaction by methyl iodide. By adding benzonitrile as an extra electron acceptor, the bond rupture may be controlled and the selectivity of the reaction has been improved in favour of the thiophenyl ether but at the expense of the overall reaction rate [118].

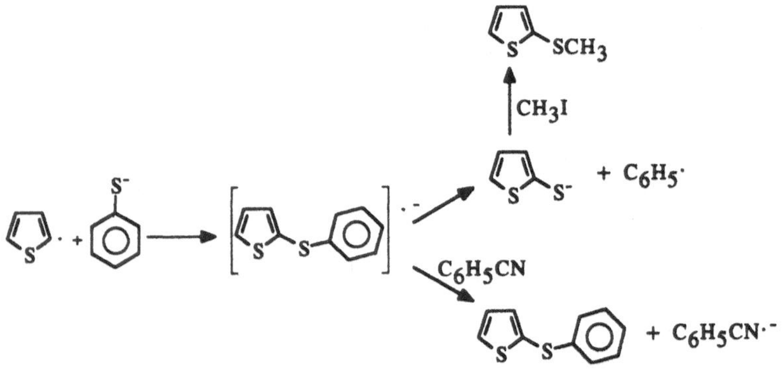

[reaction scheme: thiophene SCH₃ product via CH₃I]

CH_3I

[thiophene anion] + [benzene] $\longrightarrow$ [radical anion intermediate] $\longrightarrow$ [thiophene-S⁻] + C_6H_5·

C_6H_5CN

[thiophene-S-phenyl product] + C_6H_5CN·$^-$

Intramolecular examples of photostimulated $S_{RN}1$ reactions implying excited anions with a demonstrated chain mechanism have been reported as in the case of anilide anions and N-acyl benzylamines used as precursors of oxindoles and isoquinolines [129].

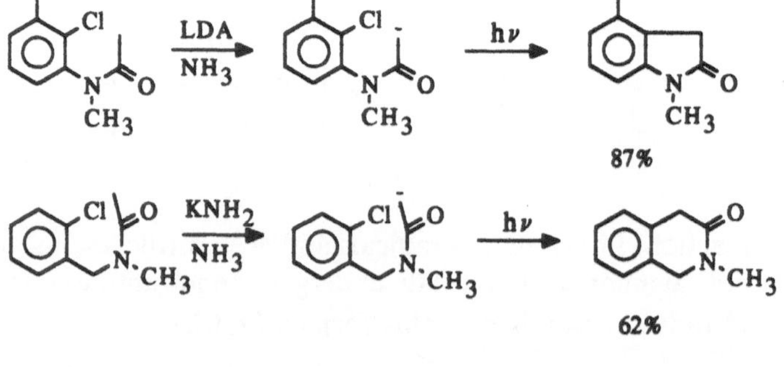

[reaction scheme 1]

$\xrightarrow[NH_3]{LDA}$ $\xrightarrow{h\nu}$

87%

[reaction scheme 2]

$\xrightarrow[NH_3]{KNH_2}$ $\xrightarrow{h\nu}$

62%

4.2 Carbanion Reactions

Photoinduced electron transfer from carbanions were observed by Tolbert who started working in this area with "the conviction that the use of low energy chromophores and visible light should minimize troublesome oxidation-reduction reactions" [130]. His objective was successfully achieved in certain cases [6], but a substantial part of his work was however concerned with the electron donating propensity of excited carbanions.

When considering the sequence $CH_3^-, C_6H_5CH_2^-, (C_6H_5)_2CH^-, (C_6H_5)_3C^-$, the last one is the most bathochromic of the series and the most stabilized one. This must be accompanied by a weakest oxidability. Nevertheless, a photoinduced electron transfer from photoexcited triphenylmethyl anion to DMSO was shown to occur, leading to C–S bond rupture and formation of triphenylethane. The mechanism, consisting of Eqs (15) to (18), resembles the $S_{RN}1$ but instead of regenerating the radical anion of the electron accepting reactant, recovering of the triphenylmethyl anion occurs from the triphenylmethyl radical [131, 132].

$$Ph_3C^{-*} + (CH_3)_2SO \longrightarrow Ph_3C^{\cdot} + (CH_3)_2SO^{\cdot -} \tag{15}$$

$$(CH_3)_2SO^{\cdot -} \longrightarrow CH_3^{\cdot} + CH_3SO^- \tag{16}$$

$$CH_3^{\cdot} + Ph_3C^- \longrightarrow Ph_3CCH_3^{\cdot -} \tag{17}$$

$$Ph_3CCH_3^{\cdot -} + Ph_3C^{\cdot} \longrightarrow Ph_3CCH_3 + Ph_3C^- \tag{18}$$

The tetraphenylcyclopentadienide and the fluorenide carbanions were photodimethylated by DMSO in a quite similar way [133].

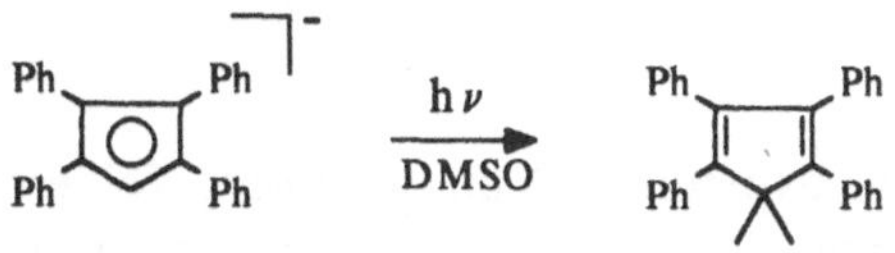

In these non chain $S_{RN}1$ type reactions occurring in DMSO, the process may be quenched by molecules accepting an electron from the DMSO radical anion [134]. This $S_{RN}1$ type-DMSO mechanism was clearly distinguished from other single electron transfer reactions runned in other solvents with the excited triphenylmethyl anion. When the photolysis is runned in THF and in the presence of acceptors like bromobenzene three products, among which a solvent adduct, are formed [135]. The product distribution between tetraphenylmethane and p-biphenylyldiphenylmethane arises from the reaction between a triphenylmethyl anion and a phenyl radical.

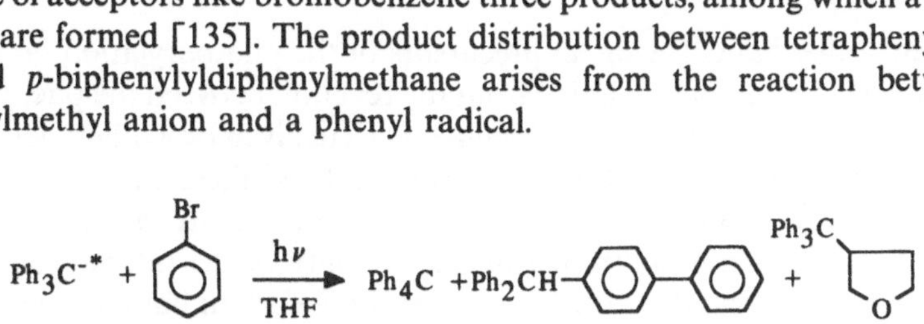

Tolbert was able to demonstrate that an ambident anion (like the triphenyl-methyl one able to react in the *para* or α position) will add a radical at the more basic site instead of giving the most stable radical anion adduct. This was mainly demonstrated from a very elegant experiment performed on the diphenyl 1,3-indenyl anion photoarylated by bromobenzene [136].:

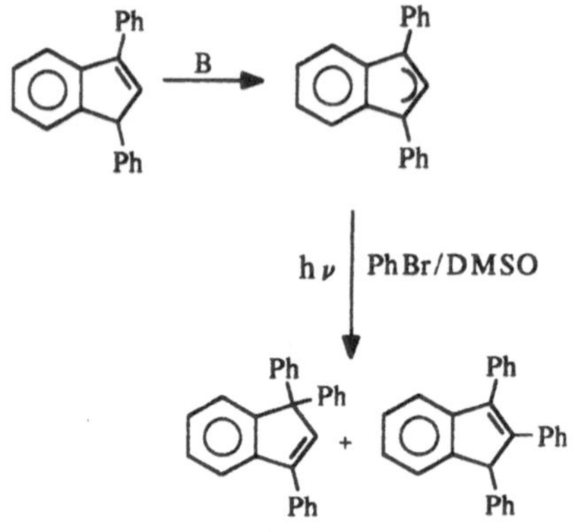

$$ \text{(indenyl anion, Ph at 1,3)} \quad \xrightarrow{\text{B}} \quad \text{(indenyl anion)} $$

$$ \text{h}\nu \;\Big|\; \text{Ph Br/DMSO} $$

$$ \text{(a)} \quad + \quad \text{(b)} $$

(a) 50%   (b) 9%

The conjugated base of 1,3-diphenylindene and -isoindene is the same molecule obtainable by proton abstraction from a (c) or (d) precursor:

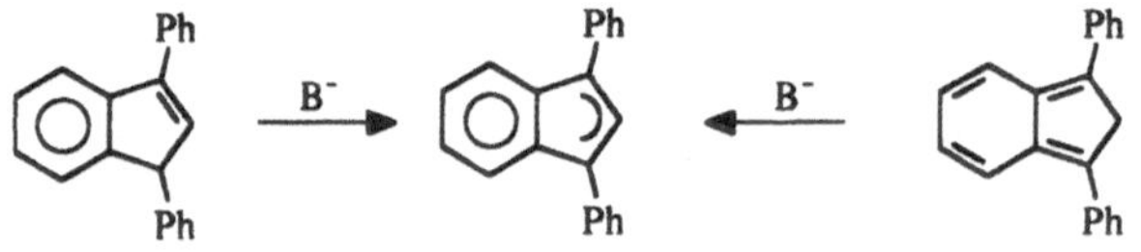

$$ \text{(c)} \quad \xrightarrow{\text{B}^-} \quad \text{(anion)} \quad \xleftarrow{\text{B}^-} \quad \text{(d)} $$

(c)   (d)

Referring to other experimental observations, Tolbert underlined that the more basic site of the conjugated anion corresponds to the site of proton abstraction from precursor (c).

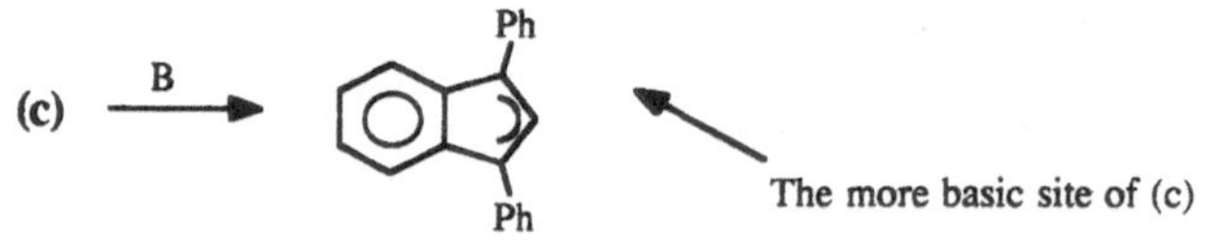

$$ \text{(c)} \quad \xrightarrow{\text{B}} \quad \text{(anion)} \quad \nwarrow $$

The more basic site of (c)

On the other hand, the direct precursors of the photoarylation products (a) and (b) are radical anions resulting from the reaction between the phenyl radical and the hereabove indenyl anion. Arguments were given by Tolbert demonstrating that the precursor of (b), is the more stable of these radical anions. His conclusions were therefore that the selectivity of the attack of a radical on ambident anions is not controlled by the stability of the intermediate radical anion but by the strength of the incipient carbon–carbon bond as measured by the basicity of the reaction sites. This conclusion was also reached from competition experiments [132].

The studies by Tolbert on indenyl carbanions were extended to the 2 halogenated anions. In this interesting cases halide ejection was shown to override the electron transfer chemistry and an hypovalent intermediate of the carbene or allene type is formed leading to C-H insertion products [137, 138, 139].

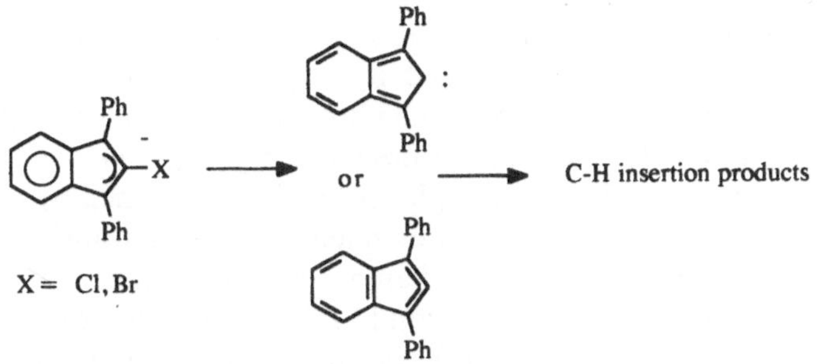

1,3-Diphenylallyl [140, 141] or azaallylcarbanions [142] undergo photo-isomerization ($E, E \Leftrightarrow E, Z$) which may be a photoinitiated electron transfer process. It has been suggested that this type of isomerization may follow an electron transfer from the excited anion to its metallic counterion leading to a radical-metal pair in which the E/Z conversion takes place before the back electron transfer [4, 32]. Furthermore, the rotational barrier in the radical has been shown to be lower than in the corresponding anion [143]. Strengthening the idea of an electron transfer mechanism are the variations of the photo-isomerization equilibria between tight and loose ion pairs [142]. In another study devoted to the 1,3-diphenylallyl anion, the photoisomerization was found to be related to the anion itself and to the presence of a trace of diphenylpropene impurity acting as an electron acceptor catalyst [144].

In 1,2,3-triphenylallyl carbanions, electron transfer photochemistry is found in the presence of acceptors like stilbene as demonstrated by the sensitized *cis-trans* isomerization of this molecule according to the following pathway [145].

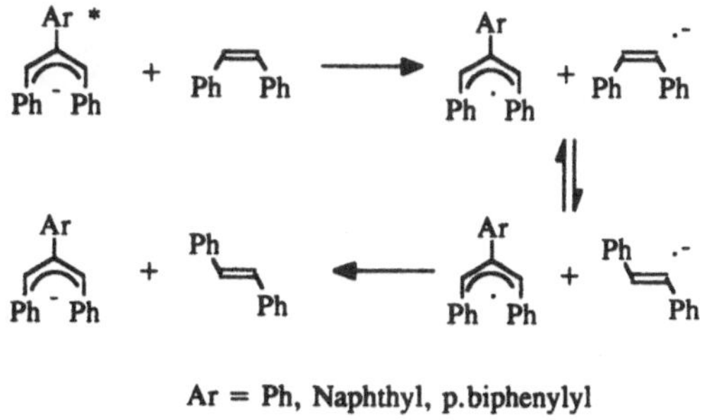

Ar = Ph, Naphthyl, p.biphenylyl

In the triarylallyl carbanion study and in the absence of accepting stilbene, E/Z photoisomerization of the irradiated carbanion was observed and the effect of changing the substituent at position 2 was examined. Starting from the zero coefficient at carbon-2 in the non bonding MO of the allyllic system, Tolbert considers that the substituent at the 2 position does not affect the energy of the MO. If an electron transfer mechanism governs the reactivity, the substituent on this position will not modify the process in a significant way. The experimental results were very dependent on the C-2 substitution and this reaction was therefore considered as relevant of the intrinsic photochemistry of the anion [145]. This view has been confirmed in other studies on 1,3-diphenylallyl carbanion: the kinetic parameters of the photoisomerizations were found to be inconsistent with an electron transfer mechanism [146, 147].

The choice between the intrinsic photochemistry governed by orbital topological control or a photochemistry subsequent to electron transfer must always be done in the various examples of carbanions photoreactions. Among these are the photocyclization of phenalenyl anion [148], the photorearrangement in 8,8-dimethylcyclooctatrienyl anion [149], the ring photoopening of a cyanocyclopropanyl [150] or diphenylcyanocyclopropanyl anions [151].

Cyclic conjugated carbanions have also been studied and rarely found to give an electron transfer photochemistry in the absence of an electron acceptor. One example was however found in the case of cyclooctadienyl carbanion irradiated in THF and in the presence of chlorobenzene or *trans*-stilbene. Dimeric radical products and phenylated cyclooctadiene were formed in the first case while *trans-cis* isomerization of stilbene was observed in the second one [152].

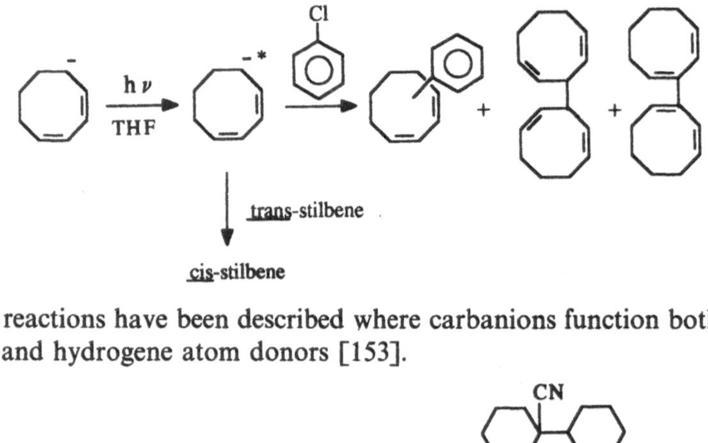

New reactions have been described where carbanions function both as one electron and hydrogene atom donors [153].

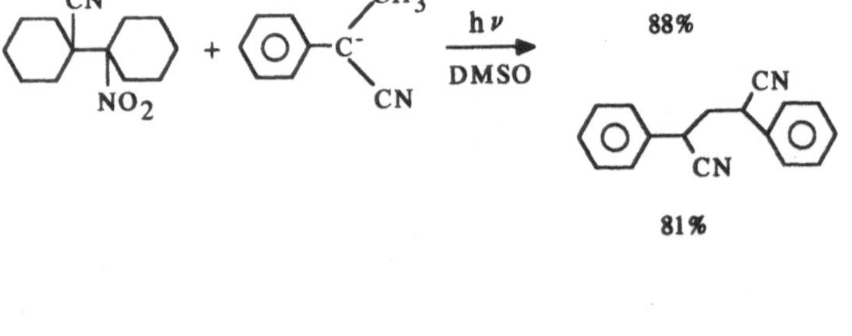

This reaction gives high product yields and works with a methylbenzylcyanide, 9-methylfluorenide anions as well as with the anion of α-methylphenylacetic ester or propiophenone. No reaction is however found with the non methylated anions of phenylacetonitrile, fluorene, phenylacetic ester or acetophenone. The postulated mechanism implies an hydrogen abstraction from an anion leading to a radical anion intermediate:

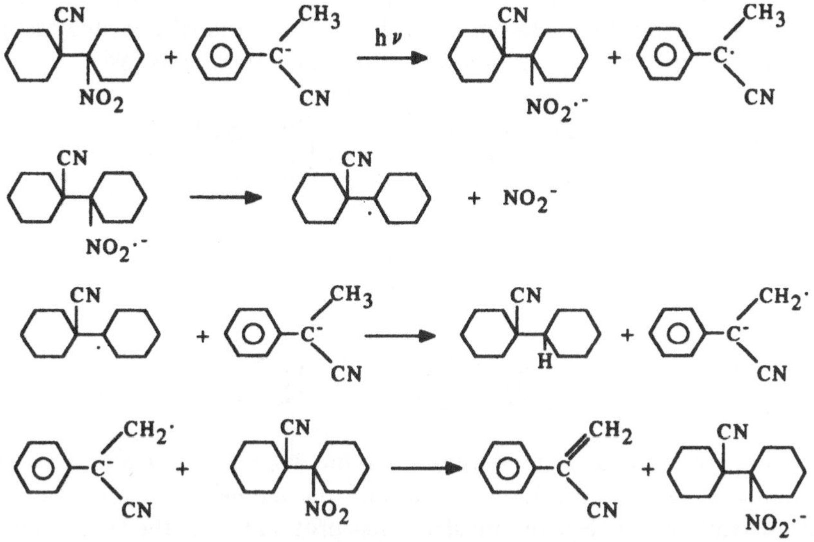

In experiments devoted to the photodecarboxylation of benzannelated acetic acids, the corresponding carbanions were shown to be formed [154]. During photolysis in water (pH > 7) and in the presence of p-nitrobenzoic acid, ESR signals assignable to the p-nitrobenzoate radical anion (radical dianion) were observed. This was attributed to an electron transfer from the carbanion. As the authors do not know if the carbanions are adiabatically formed, the question of an electron transfer from an excited or from a ground state carbanion remains open in this case.

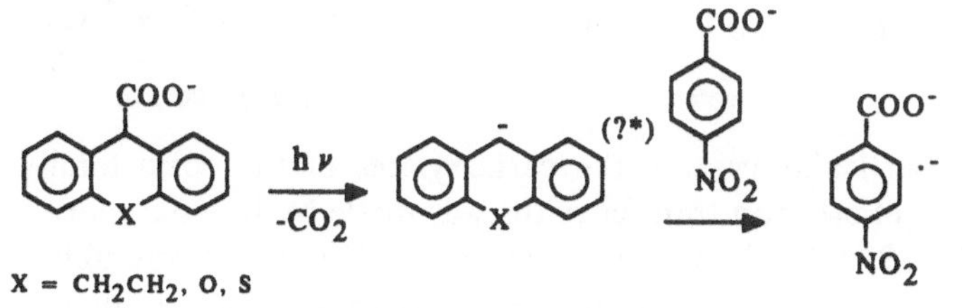

4.3 Oxyanion Reactions

When considering aromatic oxyanions involved in photoinduced electron transfer reactions, a cautious interpretation of the reaction mechanism is to be made because the electron donating tendency due to the anionic status is often overpowered by the electron accepting propensity of the conjugated system. The direction of electron transfer towards or from excited conjugated organic oxyanions will of course depend very much on the quenching partner. For instance, in the presence of the highly accepting methyl viologen (MV^{2+}), the excited rhodizonate dianion is found to induce $MV^{·+}$ formation [155]. On the other hand, it is known since years that eosin, while being an anion, oxidises phenol in aqueous solution [156] and in a more recent study about the oxidability of naphtholate anions it has been shown that the excited resorufin anion acts as an electron acceptor versus naphtholate anions [96].

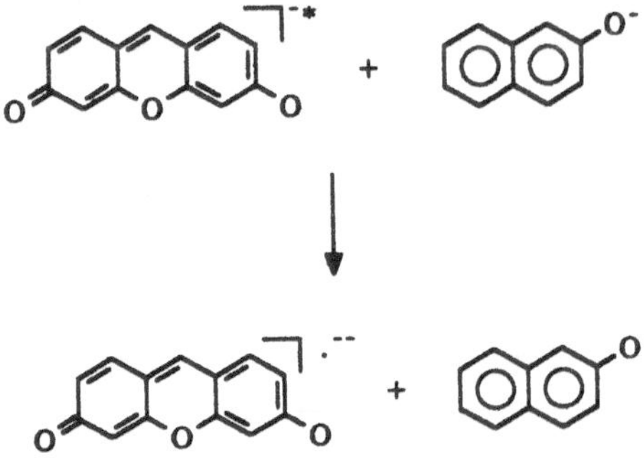

In a study devoted to the fluoresceinamine fluorescence it was shown that this molecule exhibits no fluorescence until the amine's electron lone pair is made unavailable for electron transfer. Thus, protonation of the ammonium ion restores the fluorescence of the molecule: in the deprotonated form, an intra-molecular electron transfer fluorescence quenching occurs, the resorufin moiety of the molecule behaving here as an acceptor [157].

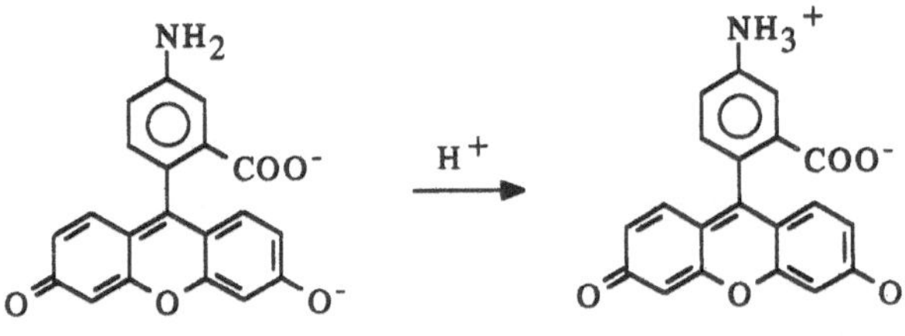

Non fluorescent   Highly fluorescent

In the particular case of the carboxylates of the eosin family, a careful examination of electron transfer paths was made by K. Tokumaru and cowor-kers [158, 159, 160]. They found that eosin Y (EY^{--}) was able to sensitize hydrogen evolution from water under visible light irradiation in a system made of an amine donor, methyl viologen (MV^{2+}) mediator and redox catalyst such as colloidal platinum. In a water-ethanol solution of EY^{--}, in the presence of MV^{2+} and triethanolamine (TEOA), a maximum quantum yield of 0.3 was measured for the $MV^{·+}$ radical cation formation. Kinetic measurements of the EY^{--} triplet quenching constant by MV^{2+} ($k_q = 3.10^9 \ M^{-1} s^{-1}$) and TEOA

$(k_q = 3.10^6 \, M^{-1} s^{-1})$ were necessary to show that electron transfers occur in two opposite directions, from and to eosin triplet. In this three component system, the $MV^{\cdot +}$ formation can be explained by a mechanism which can be switched from one electron transfer scheme (steps (19–20)) to another one (steps(21–22)) depending on the relative concentrations of MV^{2+} and TEOA. In the first one, the triplet dye gives electron to viologen while in the second one an electron is accepted from TEOA. A non fluorescing charge transfer complex between eosin and viologen explaining a rapid decrease of quantum yield at higher viologen concentrations was also demonstrated in this case.

$$EY^{--*3} + MV^{++} \longrightarrow EY^{\cdot -} + MV^{\cdot +} \tag{19}$$

$$EY^{\cdot -} + TEOA \longrightarrow EY^{--} + TEOA^{\cdot +} \tag{20}$$

$$EY^{--*3} + TEOA \longrightarrow EY^{\cdot ---} + TEOA^{\cdot +} \tag{21}$$

$$EY^{\cdot ---} + MV^{++} \longrightarrow EY^{--} + MV^{\cdot +} \tag{22}$$

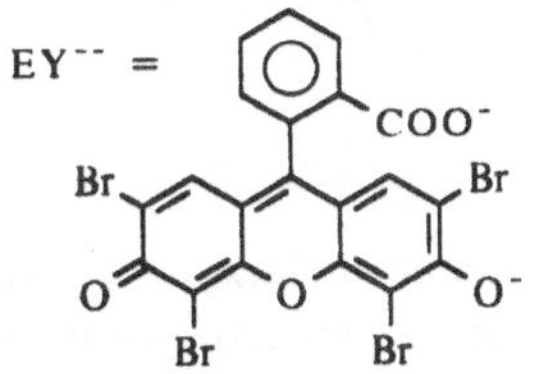

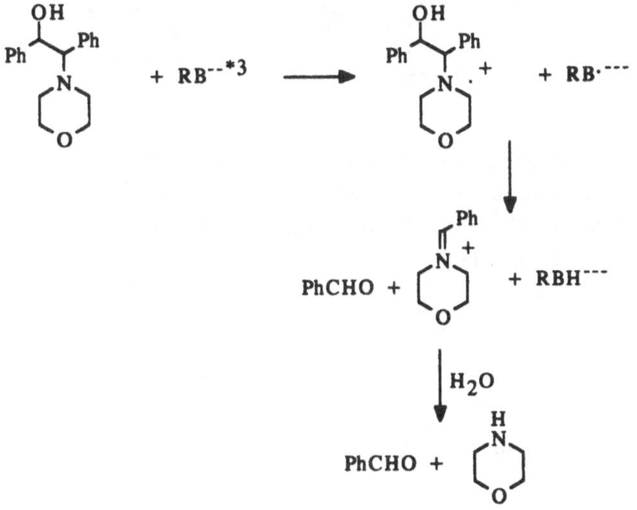

Belonging to the same dye family, Rose Bengal (RB^{--}) has also been used to sensitize the photoreduction of water via both a reductive or an oxidative quenching step: the quenchers of the RB^{--*3} were respectively EDTA or MV^{2+} [161]. Rose Bengal was also reported to induce the photofragmentation of β-amino alcohols yielding the corresponding aldehyde and amine. This was proposed by Whitten to proceed through the intermediacy of sensitized singlet oxygen and superoxide ion formation [162]. An alternative mechanism involving electron transfer from the amino alcohol to the RB^{--*3} is now supported by new evidence [163].

Energy transfer from triplet dyes to the 9-anthracene carboxylate anion (AC^-) was used to sensitize MV^{2+} reduction [164]. Here, the dye is replaced by AC^- while the triplet state of the carboxylate is formed via an efficient energy transfer from eosin or erythrosin. In this system, active under visible irradiation and without sensitizer destruction, both efficiencies of the energy transfer and of the electron transfer were close to unity.

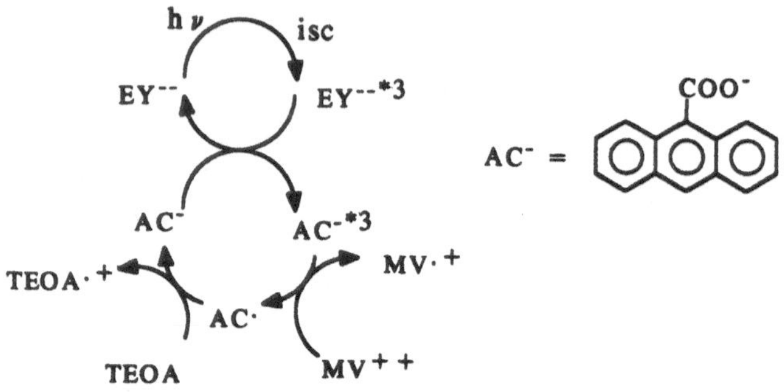

$$h\nu \quad isc$$
$$EY^{--} \quad EY^{--*3}$$
$$AC^- \quad AC^{-*3}$$
$$TEOA^{\cdot+} \quad MV^{\cdot+}$$
$$AC^{\cdot}$$
$$TEOA \quad MV^{++}$$

$$AC^- =$$

Reactions using the anthracene carboxylate as the primary sensitizer are also known as for instance in the water reduction to hydrogen: AC^- triplet is quenched by MV^{2+}. The obtained viologen radical cation initiates hydrogen formation mediated by platinum/polyvinyl alcohol. The AC^- anion is regenerated from the corresponding radical by electron transfer from EDTA and the H_2 production quantum yield is close to unity [165].

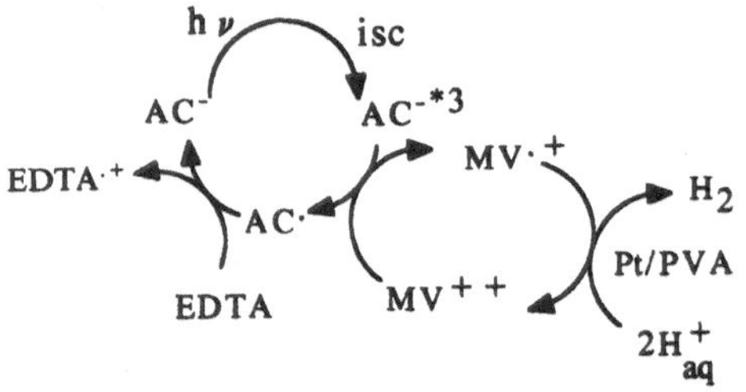

$$h\nu \quad isc$$
$$AC^- \quad AC^{-*3}$$
$$MV^{\cdot+}$$
$$EDTA^{\cdot+} \quad H_2$$
$$AC^{\cdot} \quad Pt/PVA$$
$$EDTA \quad MV^{++} \quad 2H^+_{aq}$$

Another H_2 production system was constructed using the capacity of fluorescein anion excited state to accept an electron from a reducing agent [166]. Quantum yields were in the range of 0.1 for this system working in absence of an electron relay.

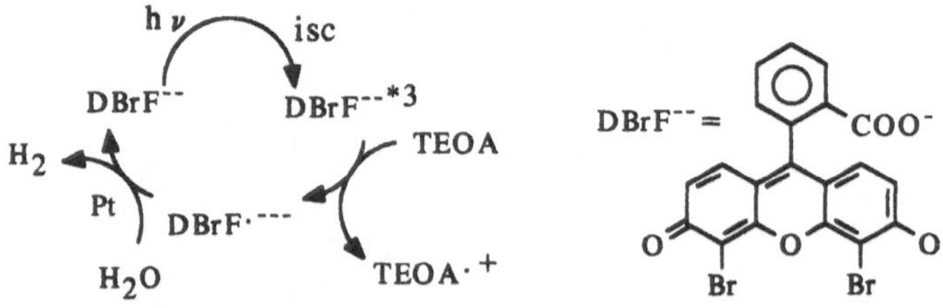

$$h\nu \quad isc$$
$$DBrF^{--} \quad DBrF^{--*3}$$
$$H_2 \quad TEOA$$
$$Pt \quad DBrF^{\cdot---}$$
$$H_2O \quad TEOA^{\cdot+}$$

$$DBrF^{--} =$$

In this family of the xanthene dyes, excitation of oxygen free aqueous solution of the dye alone is sufficient to induce the formation of the semioxidized and semireduced form of the dianionic starting material and the electron transfer steps always concern the triplet dye X^{--*3} [167–173]. Electron exchange may occur between triplet and ground state or between two triplet excited states (steps 23 and 24).

$$X^{--*3} + X^{--} \longrightarrow X^{\cdot---} + X^{\cdot-} \tag{23}$$

$$2\,X^{--*3} \longrightarrow X^{\cdot---} + X^{\cdot-} \tag{24}$$

This had been proposed, namely for a Rose Bengal derivative because the yield and the rate of formation of the radicals are dependent on the light intensity. A more recent analysis of the process performed in a micellar environment seems to agree with an alternative path (steps 25 and 26) still explaining a biphotonic behaviour [174]: a more reactive but unidentified intermediate would be formed by absorption of a photon by the triplet state.

$$X^{--*3} \xrightarrow{\text{h}\nu} Y \tag{25}$$

$$Y + X^{--} \longrightarrow X^{\cdot---} + X^{\cdot-} \tag{26}$$

New anionic photoreductions initiated by excited β-naphtholate anions have been presented recently. According to measurements of electron transfer quenchings of the β-naphtholate excited singlet state [96], it was shown that this powerful reducing species is able to reduce all the benzenic compounds bearing two or more chlorine substituents and all naphthalenes and biphenyls bearing at least monochlorinated. This means of course that this process is applicable to the polychlorobiphenyls, well known for their undesired persistence in the environment. When applied to chloronaphthalene or monochlorobiphenyl, this anionic photodechlorination presents characteristics comparing favorably with other sensitized processes [175].

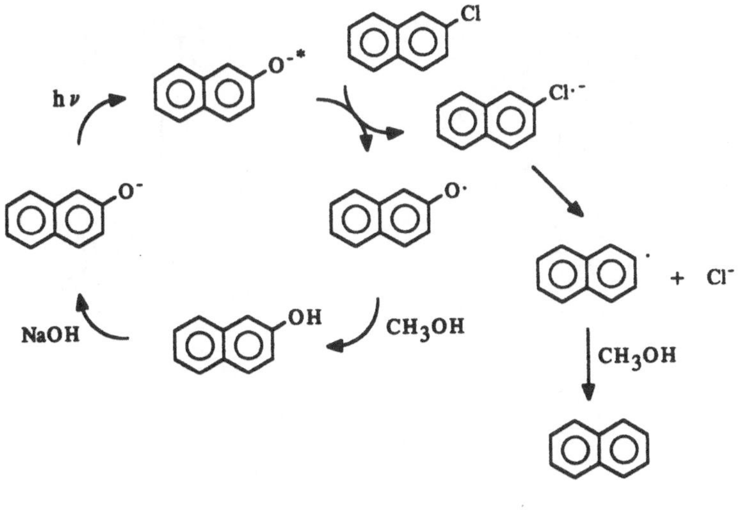

The bathochromic absorbing sensitizer allows selective excitation relative to most of the chloroaromatics. The naphtholate anion absorption spectrum fits almost exactly with black light phosphor lamps, allowing very efficient light harvesting. The quantum yield of 0.3 is better than for many direct photo-dechlorinations. The naphtholate anion is (at least partially) catalytically regenerated and the reaction can proceed to complete disappearance of the chlorobiphenyl while 100% of the disappeared organic chlorine is recovered as sodium chloride.

The naphtholate anion photodechlorination also works in DMF as solvent giving high reduction yields. It is worth mentioning that, in the same DMF solvent, when a mixture of phenylthiolate and haloacetophenones or halobenzophenones are irradiated, the dominating reaction path (> 90%) is the $S_{RN}1$ replacement of the halogen by a phenylthio substituent [176]. Being in the same solvent, the different behaviour must be related to the increased nucleophilicity of the thioanion when compared to the naphtholate.

In an intramolecular example of photodechlorination implying a phenolate anion, a triplet sensitizer and triethylamine, a cyclization of perchlorinated o-phenoxyphenol to octachlorodibenzodioxin is observed. An intramolecular electron transfer from the polychlorinated phenolate moiety to the perchlorophenyl residue is considered together with an intermolecular electron transfer from external triethylamine followed by an intramolecular $S_{RN}1$ path for the ring closure [177].

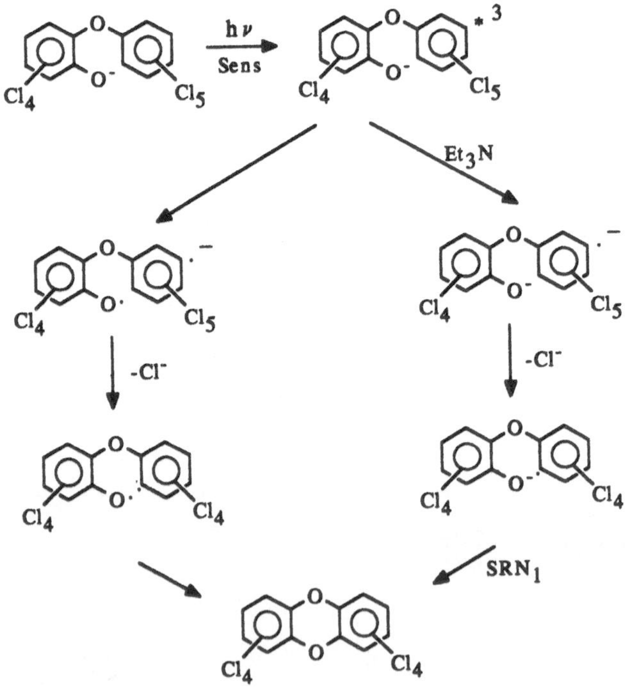

Excited β-naphtholate has also been used in an anionic photodeprotection of tosylamides. This very mild photodetosylation gives amine and sulphinic acid in the presence of sodium borohydride as a coreductant [178]. This process is even more effective than the photodechlorination described previously since a limiting quantum yield of 0.6 was measured together with a quantitative amine recovery and an almost unchanged naptholate sensitizer.

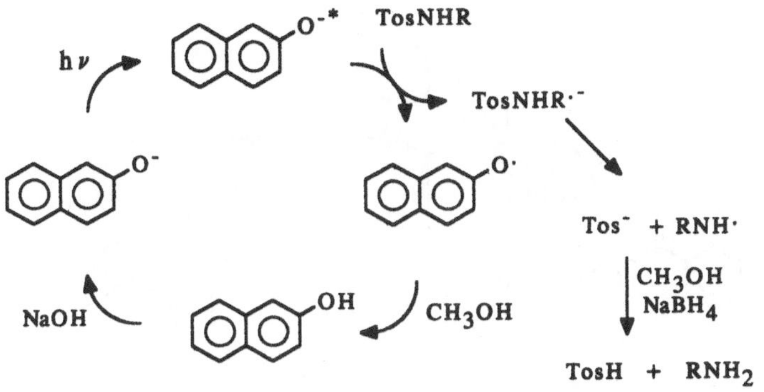

4.4 Radical Anions

Irradiation with visible light of pyrene and perylene anion radicals produced during a cyclic voltammetric experiment leads to an enhancement of the peak current and photoinduced electron transfer to chlorobenzene as acceptor has been shown to occur directly or via the dianion [179, 180].

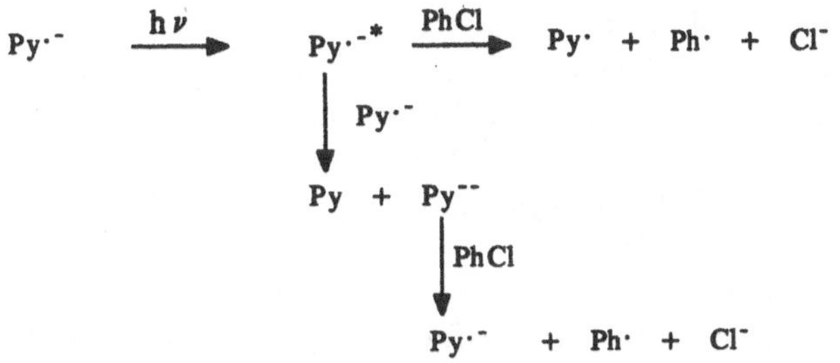

Excited radical anions may be created in a medium which is only slightly reducing, which is not always the case in the presence of anions or by direct electrochemistry. If, for instance, o-dibromobenzene is electrochemically reduced, benzene is the product, obtained via bromobenzene radical and benzyne, since the second electron uptake cannot be avoided [181].

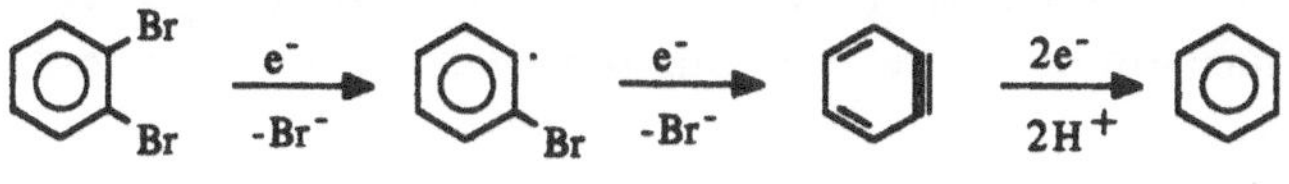

Anthraquinone radical anion has been electrochemically generated and photoexcited in the presence of anthracene and o-dibromobenzene or related haloaromatics [182]. The main product was no longer benzene but phenylated anthracene. Since photoexcited anthracene as well as electrogenerated anthracene or anthraquinone radical anion does not react with dibromobenzene, the following mechanism implying electron transfer from the excited anthraquinone radical anion has been proposed.

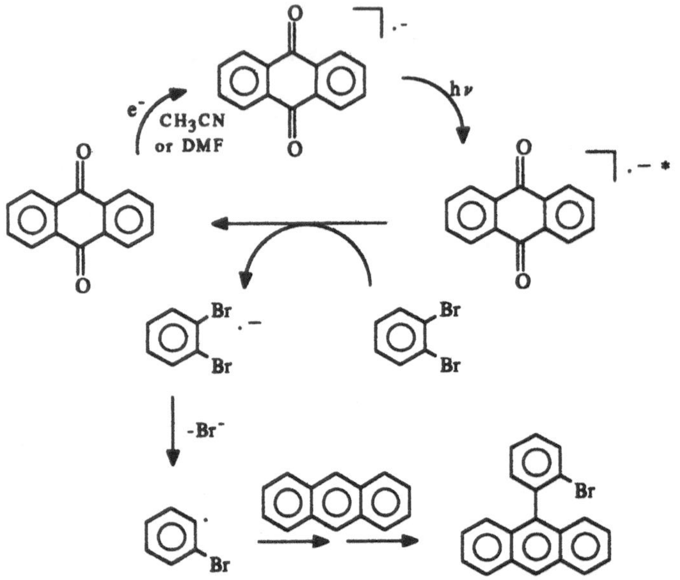

In another example rather similar electrochemically generated anion radicals of anthracene and 9,10-diphenylanthracene were excited by visible light and found to give a greater than 10-fold increase in the rate of reductive dechlorination of 4 chlorobiphenyl. The kinetic results suggested a fast light-assisted reduction pathway implying either the photoexcited anion radical or an electron ejected from it [183].

The excited radical anions of anthraquinone and 9,10-dicyanoanthracene were found to be fluorescent and the fluorescence quenching of these species by added electron acceptors has been studied [94, 95].

5 Anions Used as Electron Transfer Quenchers

$S_{RN}1$ reactions in which anions are not the initial excited species will of course take their place among these reactions but we shall not comment again about these systems.

Aqueous solutions of iminium perchlorates were irradiated in the presence of carboxylate anions and alkylated adducts were formed [184, 185]. This was shown to occur through electron transfer quenching of the iminium excited singlet, followed by an efficient decarboxylation of the resulting acyloxy radical. When α-hydroxy carboxylates are used, a reduction product is also formed and probably linked to the formation of a ketyl radical whose reducing properties are known.

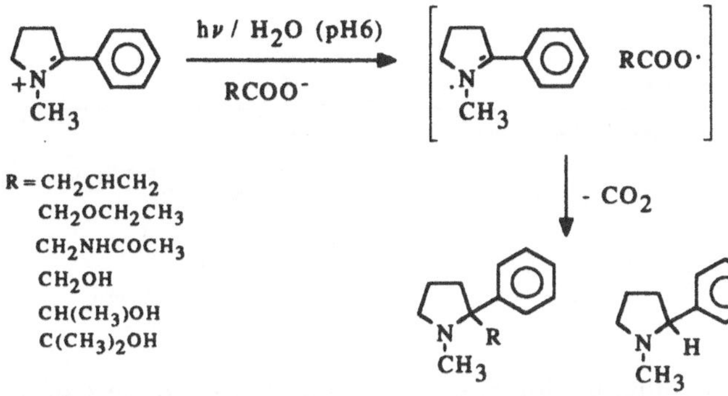
hν / H_2O (pH6)
$RCOO^-$
$RCOO\cdot$
CH_3
- CO_2
R = CH_2CHCH_2
$CH_2OCH_2CH_3$
$CH_2NHCOCH_3$
CH_2OH
$CH(CH_3)OH$
$C(CH_3)_2OH$
R
H

In the photoreactions between halogenated aromatics and sodium borohydride, reductive dehalogenation was found to occur and mechanism proposals have been made [186, 187]. Excited arene itself has been supposed to react with BH_4^- by a direct hydride transfer [188]. The photolysis of the aromatic C–X bond leading to aryl radical is proposed to be followed by hydrogen atom abstraction from the hydride [189]. Finally, an electron transfer mechanism from BH_4^- to excited arene [190] is also postulated. A single electron transfer-hydrogen atom transfer sequence was supported in the case of halocyanonaphthalene and the same mechanism found to work when alkyltriphenyl borate is used as reducting anion [191]. In the recently studied photoreduction of diphenyl 2-bromoethylene,, the same sequential process is invoked but direct hydride transfer quenching is still considered [192].

Alkyltriphenylborates were introduced as a new class of photooxidizable reagents and photoreduction of cyanoaromatics has been studied: high yields of alkylated products were formed through a mechanism involving a boranyl radical in which a carbon–boron bond is homolytically cleaved [193, 194, 195]. Electron transfer from tetraphenylborate to anthraquinone excited state has also been studied [196].

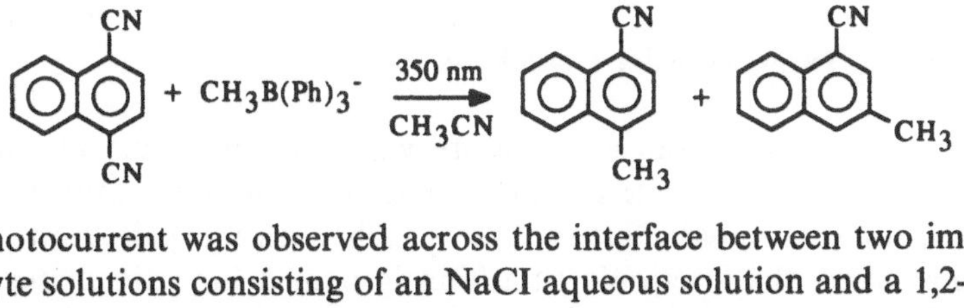
CN
CN
+ $CH_3B(Ph)_3^-$
350 nm
CH_3CN
CN
CH_3
+
CN
CH_3

A photocurrent was observed across the interface between two immiscible electrolyte solutions consisting of an NaCI aqueous solution and a 1,2-dichloroethane solution of tetrabutylammonium tetraphenyl borate, when the organic solution contains a quinone (Q) [197]. This photoelectrochemical effect is due to the transfer of radical anions of quinones across the interface. The tetraphenyl borate gives electrons to the triplet state of quinones while the phenylboranic acid arising through the formation of diphenylborinate anion is detected in the dichloroethane solution.

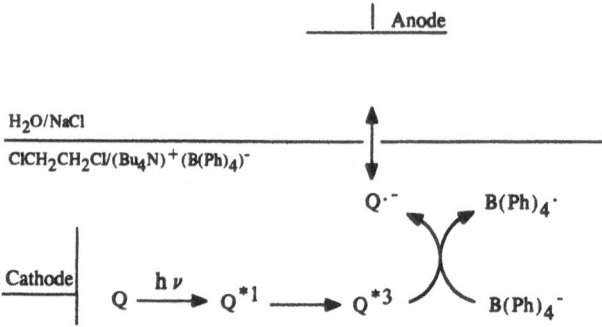

Steady state illumination of polyaromatics in the presence of triphenylstannyl anion in THF gave rise to the aromatic radical anion [198]. In the case of perylene and tetracene the radical anions persisted for a long time. In fact, their decay is modulated by the relative value of the reduction potential of the aromatic and of the formed distannane.

$$Pe \xrightarrow{h\nu} Pe*$$

$$Pe* + (Ph)_3Sn^- \longrightarrow Pe^{\cdot-} + (Ph)_3Sn^{\cdot}$$

$$2\,(Ph)_3Sn^{\cdot} \longrightarrow (Ph)_3SnSn(Ph)_3$$

$$Pe^{\cdot-} + (Ph)_3SnSn(Ph)_3 \longrightarrow Pe + (Ph)_3Sn^{\cdot} + (Ph)_3Sn^-$$

$$Pe^{\cdot-} + (Ph)_3Sn^{\cdot} \longrightarrow Pe + (Ph)_3Sn^-$$

6 Other Types of Photoinduced Electron Transfer Anionic Precursors

6.1 Anions in Charge Transfer Complexes

Charge transfer band absorption occurs whenever inter- or intramolecular complexes between donor and acceptor are formed. In these complexes, anions may be, among the most effective donors. When an ion pair shows such an absorption band, the electronic transition is named ion-pair charge-transfer (IPCT). Ion pairs of coordination compounds are often concerned by IPCT and these processes have been recently reviewed [199]. In this paper, anions belonging to ion pairs of coordination compounds will not be considered.

According to Mulliken [200], ion radical pairs arise from excitation of a CT band: this corresponds to a vertical optical electron transfer from the donating to the accepting moiety of the complex. The CT ion pair so obtained is considered to be akin to a contact ion pair [201]. A radical pair is thus expected from IPCT.

Among CT band promoted electron transfers, some cases are known for anionic donors as for example when mixtures of methyl viologen (as its tetrafluoroborate) and dithiophosphate anions are irradiated by visible light in acetonitrile [202]. In this system, the dark back electron transfer is prevented by the dimerization of the obtained thio radicals.

$$[MV^{++} \cdots (EtO)_2PS_2^-] \xrightarrow{h\nu} MV^{\cdot+} + (EtO)_2PS_2\cdot$$

$$2\,(EtO)_2PS_2^{\cdot} \longrightarrow Dimer$$

In viologen-carboxylate mixtures, the charge transfer interaction between the bipyridinium ion and carboxylate is the dominant activating influence: light of wavelength longer than can be absorbed by either oxidant or reductant is active and $MV^{\cdot+}$ is formed. Quantum yields between 0.3 and 1 were found [203] and its dependence on the pH of the solution was linked to the pH effect on the CTC concentration [204].

$$MV^{++} + RCOO^- \xrightarrow{h\nu_{CT}} MV^{\cdot+} + R^{\cdot} + CO_2$$

Oxalate viologen ion pair complexes have been examined in detail under pulsed laser flash and continuous photolysis [205]. As the photolysis leads to the oxidation of oxalate dianion, the strongly reducing $CO_2^{\cdot-}$ is formed after decarboxylation and a second MV^{2+} is reduced, generating another equivalent of $MV^{\cdot+}$. Malonate, succinate, glutarate, polyacrylate and polymethacrylate were also tested and found to be effective as CTC donors for $MV^{\cdot+}$ formation. In the case of malonate, decomposition of the carboxylate was not accompanied by MV^{2+} consumption. This was attributed to the efficiency of a back electron transfer step following immediately the decarboxylation [206].

$$[MV^{++},\ ^-O_2CCH_2CO_2^-] \xrightarrow{h\nu_{CT}} [MV^{\cdot+},\ ^-O_2CCH_2CO_2\cdot]$$

$$[MV^{\cdot+},\ ^-O_2CCH_2CO_2\cdot] \longrightarrow [MV^{\cdot+},\ ^-O_2CCH_2\cdot] + CO_2$$

$$[MV^{\cdot+},\ ^-O_2CCH_2\cdot] \longrightarrow MV^{++} + ^-O_2CCH_2^-$$

$$\downarrow H_2O$$

$$CH_3CO_2^- + OH^-$$

Starting from one of the cases presented above (when $RCOO^-$ is benzilate anion) a two compartment photoelectrochemical cell has been presented [207].

In the illuminated compartment benzophenone is obtained as the oxidation product, the viologen radical cation serving as the photocurrent source. In the dark compartment an electron acceptor is reduced at the cathode. The system was found to drive a net chemical reaction with an irreversible acceptor (Q = tropylium or dithiolium) in the cathodic compartment. The crucial point in this system is that the back electron transfer between $MV^{\cdot+}$ and benziloxy radical is prevented by a rapid and irreversible decomposition of the benziloxy radical.

Illuminated compartment

$$[MV^{++}\cdots\cdot Ph_2C(OH)COO^-] \xrightarrow{h\nu} MV^{\cdot+} + Ph_2C(OH)COO^{\cdot}$$

$$Ph_2C(OH)COO^{\cdot} \longrightarrow Ph_2C^{\cdot}(OH) + CO_2$$

$$Ph_2C^{\cdot}(OH) + MV^{++} \longrightarrow Ph_2CO + MV^{\cdot+} + H^+$$

$$MV^{\cdot+} \longrightarrow MV^{++} + e^-$$

Dark compartment

$$Q + e^- \longrightarrow Q^{\cdot-}$$

In a solid salt of viologen $MV^{2+}(TFPB^-)_2$ where the anion is not the usual chloride but tetrakis[3,5-bis(trifluoromethyl)phenyl]borate ($TFPB^-$), an IPCT absorption is found. When excited in this band in vacuo or under argon, $MV^{\cdot+}$ is observed to accumulate and to reversibly disappear when heated or dissolved [208, 209]. This system has also been observed in degassed organic solutions [210] showing that the extraordinarily bulky and stable $TFPB^-$ anion works as a reversible electron donor. The MV^{2+}-$TFPB^-$ ion pair is also the only example for which ion pair luminescence (optical back electron transfer) is observed at room temperature [211].

$$TFPB^- = \begin{bmatrix} \begin{array}{c} CF_3 \\ \bigcirc \\ CF_3 \end{array} \ B^- \end{bmatrix}_4$$

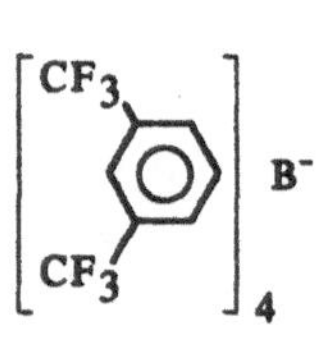

These exceptional characteristics have been used later in several interesting applications. An elastic polymer film containing the same MV^{2+} salts as part of the main chain was found to be photochromic and results indicate that data can be written optically ($\lambda > 365$ nm) and erased thermally [212]. The photochromism was also observed in monolayer assemblies (Langmuir-Bloodgett films) made of mixtures of N,N'–hexadecyl-4,4'-bipyridinium salts of $TFPB^-$ and arachidic acid [213, 214, 215].

When the less stable tetraphenylborate is used, decomposition of the oxidized anion into triphenelborane is observed and the viologen precursor is not reversibly restored. As in the case of the benzilate MV^{2+} example, a photo-electrochemical cell based on this system has been presented [216].

Electron transfer substitution at saturated carbon has been performed through photochemical initiation involving a charge transfer complex between the nucleophilic anion and a neutral substrate: in HMPA, nitrite anion was displaced by azide on α,p–dinitrocumene [217].

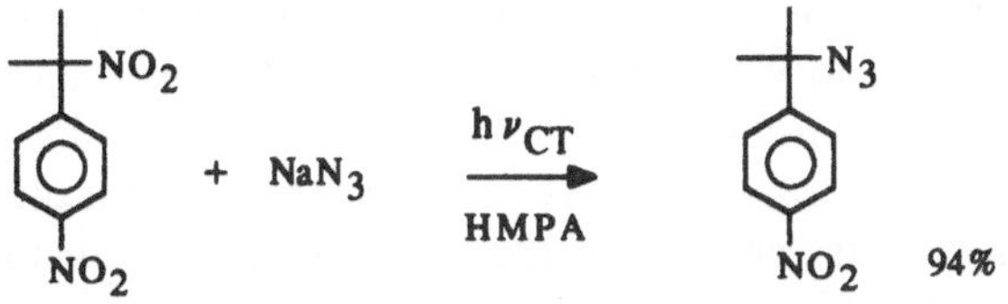

In DMF, with carbanions (malonates or nitronates) as nucleophiles and phenylthio or phenylsulphonyl leaving groups, a chain radical substitution takes place at an allylic carbon. This $S_{RN}1$ like reaction was found to be activated by visible light, the only possible absorbing species being a charge transfer complex between the anion and the substrate [218].

$$L = C_6H_5S$$
$$C_6H_5SO_2$$

$$Nu^- = {}^-CEt(COOEt)_2$$
$${}^-CH(COOEt)_2$$
$${}^-CMe_2NO_2$$
$${}^-CHMeNO_2$$

A photoreduction affording bond cleavage products was also initiated through charge transfer band excitation: in DMF, with thiolate anion, N-tosylsulfilimines were cleaved leading to sulfide, disulfide and tosylamide [219].

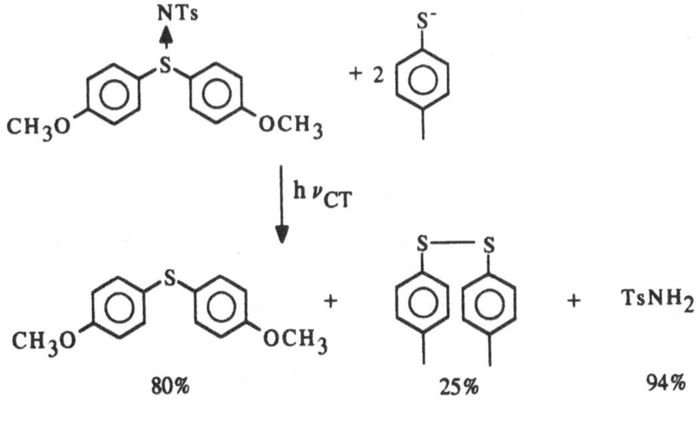

6.2 Intra Ion Pair Photoelectron Transfer

The design of photopolymerization initiators sensitive to visible light was based on the intra ion pair electron transfer photoinduced from borate anion to an excited counter cation [195]. In non polar solvents cyanine or pyrylium borates form tight ion pairs whose absorption spectra are insensitive to the nature of the organic borate anions showing that excitation affords a locally excited state of the cyanine cation. On the other hand, their fluorescence efficiency is quite dependent on the identity of the borate anion and provides a direct measure of the rate of electron transfer from the borate to the excited cyanine. Very high intra ion pair electron transfer rates (100 times greater than the diffusion limited rate for a bimolecular reaction in the same solvent) were measured in these systems [220, 221]. However, this is not necessarily always true and steric factors need to be considered. Slow electron transfer rates are observed in a thermodynamically favourable case where a "penetrated" cyanine borate ion pair is formed: with phenyltriptycyl groups in the borate anion, a cavity accomodates penetration of the cyanine cation and this structural singularity seems to explain the observed properties [222]. When the electron transfer works radicals are formed and this property makes cyanine or pyrylium borates useful polymerization initiators having sensitivity throughout the entire visible spectral region [93, 223]. This is thus an alternative route for IPCT which does not occur here, the electron transfer being a classical quenching process.

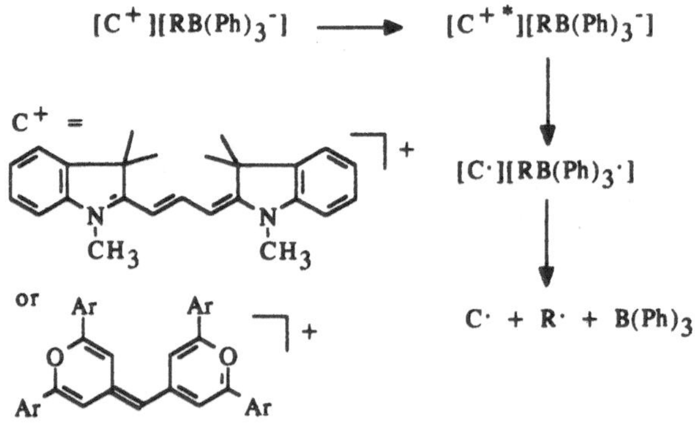

Another similar example, also used in polymerization initiation, is the case of the iodonium salts of Rose Bengal (RB^{2-}) [224, 225]. Methylene chloride solutions of these salts bleach in a few seconds in room light through an electron transfer photoinduced from the excited RB^{2-} to the iodonium cation: the resulting phenyl radicals were reported to initiate polymerization of acrylate.

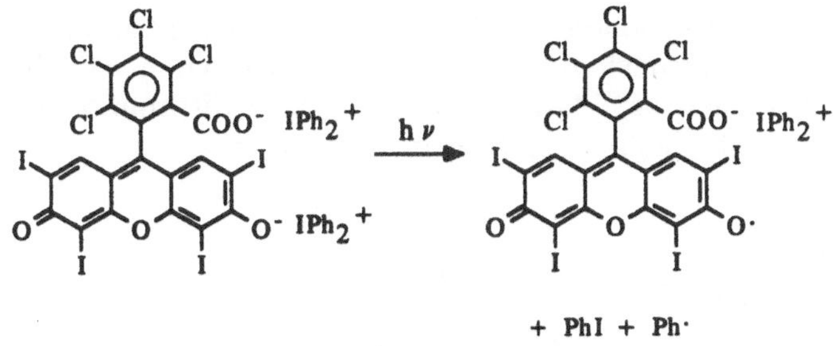

+ PhI + Ph·

RB^{2-} has been replaced by the 9,10-dimethoxyanthracene-2-sulfonate and the products of the irradiation of the diphenyliodonium salt in degassed acetonitrile are the parent sulfonic acid, iodobenzene and benzene [226].

With p-tolyldiazonium salts (4-MeC$_6$H$_4$N$_2^+$X$^-$), in the presence of oxidizable anions like B(Ph)$_4^-$ or oxalate or bromide, electron transfer quenching of the diazonium excited state gives aryl radical. Ion pairing, when forced by the use of dichloromethane as solvent, favours this process [227].

6.3 Electron Transfer from Excited Charge Transfer States

The anomalous fluorescence of p-N,N'-dimethylaminobenzonitrile (DMABN) in polar solvents has been attributed to a twisted intramolecular charge transfer state (TICT) [228]. The subject has been reviewed [229, 230]. TICT perpendicular states are considered as radical-cation-radical anion pairs resulting from a total intramolecular electron transfer from the donor to the acceptor end. Since these transients are able to fluoresce, they are excited states and their fluorescence corresponds to the back electron transfer leading to an unstable conformer of the ground state. In the TICT state the zwitterionic charges are stabilized by electrostatic interactions. As excited ions are very efficient electron transfer reagents, the feasibility of electron transfer reactions starting from a TICT state has been examined in the case of the N-(α-naphthyl) carbazole [231].

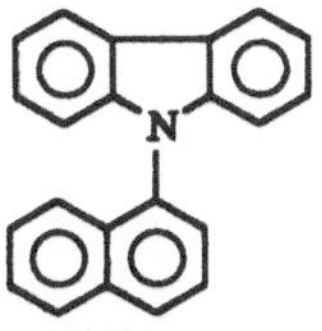

NC

The anionic and cationic ends of the NC's TICT state were found to be quenched by electron donors and acceptors respectively. The quenching rate constants were analyzed with the help of the Marcus model and found consistent with the expected value for electron transfer mechanism. This was confirmed by photochemical reactions running through a radical cation or a radical anion intermediate. The isomerization of quadricyclane to norbornadiene was used as a check for the reactivity of the radical-cation end of the NC's

TICT state. Radical anionic sensitization was also performed, taking the dechlorination of pentachlorobenzene as the test reaction. The three isomeric tetrachlorobenzenes were formed with the same distribution as previously observed in another anionic dechlorination process.

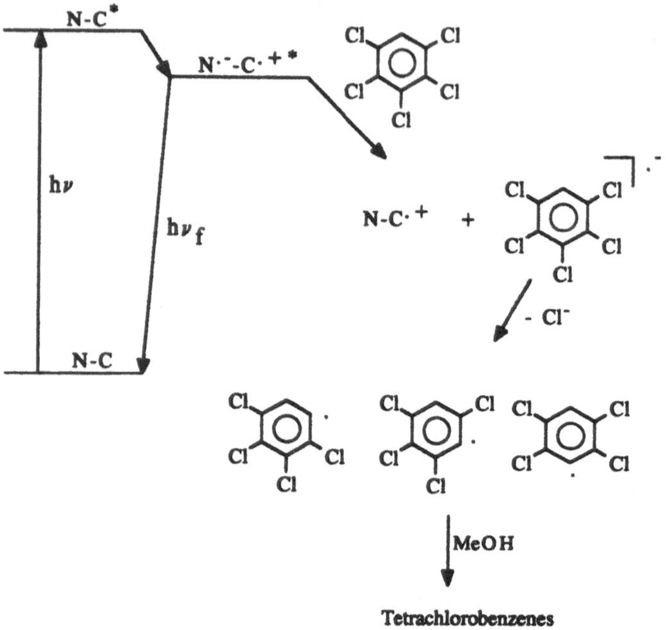

Tetrachlorobenzenes

TICT compounds may thus be useful as anionic-like photoreducers. This reactivity is, however, modest and will, of course, depend on the energy gap between the initial locally excited state and the TICT state itself. As far as the reactivity is concerned, these states may be considered as a contact ion pair in the excited state.

7 Conclusion

The use of anions as electron transfer sensitizers is a more delicate matter to master than the use of neutral donors. Of course, there are advantages like the increased donating capacity due to the charge, but the reactivity may be very strongly influenced by the experimental conditions. Solvent changes lead to more or less ion pairing effects as well as to changes in the excitation wavelength. This will modify the thermodynamics of the reaction and of course the reactivity: a 10 Kcal change in the normal region of a Marcus graph will affect the rate constants by more than two orders of magnitude.

The manipulation of organic anions, especially carbanions, often requires an inert atmosphere and the presence of very strong bases. These experimental conditions are not very mild and to achieve some clean photochemistry under these circumstances is not an easy task. However, many examples have been shown of useful processes starting from excited anions and $S_{RN}1$ is certainly among them. Sometimes, advantage was taken of the bathochromic transitions offered by the anionic sensitizer and visible light activation may be achieved in certain cases.

When part of a delocalized aromatic molecule, the anion, after excited state electron transfer, gives rise to a not too reactive radical so that if one plays with the sequence of hydrogen abstraction from solvent followed by proton abstraction from the base, the initial anion may be restored and the solvent usefully replaces the sacrificial donor. This is the case with naphtholate as anion, certainly an interesting anionic sensitizer to test in other systems. Since the solvent is by definition at high concentration, its recovering role is played with a maximum efficiency: reaction between the solvent and one of the radicals arising from the electron transfer is a good way to circumvent the back electron transfer.

As far as the use of the Marcus model is concerned, the fluorescence quenching of excited anions has demonstrated two important points. The inverted region probably starts in a more exothermic region than previously expected. With molecules presenting a large Stokes shift, the available excitation energy must be estimated carefully and is not necessarily located at the center of gravity of the absorption and emission energies.

Acknowledgments: This review has been written as part of our research program on photoreductive anionic reactions. This research is funded by the F.N.R.S. (Fonds National de la Recherche Scientifique).

8 References

1. Holy NL (1974) Chem Rev 74: 243
2. Holy NL, Marcum JD (1971) Angew Chem Internat Ed 10: 115
3. Fox MA (1979) Chem Rev 79: 270
4. Tolbert LM (1986) Org Photochem 6: 177
5. Krogh E, Wan P (1990) Photoinduced electron transfer I, Top Curr Chem Springer Berlin Heidelberg New York, p 93
6. Tolbert LM (1963) Acc Chem Res 19: 268
7. Zaugg HE, Schaefer AD (1965) J Am Chem Soc 87: 1857
8. Garst JF, Klein RA, Walmsley D, Zabolotny ER (1965) J Am Chem Soc 87: 4080
9. Garst JF, Richards WR (1965) J Am Chem Soc 87: 4084
10. Jortner J, Ottolenghi M, Stein G (1963) J Am Chem Soc 85: 2712
11. Ullman EF, Badab E, Sung M (1969) J Am Chem Soc 91: 5792
12. Porter G, Suppan P (1964) Pure Appl Chem 9: 499
13. Porter G, Suppan P (1965) Trans Faraday Soc 61: 1664

14. Bjerrum N (1926) Kgl Dan Selsk Mat Fys Medd 7: 9
15. Szwarc M (1968) Carbanions living polymers and electron transfer processes, Interscience, New York, pp 101, 207, 218.
16. Winstein S, Clippinger E, Fainberg AH, Robinson GC (1954) J Am Chem Soc 76: 2597
17. Sadek H, Fuoss R (1954) J Am Chem Soc 76: 5897
18. Marcus Y (1985) Ion solvation. J Wiley, Chapter 8
19. Smid J (1972) In: Szwarc M (ed) Ions and ion pairs in organic reactions, Wiley Interscience, vol 1, chap 3
20. Hogen Esch TE (1977) Adv Phys Org chem 15: 153
21. Carter HV, McClelland BJ, Warhurst E (1960) Trans Faraday Soc 56: 455
22. Hogen Esch TE, Smid J (1966) J Am Chem Soc 88: 307
23. Kaufman MJ, Gronert S, Bors DA, Streitwieser Jr A (1987) J Am Chem Soc 109: 602
24. Soumillion JPh, Vandereecken P, Van Der Auweraer M, De Schryver FC, Schanck A (1989) J Am Chem Soc 111: 2217
25. Bayliss NS, McRae EG (1954) J Am Chem Soc 58: 1002
26. Aten AC, Dieleman J, Hoijtink GJ (1960) Discuss Faraday Soc 29: 182
27. Buschow KHJ, Dieleman J, Hoijtink GJ (1965) J Chem Phys 42: 1993
28. Vos HW, MacLean C, Velthorst NH (1976) J Chem Soc Faraday Trans II, 72: 63
29. Vos HW, Rietveld GGA, MacLean C, Velthorst NH (1976) J Chem Soc Faraday Trans II, 72: 1636
30. Hogen Esch TE, Plodinec MJ (1978) J Am Chem Soc 100: 7633
31. Hogen Esch TE, Plodinec MJ (1976) J Phys Chem 80: 1090
32. Fox MA, Voynick TA, (1980) Tetrahedron Letters 21: 3943
33. Plodinec J, Hogen Esch TE (1974) J Am Chem Soc 96: 5262
34. Velthorst NH (1979) Pure Appl Chem 51: 85
35. Vandereecken P, Soumillion JPh, Van Der Auweraer M, De Schryver FC (1987) Chem Phys Letters, 136: 441
36. Pearson RG, Songstad J (1967) J Am Chem Soc 89: 1827
37. Mishra AK, Sato M, Hiratsuka H, Shizuka H (1991) J Chem Soc Faraday Trans 87: 1311
38. Gould IR, Young RH, Moody RE, Farid S (1991) J Phys Chem 95: 2068
39. Ojima S, Miyasaka H, Mataga N (1990) J Phys Chem 94: 7534
40. Asahi T, Mataga N (1991) J Phys Chem 95: 1956
41. Goodman JL, Peters KS (1985) J Am Chem Soc 107: 1441
42. Simon JD, Peters KS (1984) Acc Chem Res 17: 277
43. Streitwieser Jr A (1984) Acc Chem Res 17: 353
44. Ramme R, Fisher M, Claesson S, Szwarc M (1972) Proc Roy Soc London A237: 467
45. Ramme R, Fisher M, Claesson S, Szwarc M (1972) Proc Roy Soc London A237: 481
46. Szwarc M, Levin G (1976) J Photochem 5: 119
47. Ramme G, Ehgdahl KA (1976) Acta Chem Scand A 30: 794
48. Giling L, Kloosterboer JG (1973) Chem Phys Letters 21: 127
49. Fischer M, Ramme G, Claesson S, Szwarc M (1971) Chem Phys Letters 9: 306
50. Huber W (1985) Tetrahedron Letters 26: 181
51. Huber W, Mullen K (1986) Acc Chem Res 19: 300
52. Wilhelm D, Courtneidge JL, Clark T, Davies AG (1984) J Chem Soc Chem Commun 810
53. Dvorak V, Michl J (1976) J Am Chem Soc 98: 1080
54. Smentowski FJ, Stevenson GR (1969) J Phys Chem 73: 340
55. Fox MA, Kabir-ud-Din (1979) J Phys Chem 83: 1800
56. Hohman JR, Fox MA (1982) J Am Chem Soc 104: 401
57. Vander Donckt E, Thiry P (1972) J Chem Soc Chem Commun 1293
58. Vander Donckt E, Nasielski J, Thiry P (1969) J Chem Soc Chem Commun 1249
59. Wang HC, Lillie ED, Slomkowski S, Szwarc M (1977) J Am Chem Soc 99: 4612
60. Szwarc M, Wang HC, Levin G (1977) Chem Phys Letters 51: 296
61. Levin G, Holloway BE, Mao CR, Szwarc M (1978) J Am Chem Soc 100: 5841
62. Grabner G, Kohler G, Zechner J, Getoff N (1980) J Phys Chem 84: 3000
63. Cocivera M, Tomkiewicz M, Groen A (1972) J Am Chem Soc 94: 6598
64. Ottolenghi M (1963) J Am Chem Soc 85: 3357
65. Goldschmidt CR, Stein G (1971) Chem Phys Letters 12: 339
66. Feitelson J, Stein G (1972) J Chem Phys 57: 5378
67. Klaning UK, Goldschmidt CR, Ottolenghi M, Stein S (1973) J Chem Phys 59: 1753
68. Matsuzaki A, Kobayashi T, Nagakura S (1978) J Phys Chem 82: 1201

69. Lachish U, Ottolenghi M, Stein G (1977) Chem Phys Letters 48: 402
70. Jortner J, Ottolenghi M, Stein G (1963) J Am Chem Soc 85: 2712
71. Grabner G, Kohler G, Zechner J, Getoff N (1977) Photochem Photobiol 26: 449
72. Mialocq JC, Sutton J, Goujon P (1980) J Chem Phys 72: 6338
73. Mialocq JC, Sutton J, Goujon P (1981) II Nuovo Cimento 63B: 317
74. Bradbury D, Barltrop J (1975) J Chem Soc Chem Commun 842
75. Meiggs TO, Grossweiner LI, Miller SI (1972) J Am Chem Soc 94: 7981
76. Epling GA, Lopes A (1977) J Am Chem Soc 99: 2700
77. Kohler G, Solar S, Getoff N, Holzwarth AR, Schaffner K (1985) J Photochem 28: 383
78. Mittal LJ, Mittal JP, Hayon E (1973) J Am Chem Soc 95: 6203
79. Bent DV, Hayon E (1972) J Am Chem Soc 97: 2606, 2612
80. Bryant FD, Santus R, Grossweiner LI (1975) J Phys Chem 79: 2712
81. Langmuir ME, Hayon E (1969) J Chem Phys 51: 4893
82. Kohler G, Getoff N (1978) J Chem Soc Faraday Trans 1, 74: 1029
83. McAlpine RD, Cocivera M, Chen H (1972) Can J Chem 51: 1682
84. Levin G, Claesson S, Szwarc M (1972) J Am Chem Soc 94: 8672
85. Zechner J, Kohler G, Grabner G, Getoff N (1980) Can J Chem 58: 2006
86. Das PK, Bhattacharyya SN (1981) J Phys Chem 85: 1391
87. Miedlar K, Das PK (1982) J Am Chem Soc 104: 7462
88. Rehm D, Weller A (1970) Isr J Chem 8: 259
89. Eberson L (1987) Electron transfer reactions in organic chemistry. Springer, Berlin Heidelberg New York
90. Marcus RA (1956) J Phys Chem 24: 966
91. Marcus RA (1964) An Rev Phys Chem 15: 155
92. Rehm D, Weller A (1969) Ber. Bunsen-Ges Phys Chem 73: 384
93. Chatterjee S, Davis PD, Gottscahlk P, Kurz ME, Sauerwein B, Yang X, Schuster GB (1990) J Am Chem Soc 112: 6329
94. Eriksen J, Lund H, Nyvad AI (1983) Acta Chem Scand B 37: 459
95. Eriksen J, Jorgensen KA, Linderberg J, Lund H (1984) J Am Chem Soc 106: 5083
96. Legros B, Vandereecken P, Soumillion J.Ph (1991) J Phys Chem 95: 4752
97. Weller A (1982) Z Physik Chem Neue Folge 133: 93
98. Suppan P (1988) Chimia 42: 320
99. Eberson L (1984) Acta Chem Scand B 38: 439
100. Denney DB, Denney DZ (1991) Tetrahedron 47: 6577
101. Beckwith ALJ, Palacios SM (1991) J Phys Org Chem 4: 404
102. Bunnett JF, Sundberg JE (1976) J Org Chem 41: 1702
103. Borosky GL, Pierini AB, Rossi RA (1990) J Org Chem 55: 3705
104. Cheng C, Stock LM (1991) J Org Chem 56: 2436
105. Tolbert LM, Siddiqui S (1984) J Org Chem 49: 1744
106. Bordwell FG, Clemens AH, Smith DE, Begemann J (1985) J Org Chem 50: 1152
107. Kornblum N (1978) Angew Chem Int Ed 14: 734
108. Bunnett JF (1978) Acc Chem Res 11: 413
109. Wolfe JF, Carver DR (1978) Org Prep Proceed Int 10: 225
110. Rossi RA (1982) Acc Chem Res 15: 164
111. Chanon M, Tobe ML (1982) Angew Chem Int Ed 21: 1
112. Beugelmans R, Bois-Choussy M (1988) Tetrahedron Letters 29: 1289
113. Petrillo G, Novi M, Dell'Erba C, Tavani C, Berta G (1990) Tetrahedron 46: 7977
114. Beugelmans R, Bois-Chouissy M, Tang Q (1988) Tetrahedron Letters 29: 1705
115. Pierini AB, Baumgartner MT, Rossi RA (1988) Tetrahedron Letters 29: 3429
116. Pierini AB, Baumgartner MT, Rossi RA (1991) J Org Chem 56: 580
117. Baumgartner MT, Pierini AB, Rossi RA (1992) Tetrahedron Letters 33: 2323
118. Novi M, Garbarino G, Petrillo G, Dell'Erba C (1987) J Org Chem 52: 5382
119. Santiago AN, Iyer VS, Adcock W, Rossi RA (1988) J Org Chem 53: 3016
120. Rossi RA, Pierini AB, Palacios SM (1989) J Chem Ed 66: 720
121. Santiago AN, Morris DG, Rossi RA (1988) J Chem Soc Chem Commun 220
122. Santiago AN, Rossi RA (1990) J Chem Soc Chem Commun 206
123. Bornancini ERN, Palacios SM, Penenory AB, Rossi AR (1989) J Phys Org Chem 2: 255
124. Dillender SC, Greenwood TD, Hendi S, Wolfe JF (1986) J Org Chem 51: 1184
125. Alonso RA, Austin E, Rossi RA (1988) J Org Chem 53: 6065
126. Pierini AB, Baumgartner MT, Rossi RA (1987) J Org Chem 52: 1089

127. Novi M, Petrillo G, Dell'Erba C (1987) Tetrahedron Letters 28: 1345
128. Petrillo G, Novi M, Garbarino G, Dell'Erba C (1987) Tetrahedron 43: 4625
129. Goehring RR, Sachdeva YP, Pisipatis JS, Sleevi MC, Wolfe JF (1985) J Am Chem Soc 107: 435
130. Tolbert LM (1978) J Am Chem Soc 100: 3952
131. Tolbert LM (1980) J Am Chem Soc 102: 3531
132. Tolbert LM (1980) J Am Chem Soc 102: 6808
133. Fox MA, Owen RC (1980) J Am Chem Soc 102: 6559
134. Tolbert LM, Merrick RD (1982) J Org Chem 47: 2808
135. Tolbert LM, Martone DP (1983) J Org Chem 48: 1185
136. Tolbert LM, Siddiqui S (1982) Tetrahedron 38: 1079
137. Tolbert LM, Siddiqui S (1982) J Am Chem Soc 104: 4273
138. Tolbert LM, Siddiqui S (1984) J Am Chem Soc 106: 5538
139. Tolbert LM, Islam MN, Johnson RP, Loiselle PM, Shakespeare WC (1990) J Am Chem Soc 112: 6416
140. Parkes HM, Young RN (1978) J Chem Soc Perkin Trans II, 249
141. Bushby RJ (1980) J Chem Soc Perkin Trans II, 1419
142. Young RN, Ahmad MA (1982) J Chem Soc Perkin Trans II, 35
143. Boche G, Schneider DR (1977) Angew Chem Int Ed 16: 869
144. Bushby RJ, Tytko MP (1986) J Chem Soc Chem Commun 23
145. Tolbert LM, Ali MZ (1985) J Org Chem 50: 3288
146. Parmar SS, Brocklehurst B, Young RN (1987) J Photochem Photobiol A 40: 121
147. Brocklehurst B, Young RN, Parmar SS (1988) J Photochem Photobiol A 41: 167
148. Hunter DH, Perry RA (1980) J Chem Soc Chem Comm 877
149. Staley SW, Pearl NJ (1973) J Am Chem Soc 95: 2731
150. Newcomb M, Ford WT (1974) J Am Chem Soc 96: 2968
151. Fox MA (1979) J Am Chem Soc 101: 4008
152. Fox MA, Singletary NJ (1982) J Org Chem 47: 3412
153. Kornblum N, Chen SI, Kelly WJ (1988) J Org Chem 53: 1988
154. Mc Auley, Krogh E, Wan P (1988) J Am Chem Soc 110: 600
155. Iraci G, Back MH (1988) Can J Chem 66: 1293
156. Zwicker EF, Grossweiner LI (1963) J Phys Chem 67: 549
157. Munkholm C, Parkinson DR, Walt DR (1990) J Am Chem Soc 112: 2608
158. Misawa H, Sakuragi H, Usui Y, Tokumaru K (1983) Chem Letters 1021
159. Nishimura Y, Misawa H, Sakuragi H, Tokumaru K (1989) Chem Letters 1555
160. Usui Y, Misawa H, Sakuragi H, Tokumaru K (1987) Bull Chem Soc Japan 60: 1573
161. Mills A, Lawrence C, Douglas P (1986) J Chem Soc Faraday Trans 2, 82: 2291
162. Haugen CM, Whitten DG (1989) J Am Chem Soc 111: 7281
163. Epling GA, Jackson ML (1991) Tetrahedron Letters 32: 7507
164. Usui Y, Sasaki Y, Ishii Y, Tokumaru K (1988) Bull Chem Soc Japan 61: 3335
165. Johansen O, Mau WH, Sasse HF (1983) Chem Phys Letters 94: 107
166. Hashimoto K, Kawai T, Sakata T (1983) Chem Letters 709
167. Kasche V, Lindqvist L (1965) Photochem Photobiol 4: 923
168. Mialocq JC, Hebert Ph, Armand X, Bonneau R, Morand JP (1991) Photochem Photobiol 56: 323
169. Bilski P, Dabestani R, Chignell CF (1991) J Phys Chem 95: 5784
170. Ortmann W, Kassem A, Hinzmann S, Fanghanel E (1982) J Prakt Chem 324: 1017
171. Elcov AW, Smirnova NP, Ponyaev AI, Martinova WP, Schutz R, Hartmann H (1990) J Lumin 47: 99
172. Murasecco-Suardi P, Gassmann E, Braun AM, Oliveros E (1987) Helv Chim Acta 70: 1760
173. Vintgens V, Scaiano JC, Linden SM, Neckers DC (1989) J Org Chem 54: 5242
174. Flamingi L (1992) J Phys Chem 96: 3331
175. Soumillion JPh, Vandereecken P, De Schryver FC (1989) Tetrahedron Letters 30: 697
176. Julliard M, Chanon M (1986) J Photochem 34: 231
177. Freeman PK, Srinivasa R (1986) J Org Chem 51: 3939
178. Art JP, Kestemont JP, Soumillion JPH (1991) Tetrahedron Letters 32: 1425
179. Lund H, Carlsson HS (1978) Acta Chem Scand B 32: 505
180. Carlsson HS, Lund H (1980) Acta Chem Scand B 34: 409
181. Wavzonek S, Wagenknecht JH (1963) J Electrochem Soc 110: 420
182. Nelleborg P, Lund H, Eriksen J (1985) Tetrahedron Letters 26: 1773
183. Shukla SS, Rusling JF (1985) J Phys Chem 89: 3353

184. Kurauchi Y, Ohga K, Nobuhara H, Morita S (1985) Bull Chem Soc Japan 58: 2711
185. Kurauchi Y, Nobuhara H, Ohga K (1986) Bull Chem Soc Japan 59: 897
186. Epling GA, Florio E (1986) Tetrahedron Letters 27: 675
187. Epling GA, Florio E (1988) J Chem Soc Perkin Trans 1, 703
188. Tsujimoto K, Tasaka S, Ohashi M (1975) J Chem Soc Chem Commun 758
189. Barltrop JA, Bradbury D (1973) J Am Chem Soc 95: 5085
190. Freeman PK, Ramnath N (1988) J Org Chem 53: 148
191. Kropp M, Schuster GB (1987) Tetrahedron Letters 28: 5295
192. Zupancic N, Sket B (1991) Tetrahedron 47: 9071
193. Lan JY, Schuster GB (1985) J Am Chem Soc 107: 6710
194. Lan JY, Schuster GB (1986) Tetrahedron Letters 27: 4261
195. Schuster GB (1990) Pure & Appl Chem 62: 1565
196. Compton RG, Fisher AC, Wellington RG, Bethell D (1991) Electroanalysis 3: 183
197. Kotov NA, Kuzmin MG (1990) J Electroanal Chem 285: 223
198. Aruga T, Ito O, Matsuda M (1982) J Phys Chem 86: 2950
199. Billing R, Rehorek D, Hennig H (1990) Photoinduced Electron Transfer II, Springer Verlag 151
200. Mulliken RS (1950) J Am Chem Soc 72: 600
201. Kochi JK (1988) Angew Chem Int Ed Eng 27: 1227
202. Deronzier A (1982) J Chem Soc Chem Commun 329
203. Barnett JR, Hopkins AS, Ledwith A (1973) J Chem Soc Perkin Trans II, 80
204. Tagaya H, Saito H, Suda S, Chiba K (1989) Bull Chem Soc Japan 62: 768
205. Prasad DR, Hoffman MZ, Mulazzani QC, Rodgers MAJ (1986) J Am Chem Soc 108: 5135
206. Jones GII, Zisk MB (1986) J Org Chem 51: 947
207. Deronzier A, Esposito F (1983) Nouv J Chimie 7: 15
208. Nagamura T, Sakai K (1986) J Chem Soc Chem Commun 810
209. Nagamura T, Sakai K (1989) Ber Bunsenges Phys Chem 93: 1432
210. Nagamura T, Sakai K (1988) J Chem Soc Faraday Trans 84: 3529
211. Nagamura T, Sakai K (1987) Chem Phys Letters 141: 553
212. Nagamura T, Isoda Y (1991) J Chem Soc Chem Commun 72
213. Nagamura T, Sakai K Ogawa T (1988) J Chem Soc Chem Commun 1035
214. Nagamura T, Sakai K Ogawa T (1989) Thin Solid Films 179: 375
215. Nagamura T, Isoda Y, Sakai K Ogawa T (1990) J Chem Soc Chem Commun 703
216. Sullivan BP, Dressick WJ, Meyer TJ (1982) J Phys Chem 86: 1473
217. Wade PA, Morrison HA, Kornblum N (1987) J Org Chem 52: 3102
218. Tamura R, Yamawaki K, Azuma N (1991) J Org Chem 56: 5743
219. Fujimori K, Togo H, Pelchers Y, Nagata T, Furakawa N, Oae S (1985) Tetrahedron Letters 26: 775
220. Tatikolova A, Yang X, Sauerwein B, Schuster GB (1990) Acta Chem Scand 44: 837
221. Zou C, Miers JB, Ballew RM, Dlott DD, Schuster GB (1991) J Am Chem Soc 113: 7823
222. Yang X, Zaitsev A, Sauerwein B, Murphy S, Schuster GB (1992) J Am Chem Soc 114: 793
223. Chatterjee S, Gottscahlk P, Davis PD, Schuster GB (1988) J Am Chem Soc 110: 2326
224. Neckers DC (1989) J Photochem Photobiol A, 47: 1
225. Eaton DF (1986) Adv Photochem 13: 427
226. Naitoh K, Yamaoka T, Umehara A (1991) Chem Letters 1869
227. Becker HGO, Israel G, Oertel U, Vetter HU (1985) J Prakt Chem 327: 399
228. Rotkiewicz K, Grellman KH, Grabowski ZR (1973) Chem Phys Letters 19: 315
229. Rettig W (1986) Angew Chem Int Ed Engl 25: 971
230. Grabowski ZR, Dobrowski J (1983) Pure Appl Chem 55: 245
231. Habib Jiwan JL, Soumillion JPh (1992) J Photochem Photobiol A: Chem 64: 145

PET-Reactions of Aromatic Compounds

Angelo Albini[1], Elisa Fasani[2] and Mariella Mella[2]

[1] Institute of Organic Chemistry, University of Torino v. Giuria 7, 10125 Torino, Italy
[2] Department of Organic Chemistry, University of Pavia v. Taramelli 10, 27100 Pavia, Italy

Table of Contents

The primary aim of this chapter is to provide a review of the chemistry of PET generated aromatic radical ions in solution. Extensive delocalization of the unpaired electron makes these species relatively stable. Thus it is possible to exploit both aspects of their reactivity (as radicals and as ions) and to intercept them both with electrophiles (or respectively with nucleophiles) and with radicals.

Topics in Current Chemistry, Vol. 168
© Springer-Verlag Berlin Heidelberg 1993

Other possibilities are that the original radical ions recombine or that one of them (or both) fragments either α or β to the ring yielding aryl (or respectively benzyl) radicals, which can then be trapped. Examples of all these mechanisms are given, and the synthetic potential offered by PET induced reactions of aromatics is discussed.

1 Introduction

Aromatic substrates are by far the most commonly used substrates in the rapidly expanding area of photoinduced electron transfer [1,2]. This is obviously due to the favourable location of the frontier molecular orbitals in such compounds. The same factor facilitates the formation of electron transfer donor–acceptor (EDA) complexes both in the ground state (these possibly are intermediates in some thermal reactions, e.g. selected electrophilic substitutions), and in the excited state (exciplexes).

Thus, photoinduced electron transfer may be obtained either by collision between the excited state and a quencher or by excitation of a preformed ground state complex.

$$Ar + h\nu \rightarrow Ar^* \xrightarrow{\ Q\ } Ar^{\bar{}} + Q^{\dagger} \ or \ Ar^{\dagger} + Q^{\bar{}}$$

$$(Ar \cdots Q) + h\nu' \longrightarrow Ar^{\bar{}} + Q^{\dagger} \ or \ Ar^{\dagger} + Q^{\bar{}}$$

Furthermore, on the excited surface both formation of a complex, to the stabilization of which both an exciton resonance term and an electron transfer term contribute, and full electron transfer to yield a radical ion pair may be envisaged [3]. Diffusion of the charged species leads to free solvated radical ions (FRI), but for a sizeable fraction of the systems which will be discussed in the following, the reaction takes place at the stage of the contact ion pair (CIP), and then distinction between the properties of the polar exciplex and of the radical ion pair may not be unambiguous.

$$(Ar \cdots Q)^* \qquad (Ar^{\bar{}} Q^{\dagger}) \qquad (Ar^{\bar{}})_{solv} (Q^{\dagger})_{solv}$$
$$\text{Exciplex} \qquad \text{CIP} \qquad \text{FRI}$$

In view of the extensive delocalization of the unpaired electron, the radical ions of aromatic molecules are often relatively stable, the limitation to their lifetime being rather given by back electron transfer. They are therefore ideally suited for physical studies, just as it happens for excited states, where aromatic molecules are often chosen for photophysical experiments since they show little photochemistry. Thus, systematic studies on the rate of electron transfer have been carried out using aromatic molecules [4], and aromatic substrates are in use for enhancing the quantum yield of electron transfer photoreactions through secondary electron transfer (a typical example is biphenyl, BP, which by functioning as secondary donor slows down back electron transfer between the original radical ions and allows their chemistry to show up) [5].

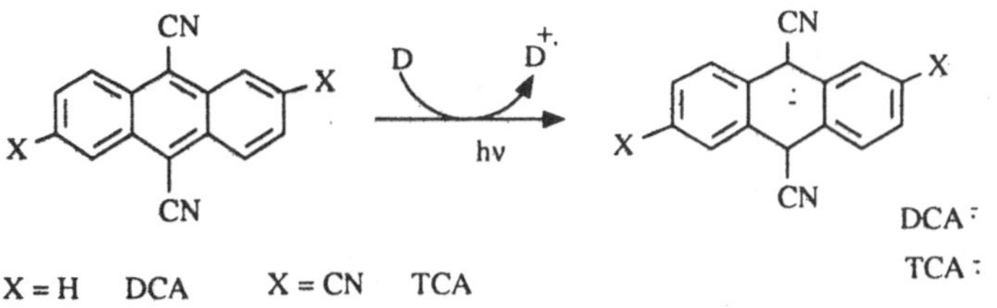

X = H   DCA   X = CN   TCA

Scheme 1.

Under proper conditions, aromatic radical ions may show unexpectedly long lifetimes. This is the case for the radical anions of 9,10-dicyanoanthracene and of 2,6,9,10-tetracyanoanthracene, which, when generated by irradiation of the arenes in the presence of triethylamine in acetonitrile containing tetraethyl-ammonium phosphate have been reported to persist for days (Scheme 1) [6].

However, aromatic radical ions have a chemistry of their own, and in many cases react efficiently, offering access to novel and useful synthetic processes. They show, in fact, a double reactivity, behaving both as radicals and as ions.

In the following, the main reactions will be discussed according to the type of mechanism operating. The reactions are classified according to the site involved in the process (the aromatic ring or a substituent) and on whether the aromatic substrate is the acceptor or the donor in the process. When there is a possible ambiguity (i.e. both the donor and the acceptor are an aromatic molecule) the classification is usually referred to the substrate absorbing the light. Heteroaromatic molecules are not included, but their PET chemistry follows a similar scheme.

2 Reactions on the Aromatic Ring

2.1 The Aromatic is the Acceptor

The occurrence of a chemical reaction involving the radical anion is linked to the following mechanisms. Recombination with the radical cation, usually resulting in aromatic substitution (Scheme 2). Addition of a cation or an electrophile giving in the end a substituted aromatic or a dihydro adduct (Scheme 3). Addition of a radical to yield an anion, which then either rearomatizes or is protonated to a dihydroaromatic; in the most usual, and interesting, instance, the radical in turn arises from the radical cation, either through a fragmentation or through an addition reaction (Scheme 4). A final possibility is fragmentation of the radical anion to yield an aryl radical; this in turn either abstracts a hydrogen atom giving the arene or adds a nucleophile (Scheme 5).

The above scheme refers to the case where the acceptor is a neutral molecule. It can be easily modified to account for different situations, e.g. electron

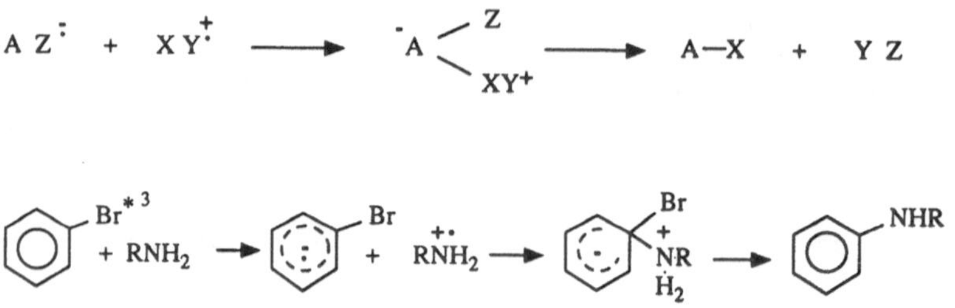

$$A\,Z^{\cdot-} + X\,Y^{\cdot+} \longrightarrow {}^{\cdot}A\!\!\begin{array}{c}Z\\XY^+\end{array} \longrightarrow A{-}X + Y\,Z$$

Scheme 2.

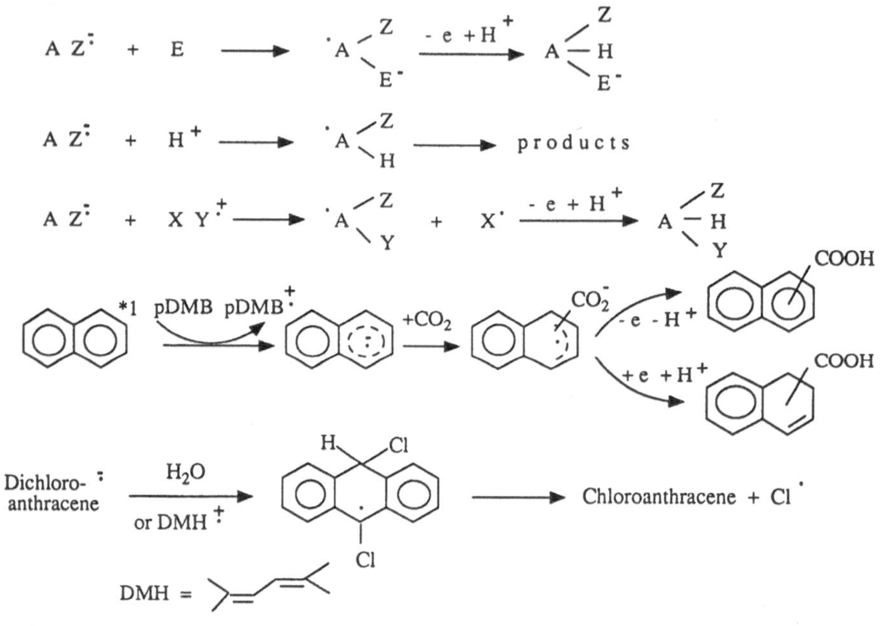

$$A\,Z^{\cdot-} + E \longrightarrow {}^{\cdot}A\!\!\begin{array}{c}Z\\E^{\cdot-}\end{array} \xrightarrow{-\,e\;+H^+} A\!\!\begin{array}{c}Z\\H\\E^{\cdot-}\end{array}$$

$$A\,Z^{\cdot-} + H^+ \longrightarrow {}^{\cdot}A\!\!\begin{array}{c}Z\\H\end{array} \longrightarrow products$$

$$A\,Z^{\cdot-} + X\,Y^{\cdot+} \longrightarrow {}^{\cdot}A\!\!\begin{array}{c}Z\\Y\end{array} + X^{\cdot} \xrightarrow{-\,e\;+H^+} A\!\!\begin{array}{c}Z\\H\\Y\end{array}$$

Dichloro-anthracene $\xrightarrow[\text{or DMH}^{\cdot+}]{H_2O}$ → Chloroanthracene + Cl$^{\cdot}$

DMH =

Scheme 3.

transfer to a diazonium salt yields a neutral radical which loses nitrogen; a similar mechanism operates with anilinium salts (Scheme 6). Of course such cases are qualitatively different from the previous ones, where pairs of oppositely charged ions are formed, while here electron transfer does not develop new charges.

2.1.1 Recombination of the Radical Ions

Photonucleophilic substitution may be preparatively useful in view of the peculiar orientation. In several cases direct attack of the nucleophile onto the excited state of the aromatic (SNAr*) occurs (and then orientation can be rationalized on the basis of the MO coefficients in the excited state), but when

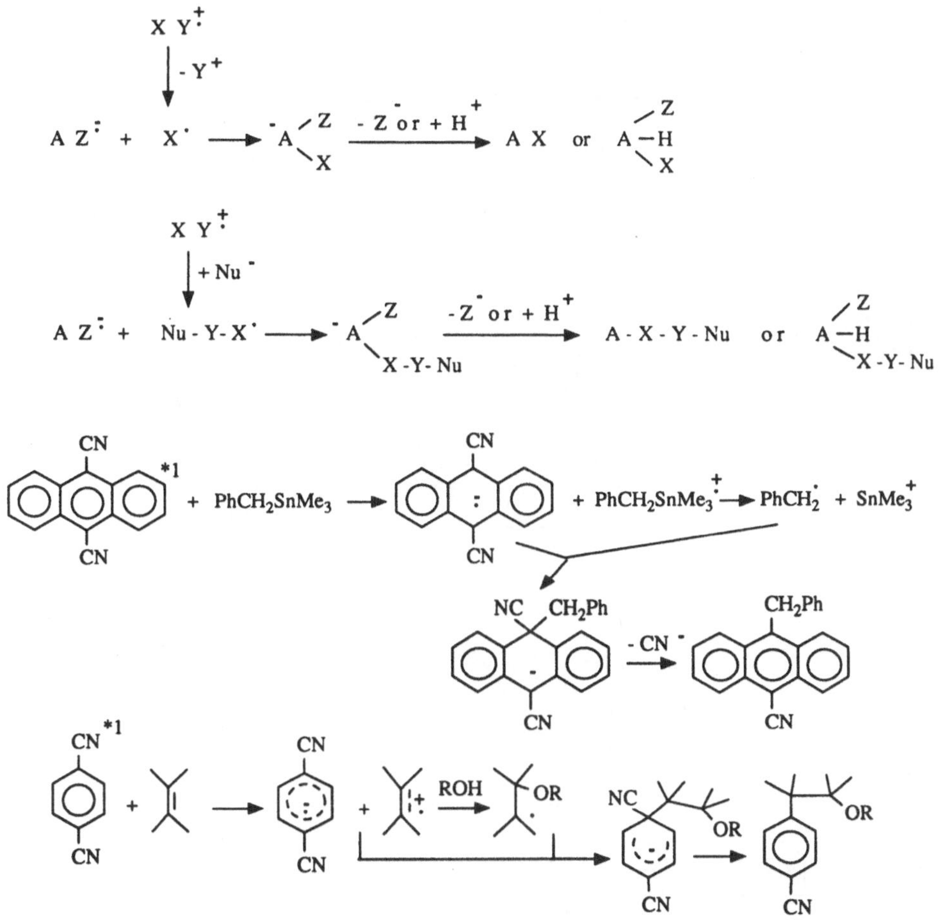

Scheme 4.

the nucleophile is a good electron donor electron transfer followed by radical ions recombination (SNAr⁻) is a viable path [7–11] (Scheme 2, for yet another path see Sect. 2.2.1). The chemical outcome is usually different, e.g. *p*-methoxy nitrobenzene undergoes substitution of the methoxy group through the electron transfer path (e.g. with hexylamine) and of the nitro group by direct attack onto the triplet (when the less efficient donor ethyl glycinate is used, Scheme 7) [12]. Roughly, it can be expected that when the "meta activation" characteristic of the excited state [7] is operating and with strong nucleophiles the SNAr* path predominates; this was the first mechanism proposed and certainly applies to those conditions. However, it has now become clear that the electron transfer mechanism operates under suitable conditions. Despite some attempts in this direction, the orientation observed in the latter case is not easily reconciled with FMO control [10]; an alternative proposal is that in this mechanism it is the stability of the σ complex which dictates the orientation [13]. Similarly, with

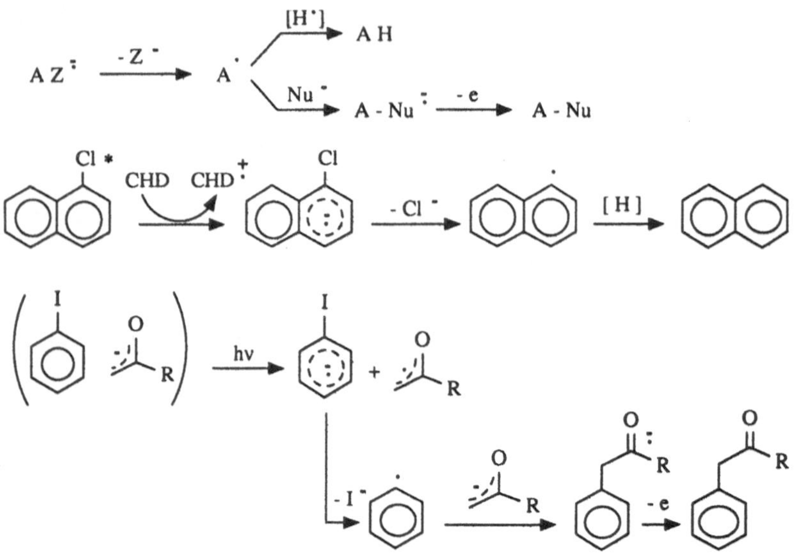

Scheme 5.

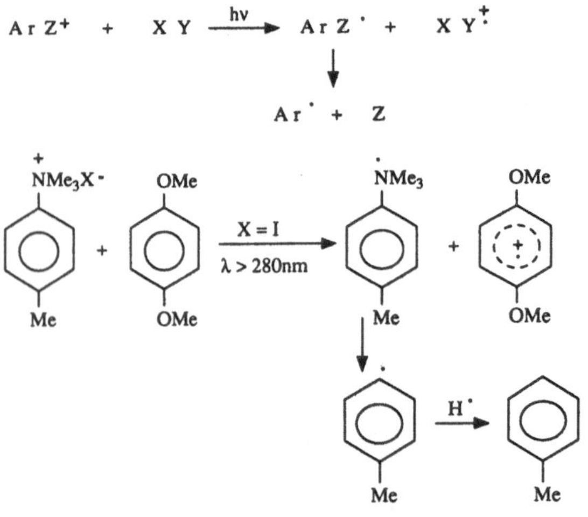

Scheme 6.

1-methoxy-4-nitronaphthalene primary amines displace the nitro group while secondary amines displace the methoxy group via an electron transfer mechanism [14], and with 4-nitroveratrole attack takes place *para* to the nitro group if electron transfer is operating (with amines) and *meta* it to if the reaction occurs on the triplet [15] (Scheme 8, the orientation depends also on the bulkiness of the amine, however) [16]. Unambiguous demonstration has been obtained only in selected cases, but this mechanistic dichotomy probably explains many other results, e.g. 3-fluoro-4-methoxynitrobenzene substitution of the fluorine atom is one of the paths with primary amines, while α-aminoacids and the hydroxide

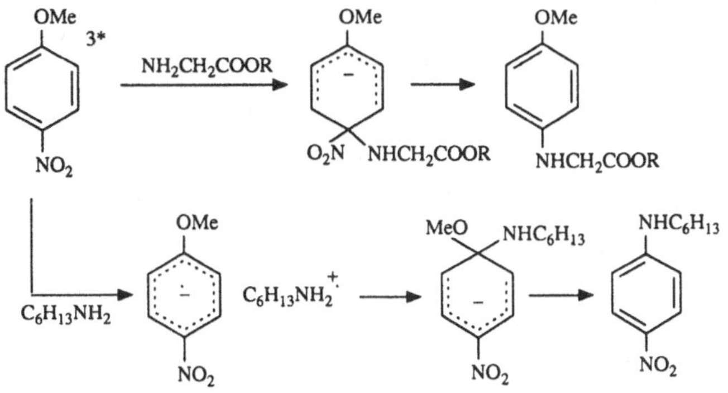

Scheme 7.

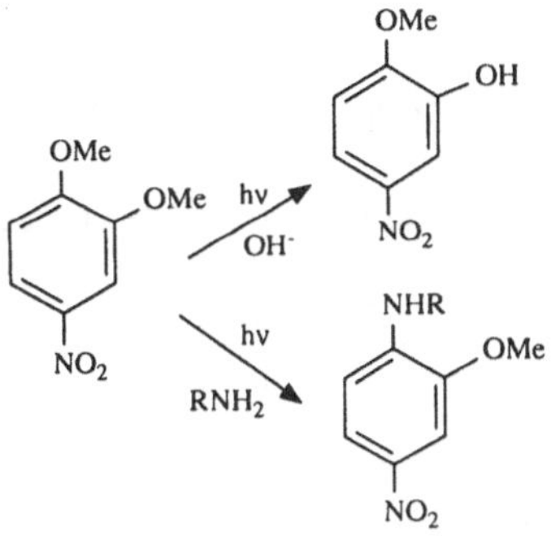

Scheme 8.

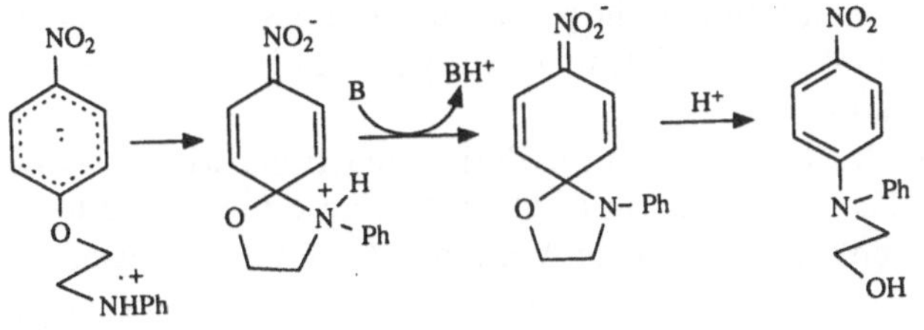

Scheme 9.

anion substitute only the two other groups [17]. Furthermore, other reactions may compete, e.g. 4,5-dinitroveratrole undergoes substitution of a methoxy group with amines with a relatively high ionization potential, but nitro group reduction when the electron transfer path operates [18]. An intramolecular version of this electron transfer induced substitution is the photo-Smiles rearrangement of 4-(2-N-phenylaminoethoxy)nitrobenzene [19–22] (Scheme 9, a reaction subject to base catalysis and inhibited by cyclodextrine complexation) [23]. The cleavage of N-phenylphenanthrenylethylamine proceeds through a similar mechanism to yield 9-methylene-9,10-dihydrophenanthrene [24–25], and the same holds for the corresponding anthracene derivative [26]. The acid

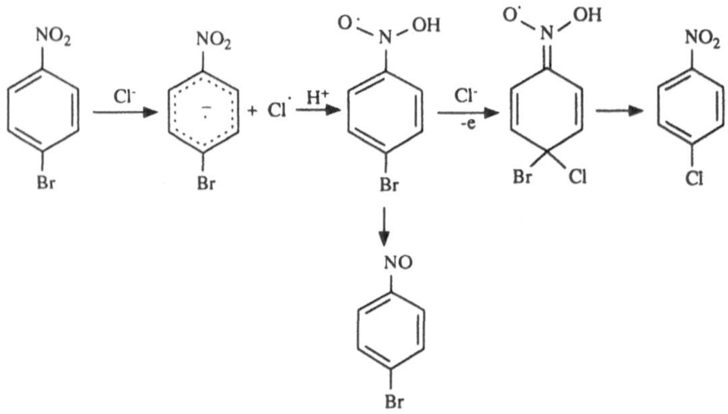

Scheme 10.

catalyzed displacement of bromine by chlorine in *p*-bromonitrobenzene follows a related path (Scheme 10) [27].

[Note. The recombination of π radical ions and cycloaddition are discussed elsewhere in this series].

2.1.2. Reaction with an Electrophile

The most typical example of a radical anion–electrophile combination is the carboxylation of aromatics (Scheme 3: naphthalene [28], phenanthrene, anthracene, pyrene [29] cyclopenta[*def*]phenanthrene [30] and, with lower yield, biphenyl [31]) with CO_2 by irradiation in the presence of donors such as amines or *p*-dimethoxybenzene (pDMB). A mixture of dihydro derivatives and re-aromatized products is usually obtained.

Although aliphatic nitriles take part in alkylation reactions with tri- and tetracyanobenzenes (Sect. 2.1.3), irradiation of DCA and an amine in aqueous acetonitrile takes a different course giving 9-amino-10-cyanoanthracene, a reaction in which the amine serves as the donor, but then the DCA radical anion reacts with acetonitrile functioning as an electrophile leading via an isomerization–elimination process to the final product [32].

Protonation is another common reaction in this category and reduction of the acceptor via protonation of the radical anion (e.g. by moisture present in the solvent) is often observed as a side process. It may become the main path either under acidic conditions or if the radical cation of the donor is a good proton donor (Scheme 3). As an example, the 1,2-dihydro derivative is the main product from 1,4-dicyanonaphthalene (DCN) when irradiated in the presence of a donor and trifluoroacetic acid [33]. Such a reaction is more important when the radical cation is the proton source. Typically, the irradiation of naphthalene with tertiary amines leads to a mixture of 1,4-dihydro- and 1,2,3,4-tetrahydro-naphthalene, as well as tetrahydrobinaphthyl and 1(1-diethylamino)ethyl-1,4-

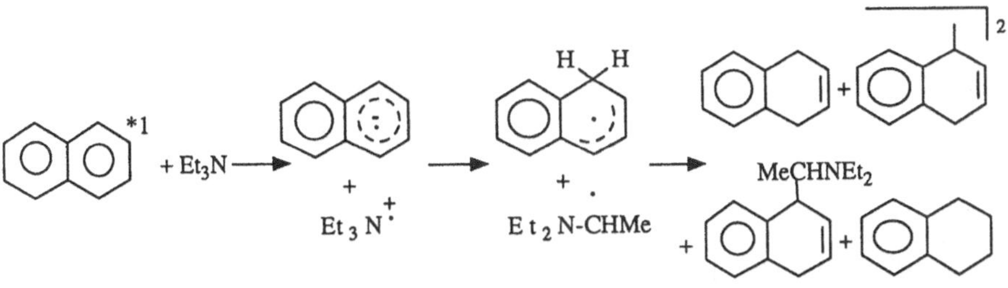

Scheme 11.

dihydronaphthalene (Scheme 11) [34]. Other photoreductions of aromatics with amines or hydroxylamines have been reported [33–40].

This class of reactions has been tagged "photo-Birch" reduction. The analogy with the Birch alkali metal reduction is not complete, however. In the thermal reaction a strong reducing agent is continuously available and protonation may occur either on the dianion or on the radical anion, and at any rate the aryl radical formed in the first step is further reduced and protonated, while in the photochemical case the reaction involves proton transfer between the oppositely charged radical ions. Thus, formation of a dihydro derivative, either by hydrogen abstraction or for the reduction of the naphthalene radical anion by the amine and protonation, is accompanied by radical coupling to yield the other products observed. The detailed mechanism of the reaction and the competition between α-C–H and N–H bonds as proton source with primary and secondary amines have been studied [37–38]. Likewise, irradiation of good acceptors, such as aromatic nitriles, and donors such as methylbenzenes and alkenes leads in some cases to electron and proton transfer resulting in reduction of the nitrile as the main path (e.g. 2-cyanonaphthalene with durene or 2,3-dimethylbutene) [41].

2.1.3 Reaction with a Radical

The most typical reaction involving radicals is alkylation of electron-poor aromatics, particularly of nitriles (Scheme 4, often the reaction is clean since in this case only the singlet reacts and there are no competitive reactions of the substituent). The scope of the reaction has been considerably enlarged recently. As sketched in Scheme 4, the radicals are formed either by cleavage of a σ bond in the radical cation, or by addition of a nucleophile onto an unsaturated radical cation. Fragmentation requires that an electrofugal group is contained in the donor. A typical example is the benzylation of 1,2,4,5-tetracyanobenzene (TCNB) when irradiated in toluene. In this case a ground state EDA complex is involved and the mechanism has been formulated as in Scheme 12 [42].

Similar reactions have been reported for a large number of substrates [43]. In polar media, proton transfer does not necessarily involve the radical anion, a relatively weak base, but rather the solvent (Scheme 13), particularly when

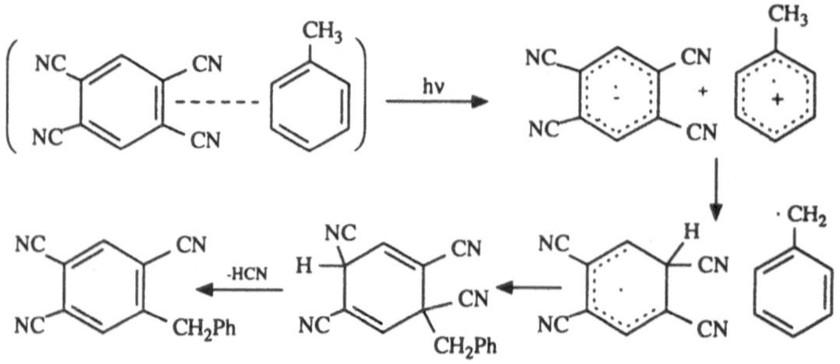

Scheme 12.

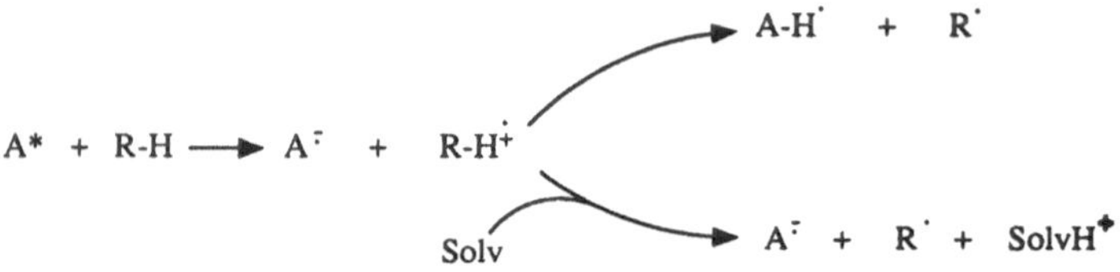

Scheme 13.

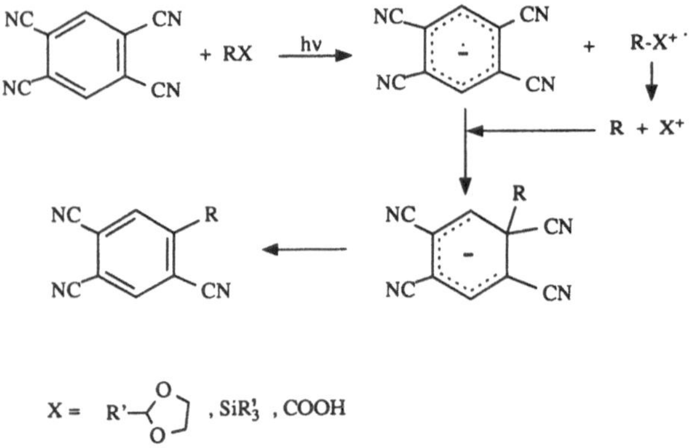

$$X = R'-\!\!\begin{array}{c}O\\ \diagdown\\ O\end{array}\!\!\Big], \ SiR'_3 \ , COOH$$

Scheme 14.

this is a nucleophile (e.g. methanol) and the electrofugal group, beside a proton, can be a silyl or stannyl cation or a carbocation. For the reaction to take place it requires that the radical cation confronts a relatively low barrier for the cleavage, in order to compete successfully with back electron transfer (see further Sect. 3.2), and that the radical anion is sufficiently stable, otherwise disproportionation may compete and the net result be reduction rather than alkylation; as mentioned above, this path often dominates with mononitriles, while alkylation is typical of polynitriles). In this way, anions are formed and the final step is rearomatization through cyanide loss with benzenenitriles (Schemes 14, 15) and protonation to yield a dihydro derivative with naphthalene and anthracene derivatives (Schemes 16, 17).

Reported examples include the alkylation of 1,2- and 1,4-dicyanobenzene with toluene [43], amines [44], alkenes [45], silanes, germanes and stannanes

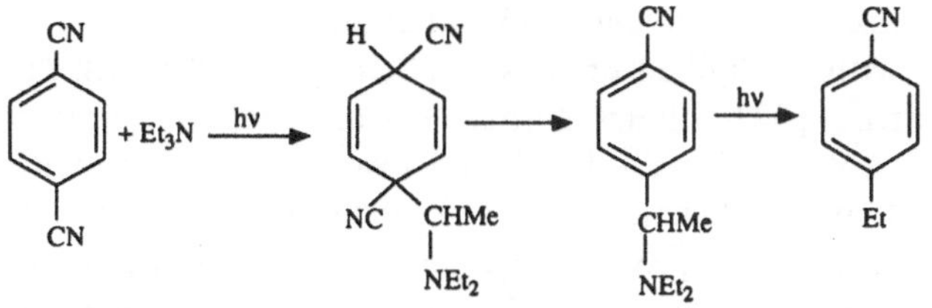

Scheme 15.

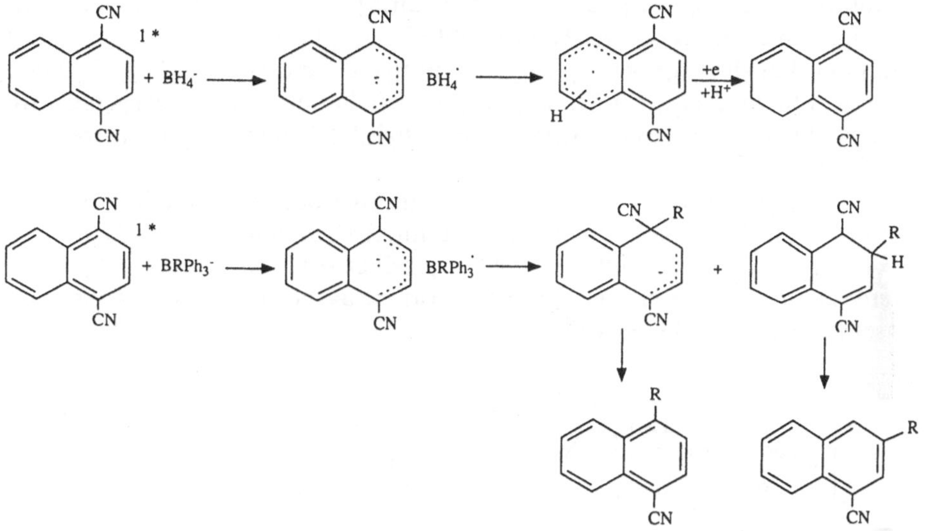

Scheme 16.

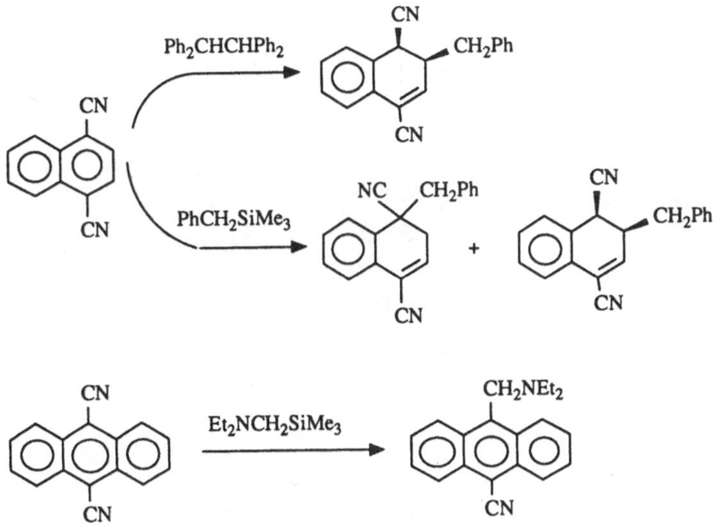

Scheme 17.

[46–8]; of 1,3,5-tricyanobenzene with toluene [49] and aliphatic nitriles [50]; of 1,2,4,5-tetracyanobenzene with methylbenzenes [42], ethers [51] (also with the related tetrahalophthalonitriles [52]), silanes, germanes, stannanes [46], acetals [53] and aliphatic nitriles [50] and acids [54]; of 1-cyanonaphthalene with alkenes [55], toluenes [41], and arylacetic acids [56]; of 1,4-dicyanonaphthalene with alkenes [57], alkylaromatics [33, 58–63], benzyl and allylsilanes and analogues [61, 64–65], alkyltriarylborates [66–67], substituted bibenzyls [59], ethers [68] and ethers of pinacols [69–71], of various cyanophenanthrenes and anthracenes with silanes [64, 72] and stannanes [73–4]. The addition of phenylcyclopropane onto positions 9 and 10 of DCA may be viewed again as a radical addition after cleavage of the phenylcyclopropane radical cation [75]. Silylation may be obtained in a similar way, e.g. by irradiation of cyanobenzenes with hexamethyldisilane [76] and with the trimethylsilyl diether of a pinacol (Scheme 18) [77].

Some reactions do require that proton transfer occurs within the originally formed radical ion pair; this is the case for the reaction between DCN and alkylbenzenes, where a different type of product is formed: noteworthy is the stereospecific formation of tetracyclic derivatives as the main products in several cases (Scheme 19) [33, 58–61]

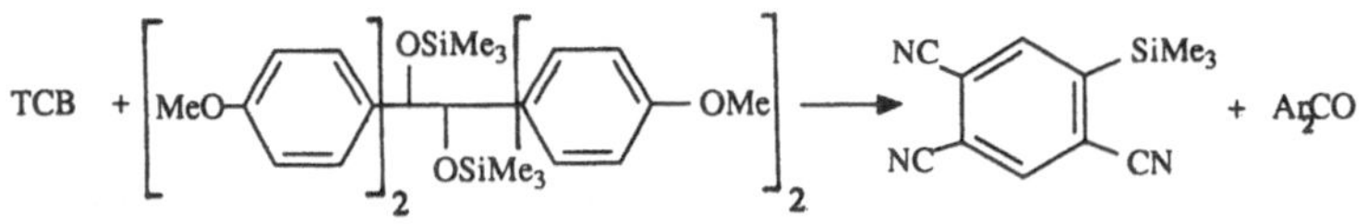

Scheme 18.

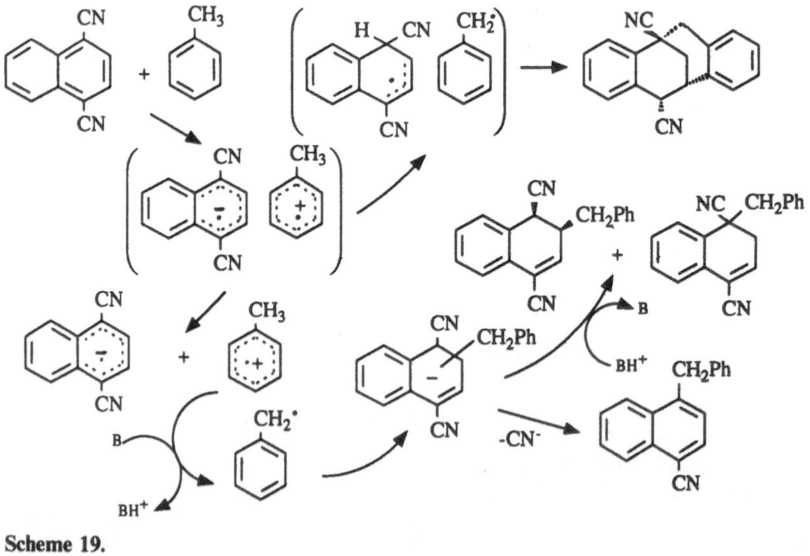

Scheme 19.

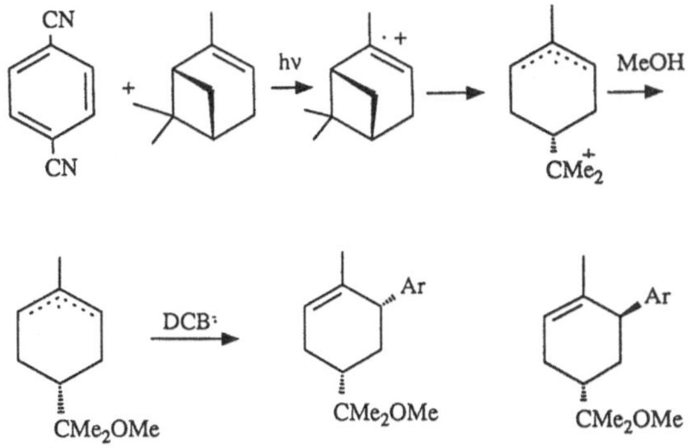

Scheme 20.

In the examples above, the attacking radical arises from the fragmentation of the radical cation, but, as already noted, in the presence of a nucleophile, radicals are also formed by addition to olefinic (Scheme 4) [45, 78–79], aromatic [80] or heterocyclic [81] radical cations, and react in the same way. For the reactions involving alkenes the acronym Nucleophile–Olefin Combination Aromatic Substitution (NOCAS process) has been proposed; in this process, the radical cation of strained alkenes may rearrange before addition (Scheme 20) [79].

2.1.4 Fragmentation

A final possibility for the radical anion is fragmentation to yield the relatively stable aryl radical and an anion. The end result is either reduction (whether via direct hydrogen abstraction or via a second electron transfer – the radical is usually easier to reduce than the original substrate – and protonation of the resulting anion) or substitution.

The PET-induced dehalogenation of aromatics has been extensively investigated. It occurs in the presence of various donors, such as aliphatic and aromatic amines [82–92], alkenes and dienes [91, 93–94], mixed hydrides [95–97], hydroxides [98], carbanions [99–101]. In the absence of an added donor, a viable mechanism involves the excimer (singlet or triplet) and its collapse to oppositely charged radical ions followed by fragmentation of the radical anion [93, 102–104] (a micellar medium has an important effect in this case [103]).

This is a useful process, since, while triplet homolysis of simple halogenated benzenes can be quite efficient, there is no such possibility for higher aromatics since the triplet energy drops much below that of the carbon–halogen bond. The PET dehalogenation, either via the singlet or the triplet, is often efficient, and may offer a method for water detoxication in view of the large use of polychlorobiphenyls and of halogenated pesticides in industry and agriculture. The process can be conveniently carried out with near UV or blue light by using a convenient sensitizer (e.g. anthracene, Scheme 21 [86]).

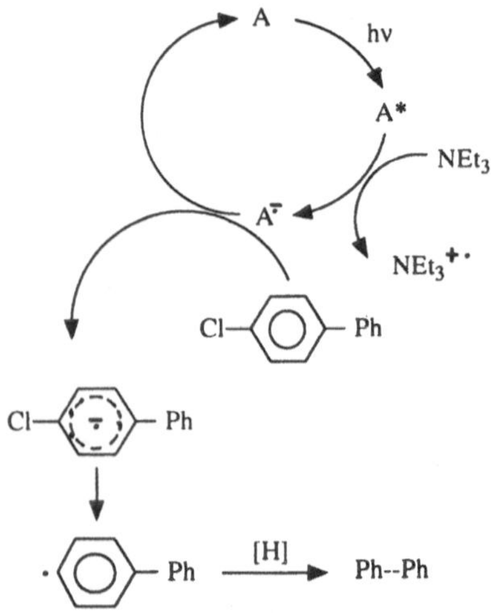

Scheme 21.

$$\mathrm{In} \xrightarrow{h\nu} \mathrm{R}^{\bullet} \qquad \mathrm{R}^{\bullet} + \mathrm{MeOH} \longrightarrow \mathrm{RH} + {}^{\bullet}\mathrm{CH_2OH}$$

$${}^{\bullet}\mathrm{CH_2OH} + \mathrm{MeO^-} \longrightarrow \mathrm{MeOH} + \mathrm{CH_2O}^{\bar{\bullet}}$$

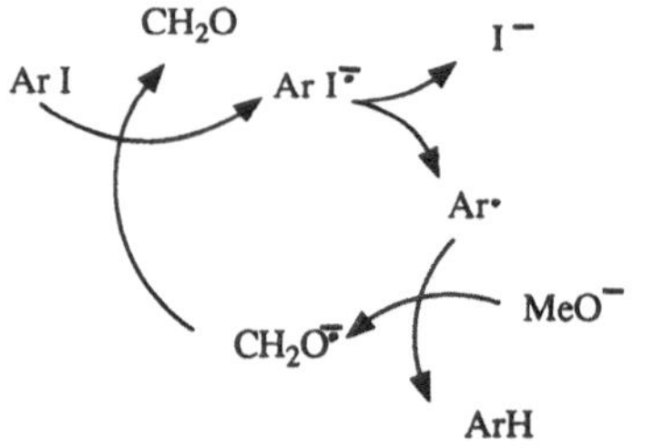

Scheme 22.

A related process involves electron transfer to an aryl halide from a photogenerated aldehyde radical anion and a propagation cycle (Scheme 22) [105–106].

Alternatively, the aryl radical may add to an aromatic substrate yielding biaryls, as in the reactions of polyhalocyanobenzenes with anisole [107–108] and of halobenzenes with *N,N*-dimethylaniline [109–110]. Related intramolecular reactions are useful for the synthesis of heterocycles [111–112].

Besides the halide anions, cyano [97, 116] and nitro [117] groups have also been found in some cases to be eliminated by irradiation in the presence of a donor.

Biaryl formation is important when cationic substrates are used, such as diazonium salts (usually in the intramolecular version, the photo-Pschorr reaction [113]) or quaternary ammonium salts [114]. In this case PET generates a neutral radical, and this expels nitrogen or an amine molecule yielding an aryl radical. Although the present discussion generally does not include ma-

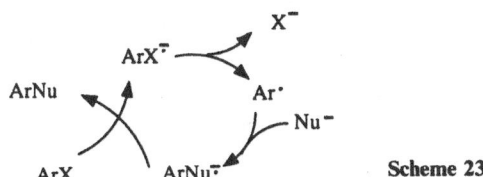

$$ArX^* + Nu^- \longrightarrow ArX^{\overline{\cdot}} + Nu^{\cdot}$$

Scheme 23.

cromolecular chemistry, one should at least mention the important role that the PET induced fragmentation of aromatic onium salts (e.g. iodonium and sulfonium salts) has for the initiation of both radicalic and ionic polymerization [115].

If the radical adds a nucleophile, a new radical anion is formed, and when oxidation ensues, the overall process is aromatic substitution (Scheme 23). When the oxidant is the substrate the reaction occurs cyclically, and the quantum yield may be far higher than 1. This is the well known $S_{RN}1$ process often observed when highly basic or nucleophilic species are used [118–120]. Mostly, these reactions do not completely fit in the general scheme presented above, since typically they involve irradiation of the nucleophile (or of a nucleophile–aromatic ground state EDA complex) rather than of the aromatic substrate, and may imply both direct electron transfer between the reagents or photoionization of the nucleophile followed by reduction of the aromatic acceptor by the solvated electron. In fact, visible light is often sufficient for initiating the reaction. This is generally carried out in very polar media, such as liquid ammonia or DMSO, but less polar solvents such as THF have been found suitable in some cases.

A typical reaction is the photostimulated substitution of aryl halides by ketone [121–131] (and much less efficiently aldehyde [124]) enolate anions (Scheme 5), both inter- [121–128] and intramolecularly [129–131]. The $S_{RN}1$ reaction with o-bromoacetophenones is a useful method for the construction of an aromatic ring (Scheme 24) [132–133], with o-halophenylalkyl ketones for macrocycles (Scheme 25) [134], with o-haloanilines for indoles [123], with o-halobenzylamines for isoquinolines [135], and several other heterocyclic syntheses are possible [136].

Similar substitutions are obtained with the carbanions of esters [129, 131], N,N-disubstituted amides [137–138] and nitriles [139–140]. The anion of cyclohexylideneacetonitrile reacts with p-bromoanisole exclusively at the γ position (Scheme 26) [141]. The ambient anions of phenols and anilines react as carbon nucleophiles, as observed in the reactions of halobenzenes with 2-naphthylamine [142], 2-naphthol [143] and various phenols (Scheme 27) [144–145].

In contrast, benzenethiolate anions react at the sulfur atom (Scheme 28) [146–149]; from 1,2-dihalobenzenes and the 1,2-dithiolate dianion, dithiin is obtained [148]. Analogous substitutions are observed with diarylphosphites and other phosphorus nucleophiles [150–152], as well as with arylarsenides

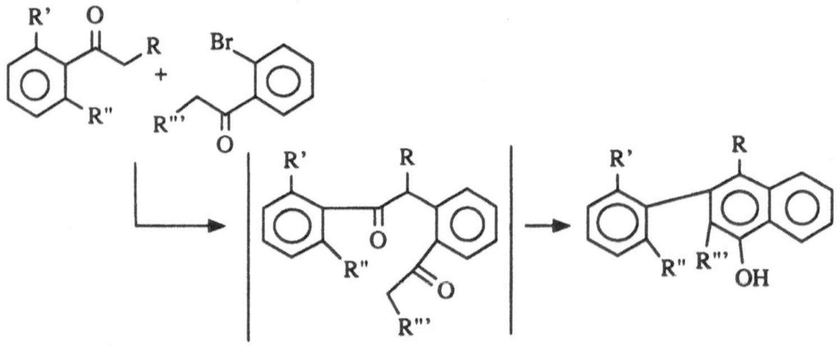

Scheme 24.

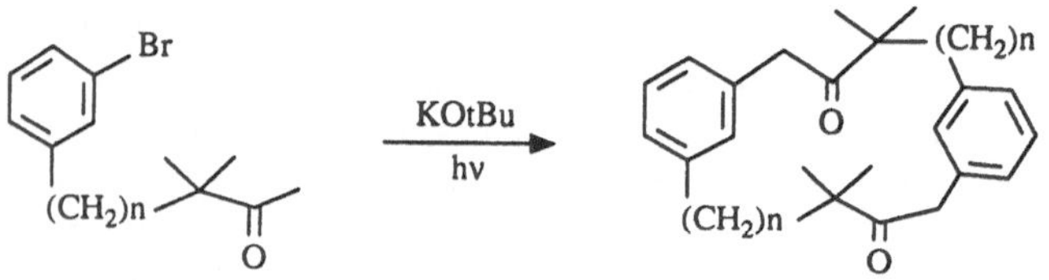

Scheme 25.

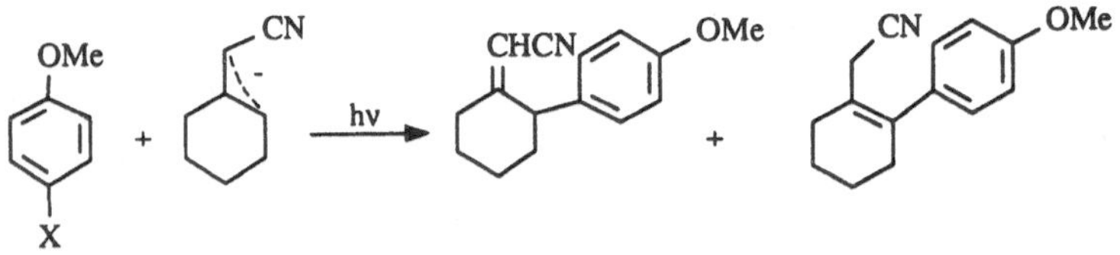

Scheme 26.

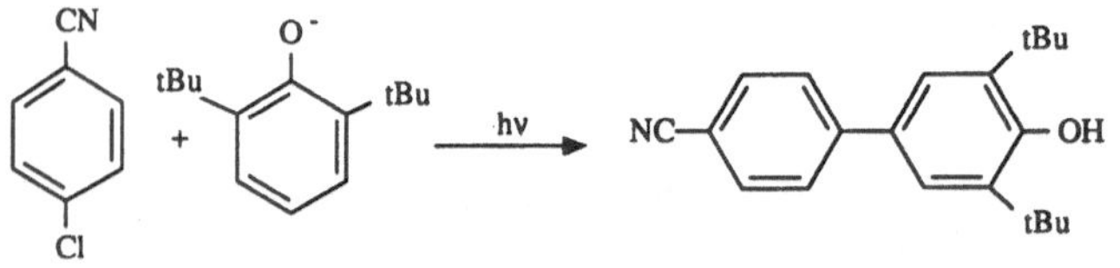

Scheme 27.

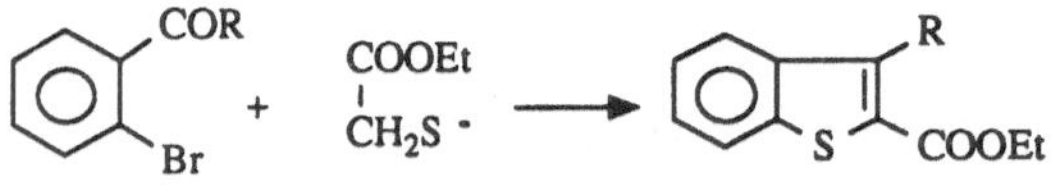

Scheme 28.

[153], selenides and tellurides [154–155]. Substitution with inorganic anions, such as sulphite [156] and cyanide [157], on halonaphthols has been found to proceed analogously. In the presence of cobalt carbonyl mono and polyfunctional benzoic acids are obtained from the corresponding halobenzenes [158–159].

Besides halogenated derivatives, the aromatic substrates liable to substitution via the $S_{RN}1$ process include compounds containing other leaving groups,

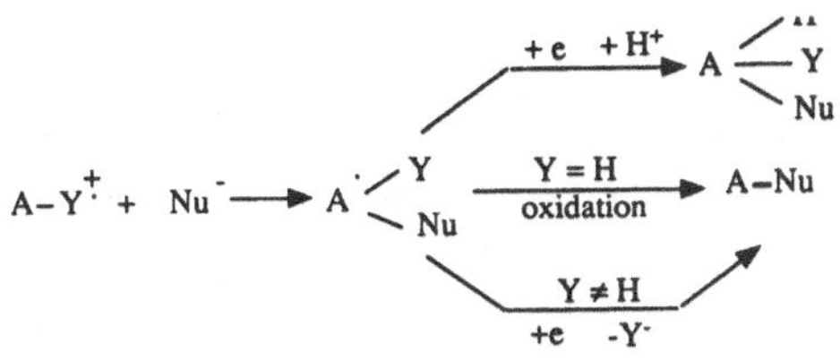

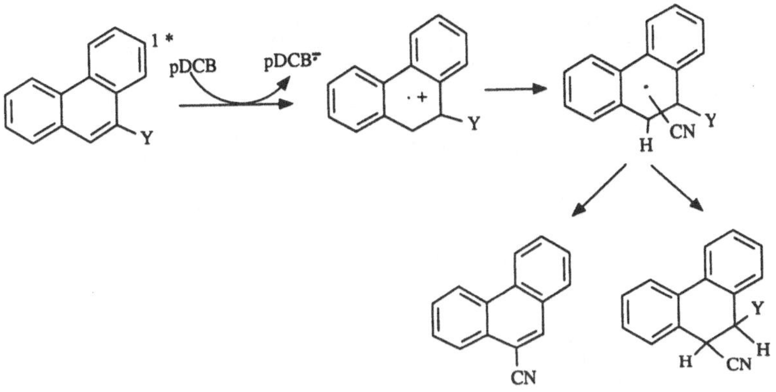

Scheme 29.

e.g. phenyltrimethylammonium and triphenylsulfonium salts, diphenyl selenide, phenyldialkyl phosphate [121] and aryl phenyl sulfides [160]. Many substituents are compatible, except those exerting a strong electron- withdrawing effect, since there the π^* MO is stabilized with respect to the σ^* MO of the Ar–X bond. However, 2-iodonitrobenzene does react, probably because in that case the nitro group is not coplanar with the ring [128].

2.2 The Aromatic is the Donor

The reactions of aromatic radical cations follow a scheme of reactivity which mirrors that of radical anions. Thus, typical reactions include addition of a charged or neutral nucleophile leading ultimately to a substituted aromatic (or its dihydro derivative, Scheme 29) and addition of a radical to yield a cation, which in turn either deprotonates or adds a nucleophile to yield respectively a substituted aromatic or a dihydro derivative (Scheme 30). The attacking radical (and in several cases also the nucleophile reacting in the second step) usually result from the fragmentation of the radical anion formed in the original electron transfer step.

2.2.1 Addition of a Nucleophile

Such a mechanism has been recognized to operate in some "anomalous" photochemical nucleophilic substitutions observed e.g. with anisoles, where

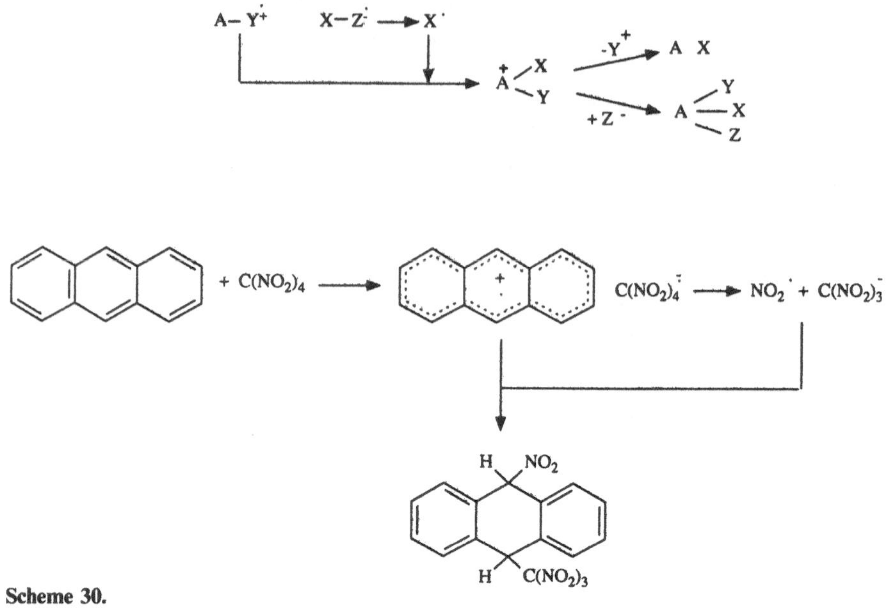

Scheme 30.

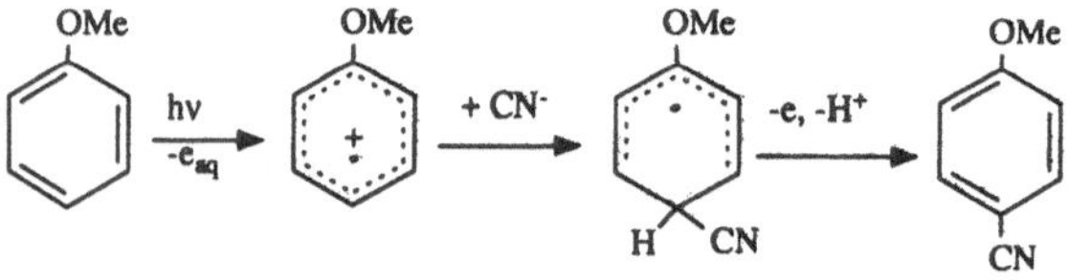

Scheme 31.

indeed the key intermediate is a radical cation [8]. This may be formed by photoionization of the substrate (Scheme 31) [161–163].

Direct photochemical cyanation of unsubstituted aromatics has also been observed, and involves PET between two molecules of the substrate (Scheme 29) [104, 164–166]. This reaction becomes preparatively more useful however, when a strong photochemical oxidant is added, e.g. 1,4-dicyanobenzene [167–168] under homogeneous and WO_3 or TiO_2 under heterogeneous conditions [169]. Addition of crown ethers or polyethylene glycol enhances the solubility of the cyanide in the organic solvent and thus gives a higher yield of the products [170–172]. Arylphosphonates can be obtained through an analogous process [173].

Generation of the radical cation of aromatic substrates in the presence of sodium boron hydride offers another path for reduction, alternative to that via the radical anion seen in Sect. 2.1.2 , and, as one may expect in view of the different mechanism, with a different regiochemistry [174–175]. Thus, e.g. irradiation of the xylenes in the presence of m-dicyanobenzene and $NaBH_4$ yields the corresponding 1,4-dihydro derivatives rather than the 2,5-dihydro derivatives obtained with dissolved metals [175].

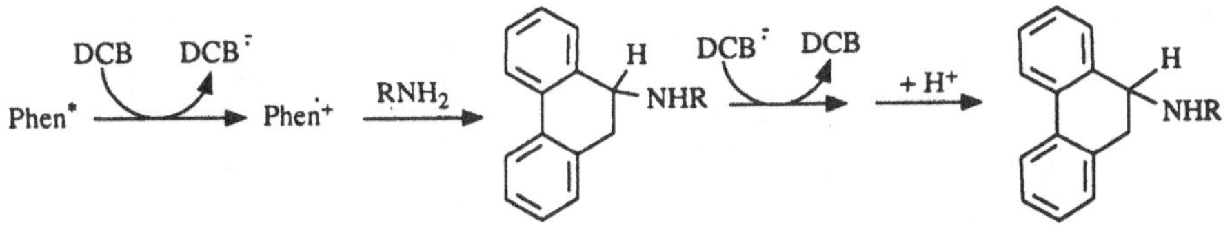

Scheme 32.

Aminodihydroaromatics are obtained when arenes are irradiated in the presence of acceptors such as cyanobenzenes and ammonia or amines (Scheme 32) [176–180]. In some cases the reaction can also be carried out in moderately polar solvents such as THF, provided that a salt is added in order to favour charge separation in the initially formed exciplex [177]. Direct irradiation of benzenes and amines leads competitively to amination and alkylation and involves some degree of charge transfer [36,181].

2.2.2 Addition of a Radical

A radical cation–radical addition appears to be the key step in the photo-chemical nitration obtained when aromatic compounds and nitrating agents NO_2Z such as nitric acid, acetyl nitrate, N-nitropyridinium, or tetranitromethane are irradiated [182]. Under these conditions a ground state EDA complex is usually present, as revealed by the colour appearing on mixing the reagents. Irradiation leads to electron transfer and cleavage of the radical ion to give NO_2 and an anion; addition of NO_2 to the radical cation gives the Wheland intermediate and fast deprotonation of the latter yields the nitrated derivative (Scheme 30) [183–5]. Side processes observed in some cases (e.g. demethylation of the methoxy group of anisoles and trans-bromination of 4-bromoanisole to afford a mixture of 4-nitroanisole and 2,4-dibromoanisole) follow the same pattern observed in the electrophilic nitration with nitric acid [186]. This and other pieces of evidence led to consider a stepwise mechanism (single electron transfer to NO_2^+ followed by radical recombination) as a general mechanism for aromatic nitration, as opposed to direct formation of the σ (Wheland) intermediate. Independently of this, photochemical nitration by selective irradiation in the charge transfer band (in the visible) is a useful process occurring with high ($\Phi \approx 0.5$) quantum yield and excellent material balance. Addition of a nucleophile (the anion resulting from the radical anion fragmentation) rather than deprotonation occurs with anthracene. Thus, with tetranitromethane 9,10-dihydro-9-nitro-10-trinitromethylanthracene is obtained.

Addition of an organic radical apparently takes place in the photochemical phenylation of naphthalene by triphenylsulfonium salts [187], a reaction which takes place also intramolecularly, with one of the aromatic substituents acting as the donor and the other one as acceptor (Scheme 33) [188]. Similarly the alkylation of aromatics by chloroacetonirile occurs [189], where the radical arises from chloride splitting. The radical can also result from the protonation of

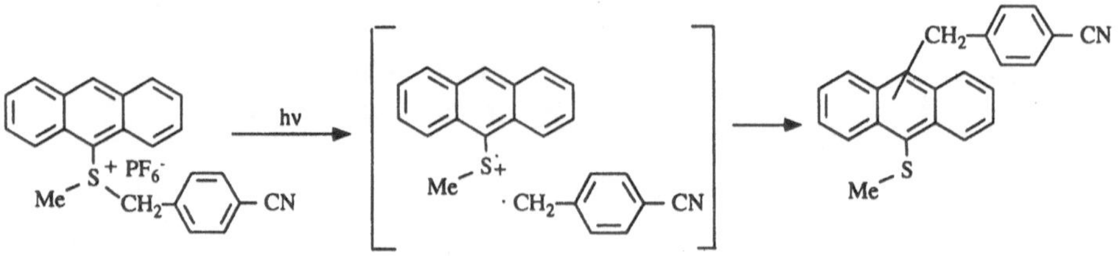

Scheme 33.

a π radical anion, as it happens in the alkylation of anisole by acrylonitrile in the presence of methanol [190]; the reaction is similar with hydroquinone and its monomethyl ether [191].

3 Reactions at the Substituent

Practically all PET-induced reactions at substituent groups involve a fragmentation of the initially formed radical ion. As one may expect, benzyl derivatives containing a nucleofugal or an electrofugal group undergo a fast cleavage to yield a benzyl radical after reduction (Scheme 34) or respectively oxidation (Scheme 35).

3.1 The Aromatic is the Acceptor

An important group of reactions pertaining to this class is the nucleophilic substitution of a chloro or a nitro group at the benzylic position of *p*-nitrocumyl derivatives; similarly to the $S_{RN}1$ ring substitution (Sect. 2.1.4) the process is accelerated by light absorption (possibly on the part of a nucleophile–substrate ground state complex) and proceeds via a chain mechanism (Scheme 36) [192–194]

Elimination of an hydroxy [195–196], acetoxy [195], and amino [197–198] group from the benzylic position via radical anion has been observed. Onium salts containing an electron rich aromatic moiety as well as an electron-withdrawing substituted benzyl group undergo elimination of the latter after intramolecular electron transfer, e.g. a *p*-cyanobenzyl radical is cleaved and then alkylates intramolecularly in various arylsulfonium salts, while the radical diffuses out of the cage and gives coupling products starting from ammonium or phosphonium salts [199] (Scheme 34). The *p*-nitrobenzyl radical is eliminated from the corresponding 9,10-dimethoxyanthracene-2-sulfonate ester [200].

Furthermore, some useful and mild methods of photochemical deprotection occur via such a pathway, in particular some selective detosylation reactions by

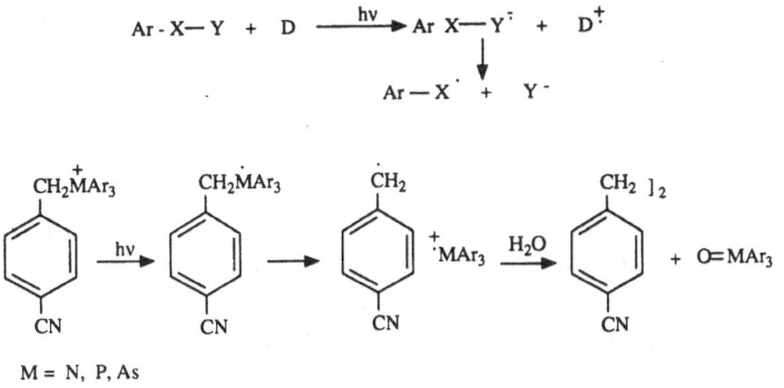
Ar-X—Y + D
hν
Ar X—Y˙⁻ + D˙⁺
Ar—X˙ + Y⁻
CH₂M⁺Ar₃
CH₂ṀAr₃
CH₂
CH₂]₂
hν
⁺˙MAr₃
H₂O
+ O=MAr₃
CN
M = N, P, As

Scheme 34.

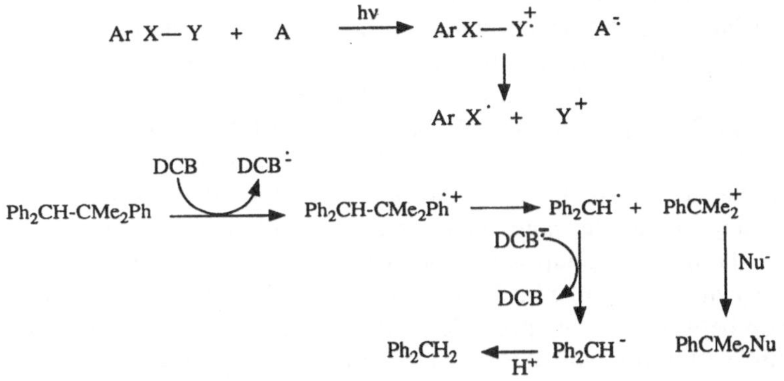
Ar X—Y + A
hν
Ar X—Y˙⁺ A˙⁻
Ar X˙ + Y⁺
DCB
DCB˙⁻
Ph₂CH-CMe₂Ph
Ph₂CH-CMe₂Ph˙⁺
Ph₂CH˙ + PhCMe₂⁺
Nu⁻
Ph₂CH₂
H⁺
Ph₂CH⁻
PhCMe₂Nu

Scheme 35.

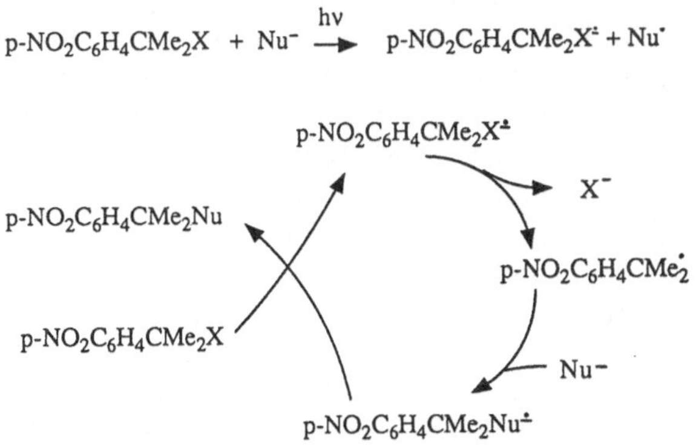
p-NO₂C₆H₄CMe₂X + Nu⁻
hν
p-NO₂C₆H₄CMe₂X˙⁻ + Nu˙
p-NO₂C₆H₄CMe₂X˙⁻
X⁻
p-NO₂C₆H₄CMe₂Nu
p-NO₂C₆H₄CMe₂˙
p-NO₂C₆H₄CMe₂X
Nu⁻
p-NO₂C₆H₄CMe₂Nu˙⁻

Scheme 36.

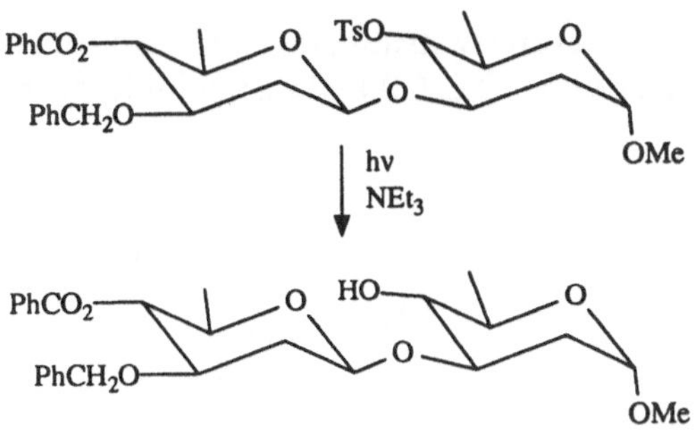
PhCO₂
PhCH₂O
O
TsO
OMe
hν
NEt₃
HO

Scheme 37.

irradiation in the presence of donors such as 1,4-dimethoxybenzene [201] or amines (Scheme 37) [202-3].

3.2 The Aromatic is the Donor

Benzyl derivatives containing an electrofugal group Y cleave when oxidized through a PET process to yield a benzyl radical (Scheme 35). In many cases the radical is then added to the radical anion of the sensitizer, and since this is usually an aromatic molecule, such reactions have been discussed among alkylations of radical anions (Sect. 2.1.3). However, the scope of these heterolytic fragmentation goes beyond that. Furthermore, interesting stereo-electronic effects have been noticed in these reactions. These are discussed in the following according to the type of bond (C-H, C-C or C-heteroatom) affected.

Very high acidity of α-protons in the radical cations of alkyl benzenes has been predicted by calculations and largely borne out in PET chemistry. Since it is expected that the C-H bond which is cleaved lies perpendicular to the plane of the aromatic molecule, the preferred conformation of the alkyl group should affect the efficiency of the cleavage. Thus as an example the reaction at the *i*-propyl and at the methyl group when DCN is irradiated in the presence of either one of the cymenes is different for each type of adduct formed, indicating that competition between the two possible modes of deprotonation depends on whether it occurs within the radical ion pair or from the free radical cation [63].

Benzylic deprotonation is often an inefficient process. It may be more important than it would appear from the end products, however, since radical cation deprotonation followed by reduction of the radical and reprotonation may regenerate the starting material. This mechanism has been proposed to explain the inefficiency of some PET alkylations [68]. In suitable models such a process has been revealed, e.g. deuterium incorporation at the bis-benzylic position in 2-(4-methoxyphenyl)-2-phenylethyl methyl ether and *cis-trans* iso-merization in 2-methoxy-1-(4-methoxyphenyl)indane (but not in the corresponding 3-methoxyphenyl derivatives) [204], as well as deconjugation of 1-phenylalkenes to 3-phenylalkenes in the presence of 1,4-dicyanobenzene, biphenyl (as a secondary donor) and a hindered pyridine as the base [205]. Deprotonation of *N,N*-dimethylaniline has likewise been observed (Scheme 38) [206-207].

Carbon-carbon cleavage has been extensively investigated. Thus bibenzyl derivatives when irradiated in the presence of 1,4-DCB and methanol form the

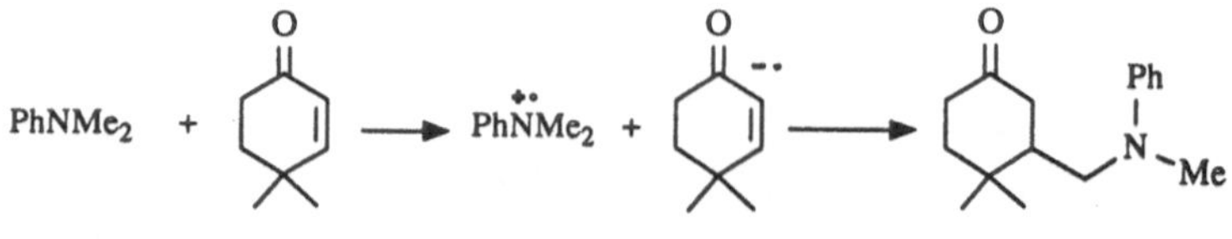

Scheme 38.

radical cations and cleave to yield a neutral radical and a cation [208–210]. The fate of the radical depends on the particular species involved: if reduction by the persistent radical anion and protonation occur the net result is ionic addition over a σ bond (Scheme 35); if the radical anion is a less efficient donor, addition rather than electron transfer takes place, and the final result is alkylation (see Sect. 2.1.3); finally, under suitable conditions (presence of a purposely added oxidant, e.g. a perchlorate [211] or when the ground state of the sensitizer is very easily reduced, e.g. when this is a quinone [212]) the radical may be oxidized to the corresponding benzyl cation and be trapped by a nucleophile. The scope of the reaction has been ascertained through systematic studies; the fragment liberated as the cation is the one corresponding to the more easily oxidized radical, and in order that the cleavage is sufficiently fast to overcome back electron transfer, the barrier has to be lower than ca. 16 kCal M^{-1}.

Another interesting model involves the irradiation of bicumenes in the presence of tetranitromethane. Both the radical cation and the radical anion of the initial diad fragment, and the observed products arise from the resulting triad and tetrad [213–215].

In a large group of the considered fragmentations, a heteroatom stabilized cation is formed. The initial observation was that 2,2-diphenylethyl methyl ether undergoes PET induced fragmentation [216]. This has been extended to various mono and diphenylethyl ethers and acetals, in particular to cyclic derivatives which open to distonic radical cation [211, 217–218]. The success of the reaction depends both on the stability of the fragments formed and on stereoelectronic factors; e.g. some cyclic derivatives do not cleave because of the poor alignment of the relevant bond (Scheme 39). Aryl substituted pinacol ethers [69–71], the nitrogen analogues of benzoin ethers [220] and α,β-bis(tertiaryamino)ethylbenzenes (Scheme 40) [221] are oxidatively cleaved by aromatic and ketone sensitizers; 9,10-dihydro-9,10-dimethyl-9,10-

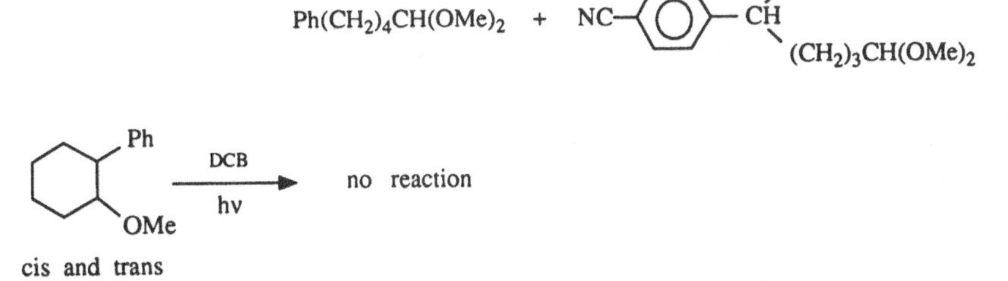

Scheme 39.

dimethoxyphenanthrene undergoes *cis-trans* isomerization through this mechanism [70–71]. The corresponding pinacols and α,β-aminoalcohols, as well as α-(*p*-dimethylaminophenyl)-phenylethyl alcohols, are cleaved by PET followed by proton transfer to the sensitizer radical anion (either an aromatic nitrile, Scheme 41, or a ketone [69, 212, 221–223]; iron (III) complexes have also been used [224]).

As one may expect, σ bond fragmentation is likewise observed with arylated small rings; thus e.g. 1,2-diarylcyclopropanes [225–226] (see also Sect. 2.1.3) as well as 2,3-diaryloxiranes [227–228] undergo easy C–C bond cleavage, and trapping by oxygen [225] (for oxygenation via radicals see also below), NO [226] and dipolarophiles [227–228] have been reported. Similar results have been obtained with arylated cyclobutanes (Scheme 42) [229–231], e.g. in a synthesis of the natural product magnoshinin (Scheme 43) [232], as well as with oxetanes (Scheme 44, notice that in the latter case the radical cation and the radical anion fragment with opposite regiochemistry) [233–234]. Other strained derivatives, e.g. the anthracene dimer, undergo an analogous fragmentation [235].

Benzylic desilylation and destannation have been often used in the alkylation of electron-poor aromatics (see Sect. 2.1.3), but are obviously useful in other contexts, a particularly well developed example being the benzylation of iminium salts for the synthesis of new heterocycles [236–238], including protoberberine alkaloids (Scheme 45) [239].

Other reactions involving the cleavage of a carbon heteroatom bond include a promising method for the deprotection of benzyl ethers by irradiation in the presence of acceptors (Scheme 46) [240–241] and the liberation of alkyl radicals (capable of initiating a polymerization) from alkyltriarylborate salts [242–243]. The PET induced decomposition of phenyldiazomethane leads to *cis*-stilbene; the reaction however appears to involve addition of the radical cation to a neutral molecule prior to nitrogen loss [244]. The detachment of a halogen after intramolecular electron transfer from the σ^* C–X bond to an electron-rich

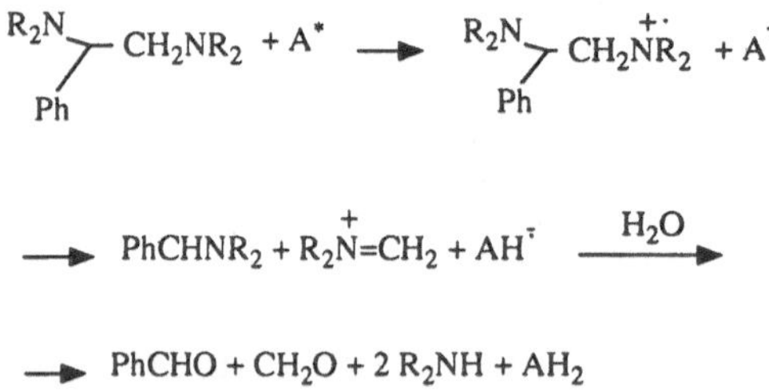

Scheme 40.

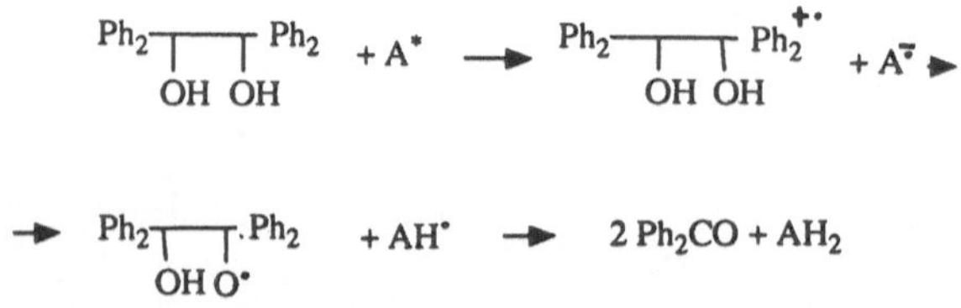

Scheme 41.

aromatic moiety in rigid models has been extensively studied [245–251]; 4-phenylbutyl iodide is converted to 4-phenylbutene, possibly again via intramolecular electron transfer and iodide loss [252].

Among reactions involving the fragmentation of a bond between two heteroatoms, one should mention the cleavage of diphenyl diselenide; the strong

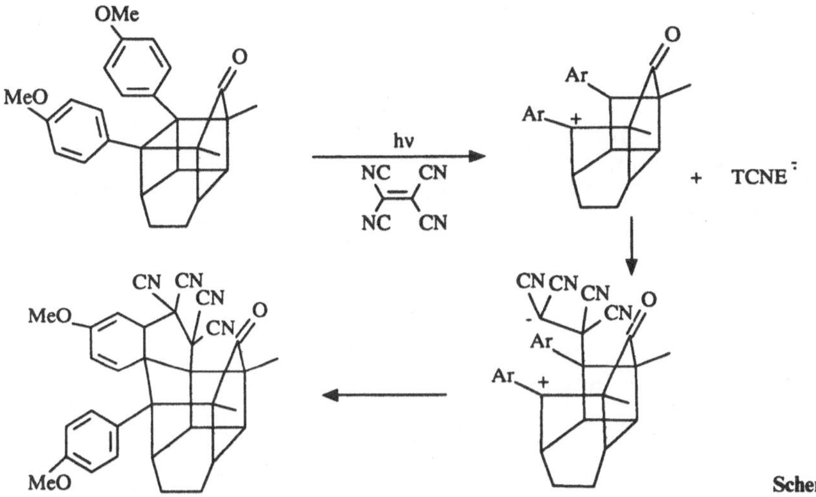

Scheme 42.

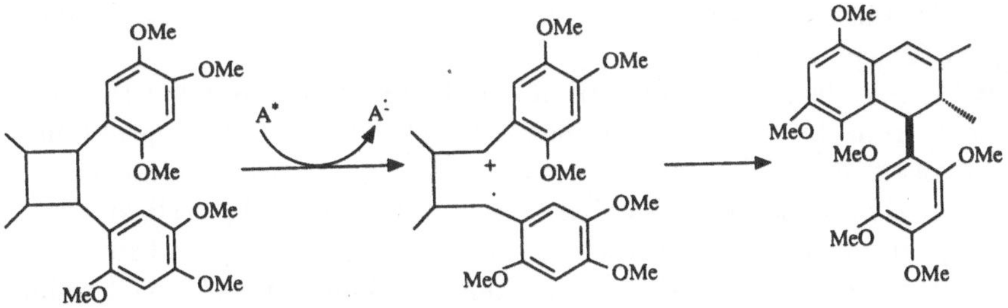

Scheme 43.

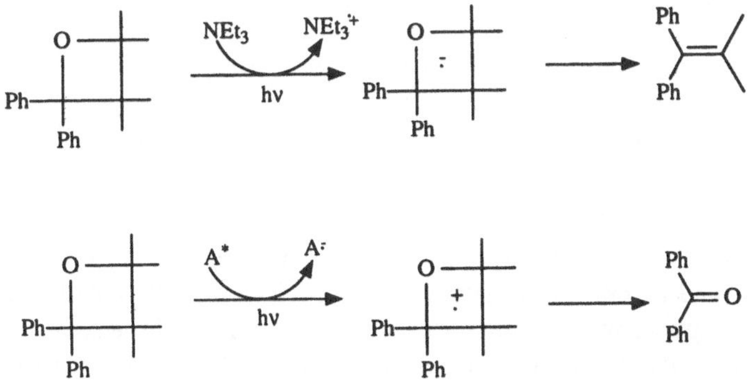

Scheme 44.

167

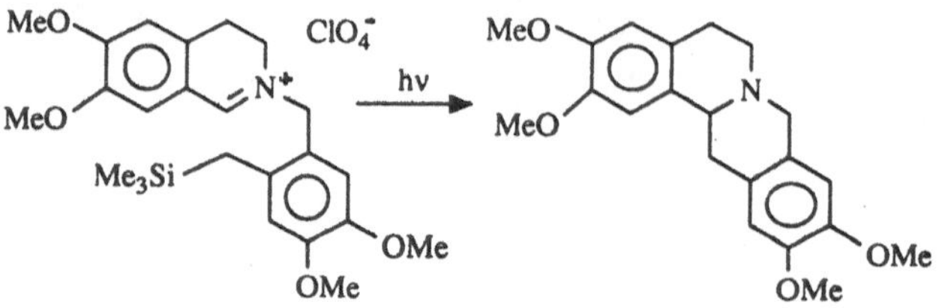

Scheme 45.

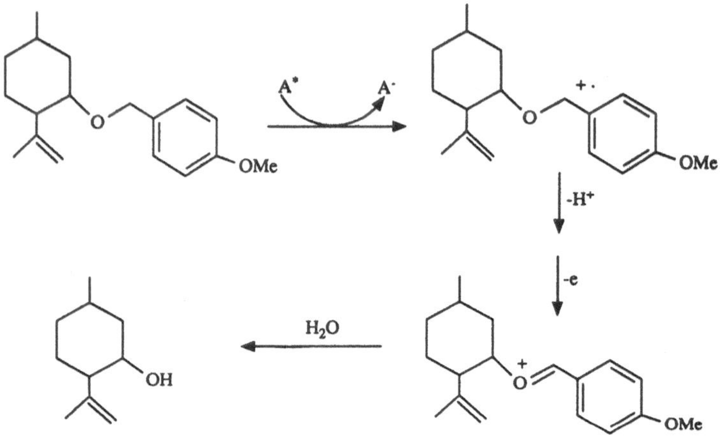

Scheme 46.

electrophile PhSe$^+$ generated in this way shows an interesting synthetic chemistry [253].

Finally, side-chain functionalization may be obtained through a PET process. Apart from the several arylations discussed in Sect. 2.1.3, oxygenation of benzyl radicals obtained from deprotonation or carbon–carbon bond cleavage is preparatively useful (the method has been recently reviewed [254]), and other functionalization are possible, notably fluorination of benzyl substrates by irradiation in the presence of titanium dioxide and fluoride ions [255]. Furthermore, the modification of a substituent, as a typical example the reduction of a nitro group [18, 27, 256–259] or an azo group [266], is in some cases obtained through a PET process.

4 Conclusions

It is hoped that the present review, though not exhaustive, also because it has been attempted to avoid extensive duplication with reviews by other Authors which are expected to appear in the same series, does give an idea of the scope of the chemistry of PET generated aromatic radical ions. The field certainly requires further mechanistic studies (e.g. in order to distinguish when electron transfer and chemical processes – fragmentation or addition – from the radical ions occur as sequential steps or concertedly, to determine the role of solvation and the assistance by additives to the observed reactions). However, in the opinion of the present authors it is even more important that an entirely new chemistry is emerging, ranging from unconventional methods of ring or side-chain functionalization to novel methods of deprotection. In this sense, the high potential of PET chemistry of aromatics will certainly be clearly demonstrated in the (hopefully near) future.

5 References

1. Albini A, Sulpizio A (1988) In: Fox MA, Chanon M (eds) Photoinduced electron transfer. Elsevier, Amsterdam, vol C, p 88
2. Amatore C, Kochi JK (1991) In: Mariano PA (ed) Advances in electron transfer chemistry. JAI Press, Greenwich, CT vol 1, p 1
3. Kavarnos GJ (1991) Top. Curr. Chem. 156: 21; Mattay J, Vondenhof M (1991) Top. Curr. Chem. 159: 219
4. Gould IR, Ege D, Moser JE, Farid S (1990) J. Am. Chem. Soc. 112: 4290
5. Schaap AP, Lopez L, Gagnon SD (1983) J. Am. Chem. Soc. 105: 663
6. Kellet MA, Whitten DG, Gould IR, Bergmark WR (1991) J. Am. Chem. Soc. 113: 358
7. Cornelisse J, Havinga E (1975) Chem. Rev. 75: 353
8. Cornelisse J, Lodder G, Havinga E (1979) Rev. Chem. Intermediat. 2: 231
9. Párkányi C (1983) Pure Appl. Chem. 55: 331
10. Mutai K, Nakagaki R (1984) Chem. Lett.: 1537
11. Berci Filho P, Neumann MG, Quina FH (1991) J. Chem. Res. (S): 70
12. Cantos A, Marquet J, Moreno-Manas M (1989) Tetrahedron Lett.: 2423
13. Cantos A, Marquet J, Moreno-Manas M, Gonzales-Lafont A, Lluch JM, Bertran J (1990) J. Org. Chem. 55: 3303
14. Bunce NJ, Cater R, Scaiano JC, Johnson LJ (1987) J. Org. Chem. 52: 4214
15. van Eijk AMJ, Huizer AH, Varma CAGO, Marquet J (1989) J. Am. Chem. Soc. 111: 88
16. Cervello J, Figueredo M, Marquet J, Moreno-Manas M, Bertran J, Lluch JM (1984) Tetrahedron Lett.: 4147
17. Pleixats R, Marquet J (1990) Tetrahedron 46: 1343
18. Marquet J, Moreno-Manas M, Vallribera A, Virgili A, Bertran J, Gonzalez-Lafont A, Lluch JM (1987) Tetrahedron 43: 351
19. Wubbels GG, Sevetson BR (1989) J. Phys. Org. Chem. 2: 177
20. Wubbels GG, Sevetson BR, Sanders H (1989) J. Am. Chem. Soc. 111: 1018
21. Wubbels GG, Celander DW (1981) J. Am. Chem. Soc. 103: 7669
22. Matai K, Nakagaki R (1985) Bull. Chem. Soc. Jpn. 58: 3663
23. Wubbels GG, Sevetson BR, Kaganove SN (1986) Tetrahedron Lett.: 3103
24. Sugimoto A, Yoneda S (1982) J. Chem. Soc., Chem. Commun.: 376
25. Sugimoto A, Sumida R, Tamei N, Inoue H, Otsuji Y (1981) Bull. Chem. Soc. Jpn. 54: 3500
26. Sugimoto A, Yamano J, Suyama K, Yoneda S (1989) J. Chem. Soc., Perkin Trans 1: 483
27. Wubbels GG, Snyder EJ, Coughlin EB (1988) J. Am. Chem. Soc. 110: 2543
28. Tagaya H, Onuki M, Karau M, Chiba K (1988) Yamagata Daigaku Kiyo Kogaku 20: 185; Chem. Abstr. 111: 96455m
29. Tazuke S, Kazama S, Kitamura N (1986) J. Org. Chem. 51: 4548
30. Minabe M, Isozumi K, Kawai K, Yoshida M (1988) Bull. Chem. Soc. Jpn. 61: 2063
31. Ito Y, Uozo Y, Matsuura T (1988) J. Chem. Soc., Chem. Commun.: 562
32. Ohashi M, Kudo H, Yamada S (1979) J. Am. Chem. Soc. 101: 2201
33. Albini A, Fasani E, Oberti R (1982) Tetrahedron 38: 1034

34. Barltrop JA (1972) Pure Appl. Chem. 32: 179
35. Onodera K, Sakuragi H, Tokumaru K (1976) Bull. Chem. Soc. Jpn. 49: 1697
36. Bellas M, Bryce-Smith D, Clarke MT, Gilbert A, Klumkin G, Krestonovich S, Marning C, Wilson S, J. Chem. Soc., Perkin Trans. 1: 2571
37. Lewis FD, Zebrowski BE, Correa PE (1984) J. Am. Chem. Soc. 106: 187
38. Lewis FD, Correa PE (1984) J. Am. Chem. Soc. 106: 194
39. Beecroft RA, Davidson RS, Goodwin D, (1983) Tetrahedron Letters: 5673
40. Lissi EA, Rubio MA, Fuentealba M (1987) J. Photochem. 37: 205
41. Albini A, Spreti S (1984) Tetrahedron 40: 2975
42. Yoshino A, Yamasaki K, Yonezawa T, Ohashi M (1975) J. Chem. Soc., Perkin Trans. 1: 735
43. Lewis FD, Petisce JR (1986) Tetrahedron 42: 6207
44. Ohashi M, Miyake K, Tsujimoto (1980) Bull. Chem. Soc. Jpn. 53: 1683
45. Borg RM, Arnold DR, Cameron TS (1984) Can. J. Chem. 62: 1785
46. Kyushin S, Masuda Y, Matsuhita K, Nakadaira Y, Ohashi M (1990) Tetrahedron Lett.: 6395
47. Mizuno K, Ikeda M, Otsuji Y (1985) Tetrahedron Lett.: 461
48. Mizuno K, Nakanishi K, Otsuji Y (1988) Chem. Lett.: 1833
49. Ohashi M, Aoyagi N, Yamada S (1990) J. Chem. Soc., Perkin Trans. 1: 1335
50. Tsujimoto K, Abe K, Ohashi M (1983) J. Chem. Soc., Chem. Commun.: 984
51. Ohashi M, Tsujimoto K, Kurukawa Y (1979) J. Chem. Soc., Perkin Trans. 1: 1147
52. Al-Fakhri K, Pratt A (1976) J. Chem. Soc., Chem. Commun.: 484
53. Mella M, Fasani E, Albini A (1992) J. Org. Chem. 57: 3051
54. Tsujimoto K, Nakao N, Ohashi M (1992) J. Chem. Soc., Chem. Commun.: 366
55. McCullough JJ, Miller RC, Wu WS (1977) Can. J. Chem 55: 2909
56. Libman J (1975) J. Am. Chem. Soc. 97: 4139
57. Arnold DR, Wong PC, Maroulis JJ, Cameron TS (1980) Pure Appl. Chem. 52: 2609
58. Albini A, Fasani E, Sulpizio A (1984) J. Am. Chem. Soc. 106: 3562
59. Albini A, Fasani E, Mella M (1986) J. Am. Chem. Soc. 108: 4119
60. Albini A, Fasani E, Montessoro E (1984) Z. Naturforsch. 39b: 1409
61. Sulpizio A, Albini A, d'Alessandro N, Fasani E, Pietra S (1989) J. Am. Chem. Soc. 111: 5773
62. d'Alessandro N, Fasani E, Mella M, Albini A (1991) J. Chem. Soc., Perkin Trans. 2: 1977
63. Albini A, Sulpizio A (1989) J. Org. Chem. 54: 2147
64. Mizuno K, Terasaka K, Ikeda M, Otsuji Y (1985) Tetrahedron Lett.: 5819
65. Mizuno K, Terasaka K, Yasueda M, Otsuji Y (1988) Chem. Lett.: 145
66. Lan JY, Schuster GB (1985) J. Am. Chem. Soc. 107: 6710
67. Lan JY, Schuster GB (1986) Tetrahedron Lett.: 4261
68. d'Alessandro N, Mella M, Fasani E, Toma L, Albini A (1991) Tetrahedron 47: 5043
69. Albini A, Mella M (1986) Tetrahedron 42: 6219
70. Reichel LW, Griffin GW, Muller AJ, Das PK, Ege S (1984) Can. J. Chem. 62: 424
71. Davis HF, Das PK, Reichel LW, Griffin GW (1984) J. Am. Chem. Soc. 106: 6968
72. Hasegawa E, Brumfield M, Yoon UC, Mariano PS (1988) J. Org. Chem. 53: 5435
73. Eaton DF (1980) J. Am. Chem. Soc. 102: 3280
74. Eaton DF (1984) Pure Appl. Chem. 56: 1191
75. Mizuno K, Ichichose N, Otsuji Y (1992) J. Org. Chem. 57: 1855
76. Kyushin S, Ehara Y, Nakadaira Y, Ohashi M (1989) J. Chem. Soc., Chem. Commun.: 270
77. Sankararaman N, Kochi JK (1989) J. Chem. Soc., Chem. Commun.: 1800
78. Arnold DR, Snow MS (1988) Can. J. Chem. 66: 3012
79. Arnold DR, Du X (1989) J. Am. Chem. Soc. 111: 7666
80. Yamada S, Kimura Y, Ohashi M (1977) J. Chem. Soc., Chem. Commun.: 667
81. Majima T, Pac C, Nakasone A, Sakurai H (1981) J. Am. Chem. Soc. 103: 4499
82. Bunce NJ, Safe S, Ruzo LO (1975) J. Chem. Soc., Perkin Trans. 1: 1607
83. Soloveichik OM, Ivanov VL, Kuz'min MG (1976) Zh. Org. Khim. 12: 859
84. Ohashi M, Tsujimoto K, Seki K (1973) J. Chem. Soc., Chem. Commun.: 1984
85. Ohashi M, Tsujimoto K (1983) Chem. Lett.: 423
86. Tanaka Y, Uryu T, Ohashi M, Tsujimoto K (1987) J. Chem. Soc., Chem. Commun.: 1703
87. Freeman PK, Srinivasan R, Campbell JA, Deinzer ML (1986) J. Am. Chem. Soc. 108: 5531
88. Freeman PK, Srinivasan R (1987) J. Org. Chem. 52: 252
89. Freeman PK, Srinivasan R (1986) J. Org. Chem. 51: 3939
90. Moore T, Pagni RM (1987) J. Org. Chem. 52: 770
91. Bunce NJ, Gallacher JC (1982) J. Org. Chem. 47: 1955
92. Chesta CA, Cosa JJ, Previtali CM (1988) J. Photochem. Photobiol. 45: 9

93. Bunce NJ, Pillon P, Ruzo LO, Sturch DJ (1976) J. Org. Chem. 41: 3023
94. Smothers WK, Schauze KS, Saltiel S (1979) J. Am. Chem. Soc. 101: 1894
95. Tsujimoto K, Tasaka S, Ohashi M (1975) J. Chem. Soc., Chem. Commun.: 758
96. Epling GA, McVicar WM, Kumar A (1988) Chemosphere 17: 2207
97. Kropp M, Schuster GB (1987) Tetrahedron Lett.: 5295
98. Kulis YY, Poletaeva IY, Kuz'min MG (1973) J. Org. Chem. U.S.S.R (engl. transl.) 9: 1242
99. Tolbert LM, Martone DP (1983) J. Org. Chem. 48: 1185
100. Fox MA, Singletary JJ (1982) J. Org. Chem. 47: 3412
101. Shukla SS, Rusling JF (1985) J. Phys. Chem. 89: 3353
102. Freeman PK, Ramnath N, Richardson AD (1991) J. Org. Chem. 56: 3643
103. Freeman PK, Lee YS (1992) J. Org. Chem. 57: 2846
104. Soumillon JP, De Wolf B (1981) J. Chem. Soc., Chem. Commun.: 436
105. Bunnett JF, Wamser CC (1967) J. Am. Chem. Soc. 89: 6712
106. Bunnett JF (1992) Acc. Chem. Res. 25: 2
107. Al-Fakhri KAK, Mowatt AC, Pratt A (1980) J. Chem. Soc., Chem. Commun.: 566
108. Al-Fakhri KAK, Pratt A (1983) Proc. R. Irish Acad. 83B: 5
109. Pac C, Tosa T, Sakurai H (1972) 45: 1169
110. Grodowski M, Latowski (1974) Tetrahedron 30: 767
111. Castedo L, Saa C, Saa JM, Suan R (1982) J. Org. Chem. 47: 513
112. Grimshaw J, De Silva AP (1982) J. Chem. Soc., Perkin Trans. 2: 857
113. Cano-Yelo H, Deronzier A (1984) J. Chem. Soc., Perkin Trans. 2: 1093; Cano-Yelo H, Deronzier A (1987) J. Photochem. 37: 315
114. Stark M, Arnold DR (1982) J. Chem. Soc., Chem. Commun.: 434
115. Timpe HJ (1990) Top. Curr. Chem. 156: 167; Fouassier JP, Burr D, Crivello JV (1989) J. Phochem. Photobiol. 49A: 317
116. Beecroft RA, Davidson RS, Goodwin D, Pratt JE (1984) Tetrahedron 40: 4487
117. Petersen WC, Letsinger WC (1973) Tetrahedron Lett.: 2197
118. Bunnett JF (1978) Acc. Chem. Res. 11: 413
119. Wolfe JF, Carver DR (1978) Org. Prep. Prog. Int. 10: 227
120. Rossi RA, de Rossi RM (1983) Aromatic Substitution by the $S_{RN}1$ Mechanism, American Chemical Society, Washington
121. Rossi RA, Bunnett J (1973) J. Org. Chem. 38: 1407
122. Bunnett JF, Sundberg JE (1975) Chem. Pharm. Bull. 23: 2620
123. Beugelmans R, Roussi G (1981) Tetrahedron 37: 393
124. Bard RR, Bunnett J (1980) J. Org. Chem. 45: 1546
125. Bunnett JF, Sundberg JE (1976) J. Org. Chem. 41: 1702
126. Alonso RA, Rossi RA (1980) J. Org. Chem. 45: 4720
127. Galli C (1988) Gazz. Chim. It. 118: 365
128. Galli C (1988) Tetrahedron 44: 5205
129. Semmelhack MF, Bargar TM (1977) J. Org. Chem. 42: 1481
130. Weinreb SM, Semmelhack MF (1975) Acc. Chem. Res. 8: 158
131. Semmelhack MF, Bargar T (1980) J. Am. Chem. Soc. 102: 7765
132. Beugelmans R, Bois-Coussy M, Tang Q (1989) Tetrahedron 45: 4203
133. Beugelmans R, Bois-Coussy M, Tang Q (1987) J. Org. Chem. 52: 3880
134. Usui S, Fukuzawa Y (1987) Tetrahedron Lett.: 91
135. Beugelmans R, Chastanet J, Roussi G (1984) Tetrahedron 40: 311
136. Beugelmans R (1984) Bull. Chem. Soc. Belg. 93: 547
137. Rossi RA, Alonso RA (1980) J. Org. Chem. 45: 1239
138. Alonso RA, Rodriguez CH, Rossi RA (1989) J. Org. Chem. 54: 5983
139. Bunnett JF, Gloor BF (1973) J. Org. Chem. 38: 4156
140. Rossi RA, de Rossi RH, Lopez AF (1976) J. Org. Chem. 41: 3371
141. Alonso RA, Austin E, Rossi RA (1988) J. Org. Chem. 53: 6065
142. Pierini AB, Baumgartner MT, Rossi RA (1987) Tetrahedron Lett.: 4653
143. Pierini AB, Baumgartner MT, Rossi RA (1988) Tetrahedron Lett.: 3429
144. Combellas C, Gautier H, Simon J, Thiebault A, Tournilhac F, Barzoukas M, Josse D, Ledoux I, Amatore C, Verpeaux JN (1988) J. Chem. Soc., Chem. Commun.: 203
145. Petrillo G, Novi M, Dell'Erba C, Tavani C (1991) Tetrahedron 47: 9297
146. Bunnett JF, Creary X (1974) J. Org. Chem. 39: 3173
147. Juillard M, Chanon M (1986) J. Photochem. 34: 231
148. Pierini AB, Baumgartner MT, Rossi RA (1987) J. Org. Chem. 52: 1089

149. Hobbs DW, Still WC (1987) Tetrahedron Lett.: 2805
150. Bunnett JF, Creary X (1974) J. Org. Chem. 39: 3612
151. Bunnett JF, Traber RP (1978) J. Org. Chem. 43: 1867
152. Schwartz SJ, Bunnett JF (1979) J. Org. Chem. 44: 4673
153. Rossi RA, Alonso RA, Palacio SM (1981) J. Org. Chem. 46: 2498
154. Pierini AB, Rossi RA (1979) J. Org. Chem. 44: 4667
155. Penenory AB, Rossi RA (1990) J. Phys. Org. Chem. 3: 266
156. Ivanov VL, Aurich J, Eggert L, Kuzmin MG (1990) J. Photochem. Photobiol. A50: 275
157. Ivanov VL, Herbst A (1988) Zh. Org. Khim. 24: 1709
158. Brunet JJ, Sidot C, Caubère P (1983) J. Org. Chem. 48: 1166 and 1919
159. Kashimura T, Kudo K, Mori S, Sugita N (1986) Chem. Lett.: 299, 483 and 851
160. Bunnett JF, Creary X (1975) J. Org. Chem. 40: 740
161. Den Heijer J, Shadid OB, Cornelisse J, Havinga E (1977) Tetrahedron 33: 779
162. Kul'bitskaya OV, Frolov AN, El'tsov AV (1979) Zh. Org. Khim. 15: 440
163. Frolov AN, Kul'bitskaya OV, El'tsov AV (1979) Zh. Org. Khim. 15: 2118
164. Lemmetyinen MJ (1983) J. Chem. Soc., Perkin Trans. 2: 1269
165. Lemmetyinen MJ, Koskikallio J, Ivanov VL, Kuz'min MG (1983) J. Photochem. 22: 115
166. Lemmetyinen MJ, Koskikallio J, Linblad M, Kuz'min MG (1982) Acta Chem. Scand. 36A: 391
167. Pac C, Nakasone A, Sakurai H (1977) J. Am. Chem. Soc. 99: 5806
168. Yasuda M, Pac C, Sakurai H (1981) J. Chem. Soc., Perkin Trans. 1: 746
169. Maldotti A, Amadelli R, Bartocci C, Carassiti V (1990) J. Photochem. Photobiol. A 53: 263
170. Suzuki N, Shimazu K, Ho T, Izawa Y (1980) J. Chem. Soc., Chem. Commun.: 1253
171. Suzuki N, Ayaguchi Y, Izawa Y (1982) Bull. Chem. Soc. Jpn. 55: 3349
172. Beugelmans R, Ginsburg H, Lecas A, Le Goff MT, Roussi G (1978) Tetrahedron Lett.: 3271
173. Yasuda M, Yamashita T, Shima K (1990) Bull. Chem. Soc. Jpn. 63: 938
174. Yasuda M, Pac C, Sakurai H (1981) J. Org. Chem. 46: 788
175. Epling GA, Florio F (1986) Tetrahedron Lett.: 1469
176. Yasuda M, Yoneshito T, Matsumoto T, Shino K (1985) J. Org. Chem. 50: 3667
177. Yasuda M, Matsuzaki T, Yamashita ·T, Shima K (1989) Chem Lett.: 551; Nippon Kagaku Kaishi: 1292
178. Yasuda M, Matsuzaki Y, Shima K, Pac C (1988) J. Chem. Soc., Perkin Trans. 2: 745
179. Meng J, Ito Y, Matsuura T (1987) Tetrahedron Lett.: 6665
180. Yasuda M, Yamashita T, Shima K, Pac C (1987) J. Org. Chem. 52: 753
181. Gilbert A, Krestonosich S, Martinez C, Rivas C (1989) Acta Cient. Venez. 40: 189
182. Kochi JK (1992) Acc. Chem. Res. 25: 39
183. Sankararaman S, Haney WA, Kochi JK (1987) J. Am. Chem. Soc. 109: 5235 and 7824
184. Masnovi JM, Kochi JK (1985) J. Org. Chem. 50: 5245
185. Masnovi JM, Hilinski EF, Rentzepis PM, Kochi JK (1986) J. Am. Chem. Soc. 118: 1126
186. Kochi JK (1990) Acta Chem. Scand. 44: 409
187. Dektar JL, Hacker NP (1989) J. Photochem. Photobiol. 46: 233
188. Saeva FD, Breslin DT (1989) J. Org. Chem. 54: 712; Saeva FD, Breslin DT, Luss HR (1991) J. Am. Chem. Soc. 113: 5333
189. Kurz ME, Lapin SC, Mariam K, Hagen TJ, Qian XQ (1984) J. Org. Chem. 49: 2728
190. Ohashi M, Tanaka Y, Yamada S (1977) Tetrahedron Lett.: 3629
191. Al-Jalal NA (1989) Gazz. Chim. Ital. 119: 569
192. Kornblum N (1975) Angew. Chem. Int. Ed. Engl. 14: 75
193. Kornblum N, Cheng L, Davies TM, Earl GW, Holy NL, Kerber RC, Kestner M, Manthley JW, Musser MT, Pinnick HW, Snow DH, Stuchal FW, Swiger RT (1987) J. Org. Chem. 52: 196
194. Wu F, Guarr TF, Guthrie RD (1992) J. Phys. Org. Chem. 5: 7
195. Ohashi M, Furukawa Y, Tsujimoto K (1980) J. Chem. Soc., Perkin Trans. 1: 2613
196. Lin CI, Singh P, Ullmann EF (1976) J. Am. Chem. Soc. 98: 6711
197. Ohashi M, Tsujimoto K, Furukawa Y (1977) Chem. Lett.: 543
198. Ohashi M, Miyake M (1977) Chem. Lett.: 615
199. Breslin DT, Saeva FD (1988) J. Org. Chem. 53: 713
200. Yamaoka T, Adachi H, Matsumoto K, Watanabe H, Shirosaki T (1990) J. Chem. Soc., Perkin Trans. 2: 1709
201. Nishida A, Hamada T, Yonemitsu O (1988) J. Org. Chem. 53: 3386
202. Binkley RW, Koholic DJ (1989) J. Org. Chem. 54: 3577
203. Hamada T, Nishida A, Yonemitsu O (1989) Tetrahedron Lett.: 4241
204. Arnold DR, Du X, Henseleit KM (1991) Can. J. Chem. 69: 839

205. Arnold DR, Mines SA (1989) Can. J. Chem. 67: 689
206. Zhang XM, Mariano PS (1991) J. Org. Chem. 56: 1655
207. Pandey G, Rani KS, Bhalerao UT (1990) Tetrahedron Lett.: 1199
208. Popielarz R, Arnold DR (1990) J. Am. Chem. Soc. 112: 3068
209. Okamoto A, Snow MS, Arnold DR (1986) Tetrahedron 42: 6175
210. Camaioni DM (1990) J. Am. Chem. Soc. 112: 9475
211. Lamont LJ, Arnold DR (1990) Can. J. Chem. 68: 391
212. Ci X, da Silva RS, Nicodem D, Whitten DG (1989) J. Am. Chem. Soc. 111: 1337
213. Maslak P, Chapman WH (1990) Tetrahedron 46: 2715
214. Maslak P, Asel SL (1988) J. Am. Chem. Soc. 110: 8260
215. Maslak P, Chapman WH (1990) J. Org. Chem. 55: 6334
216. Arnold DR, Maroulis AJ (1976) J. Am. Chem. Soc. 98: 5931
217. Arnold DR, Lamont LJ (1989) Can. J. Chem. 67: 2119
218. Arnold DR, Lamont LJ, Perrott AL (1991) Can. J. Chem. 69: 225
219. Arnold DR, Fahle BJ, Lamont LJ, Wierzchowski J, Young KM (1987) Can. J. Chem. 65: 2734
220. Kellett MA, Whitten DG (1989) J. Am. Chem. Soc. 111: 2314
221. Lee LYC, Ci X, Giannotti C, Whitten DG (1986) J. Am. Chem. Soc. 108: 175
222. Ci X, Kellet MA, Whitten DG (1991) J. Am. Chem. Soc. 113: 3893
223. Ci X, Whitten DG (1989) J. Am. Chem. Soc. 111: 3459
224. Ito Y (1991) J. Chem. Soc., Chem. Commun.: 662
225. Mizuno K, Kamiyama N, Ichinose N, Otsuji Y (1985) Tetrahedron 41: 2207
226. Ichinose N, Mizuno K, Yoshida K, Otsuji Y (1988) Chem. Lett.: 723
227. Albini A, Arnold DR (1978) Can. J. Chem. 56: 2985
228. Clawson P, Lunn PM, Whiting DA (1990) J. Chem. Soc., Perkin Trans. 1: 153 and 159
229. Ikeda H, Yamashita Y, Kabuto C, Miyashi T (1988) Tetrahedron Lett.: 5779
230. Hasegawa E, Okada K, Ikeda H, Yamashita Y, Mukai T (1991) J. Org. Chem. 56: 2170
231. Takahashi Y, Kochi JK (1988) Chem. Ber. 121: 253
232. Kadota S, Tsubono K, Makino K, Takeshita M, Kikuchi T (1987) Tetrahedron Lett.: 2857
233. Nakabayashi K, Fujimura S, Yasuda M, Shima K (1989) Bull. Chem. Soc. Jpn. 62: 2733
234. Nakabayashi K, Kojima J, Tanabe K, Yasuda M, Shima K (1989) Bull. Chem. Soc. Jpn. 62: 96
235. Masnovi JM, Kochi JK (1985) J. Am. Chem. Soc. 107: 6781
236. Lan AJY, Quillen SL, Heuckenroth RO, Mariano PS (1984) J. Am. Chem. Soc. 106: 6439
237. Borg RM, Heuckenroth RO, Lan AJY, Quillen SL, Mariano PS (1987) J. Am. Chem. Soc. 109: 2728
238. Lan AJY, Heuckenroth RO, Mariano PS (1987) J. Am. Chem. Soc. 109: 2738
239. Ho GD, Mariano PS (1988) J. Org. Chem. 53: 5113
240. Nishida A, Oishi S, Yonemitsu O (1989) Chem. Pharm. Bull. 37: 2266
241. Pandey G, Krishna A (1989) Synth. Commun. 18: 2309
242. Chatterjee S, Gottschalk P, Davis PD, Schuster GB (1988) J. Am. Chem. Soc. 110: 2326
243. Chatterjee S, Davis PD, Gottschalk P, Kurz ME, Sauerwein B, Yang X, Schuster GB (1990) J. Am. Chem. Soc. 112: 6329
244. Ishiguro K, Ikeda M, Sawaki Y (1992) J. Org. Chem. 57: 3057
245. Cristol SJ, Braun D, Scloemer GC, Vanden Plas BJ (1986) Can. J. Chem. 64: 1081
246. Cristol SJ, Dickenson WA (1986) J. Org. Chem. 51: 2973
247. Cristol SJ, Aeling EO, Heng R (1987) J. Am. Chem. Soc. 109: 830
248. Cristol SJ, Aeling EO, Strickler SJ, Ito RD (1987) J. Am. Chem. Soc. 109: 7101
249. Morrison H, Muthuramu K, Pandey G, Severance D, Bigot B (1986) J. Org. Chem. 51: 3358
250. Morrison H, Muthuramu K, Severance D (1986) J. Org. Chem. 51: 4681
251. Morrison H, Singh TV, Maxwell B (1986) J. Org. Chem. 51: 3707
252. Subbarao KV, Damodaran NP, Dev S (1987) Tetrahedron 43: 2543
253. Pandey G, Sekhar BBVS (1992) J. Org. Chem. 57: 4019
254. Lopez L (1990) Top. Curr. Chem. 156: 117
255. Wang CM, Mallouk TE (1990) J. Am. Chem. Soc. 112: 2016
256. Döpp D (1973) Top. Curr. Chem. 55: 49
257. Wubbels GG, Jordan JW, Mulls NS (1973) J. Am. Chem. Soc. 95: 1281
258. Cu A, Testa AC (1974) J. Am. Chem. Soc. 96: 1963
259. Barltrop JA, Bunce NJ (1968) J. Chem. Soc. C: 1467
260. Albini A, Fasani E, Pietra S (1986) J. Chem. Soc., Perkin Trans. 2: 681

Photoinduced Electron Transfer (PET) in Organic Synthesis

Ganesh Pandey

Division of Organic Chemistry (Synthesis), National Chemical Laboratory, Pune-411008, India.

Table of Contents

PET processes between ground and excited states of donor–acceptor pairs lead to the formation of radical ion pairs which upon disproportionation gives neutral radicals and/or ions required for synthetic purposes. A host of new reactions has been uncovered, the nature of which is governed by

Topics in Current Chemistry. Vol. 168
© Springer-Verlag Berlin Heidelberg 1993

the chemical properties and reactivity profiles of these reactive intermediates. Our objective in this chapter will be to enlist critical examples from the last ten years dealing with the radical ion reactions of synthetic importance initiated by photoexcitation of donor-acceptor pairs.

1 Introduction

Enhanced research activity during the last decade has made it increasingly evident that Single Electron Transfer (SET) plays an important role in many bimolecular reactions [1] in general and photochemical reactions in particular [2–5], as photoexcitation leads to the sufficient redox potential difference between the two interacting substrates essential for electron transfer initiation. Understanding of mechanistic details of photoinduced electron transfer (PET) processes has brought increasing emphasis on the use of these reactions in organic synthesis. This may be attributed to the unique feature of these transformations as the key reactive intermediates are radical ion species rather than the initially populated excited states. Therefore, the nature of these transformations is governed by the chemical properties and reactivity profiles of the ion radicals.

The interaction of a donor molecule (D) with acceptor (A) upon photo-excitation may result in either partial charge transfer (exciplex formation) or SET (radical ion pair formation, Eq. 1) depending upon the nature of the donor, acceptor and the solvent polarity [6]. In general, the feasibility of producing radical ions via PET can be predicted by estimating the free energy change (ΔG_{et}) associated with radical ion formation by using the well known Weller equation [7] (Eq. 2) which employs experimentally derivable parameters such as oxidation potential of the donor [$E_{1/2}^{ox}(D)$], reduction potential of the acceptor [$E_{1/2}^{red}$ (A)], energy of the excited species [$E_{0,0}$] and Coulomb interaction [E coul.] in a given solvent.

Under thermodynamically favourable PET reactions ($\Delta G_{et} < 0$) the radical ions are formed either as contact ion pair (CIP) or solvent-separated ion pair (SSIP). A closely related question is whether the primary intermediate is a SSIP or CIP. Gould and Farid [8] in their recent study have suggested that in polar solvents, such as acetonitrile, electron transfer quenching results in the formation of SSIP directly and in these solvents the fully solvated ions (SSIP) can separate to form free radical ions (FRI). Therefore, under these conditions the

$$A \ + \ D \ \xrightarrow{\ h\nu\ } \ A^{-\bullet} \ + \ D^{+\bullet} \qquad\qquad (1)$$

$$\Delta G_{et} \ = \ \Delta E_{1/2}^{ox} \ (D) \ - \ \Delta E_{1/2}^{red} \ (A) \ - \ \Delta E_{0,0} \ + \ \Delta E_{coul} \quad (2)$$

Fig. 1.

anion radicals are potentially less reactive with the cation radical than in non-polar solvents in which CIP is more important [10]. Thus, the use of polar solvents (e.g. CH_3CN and MeOH) facilitates ion-radical chemistry. Product formation in PET-reactions is often governed by the secondary processes of these initially formed ion-radicals since these serve as precursors for neutral radicals and ions (by loss of nucleofugal or electrofugal groups) required to initiate chemical reactions. Therefore, it is easy to understand the importance of investigations that focus attention on the factors governing the chemistry of ion-radicals.

Our objective in this chapter is to select the reactions of synthetic importance from the last ten years which are initiated by photoexcitation of donor–acceptor pairs.

The contents of this chapter have been organized into two main sections. Section 2, describes the photosensitized electron transfer reaction where either acceptor or donor molecule is regenerated after the initial PET processes. This includes reaction originating both from radical cations and radical anions. Section 3, deals with the intermolecular addition reaction between donor–acceptor pairs.

Over the years, excellent reviews [2–5, 10–13] have been written on the mechanistic and synthetic aspects of this subject. Therefore, our attempt will be to briefly supplement previous reviews on this topic by incorporating recent examples to put the theme of this article in total perspectives. PET reactions of carbanions and carbocations are not included here since these reactions are discussed in detail by Peter Wan in Vol. 158 of this series [14] and by T.P. Soumillion in this issue.

2 Photosensitized Electron Transfer Reactions

2.1 Reactions of Aromatic Compounds

PET generated arene radical cations and arene radical anions have found significant applications in organic synthesis. We will summarize the important examples of both types in this section.

2.1.1 Reactions of Arene Radical Cations

Two general types of reaction occur with PET generated arene radical cations; nucleophilic substitution via σ-complex (Eq. 3) and cleavage reactions at benzylic position by the loss of proton of electrofugal group (Eq. 4). In the following section we will summarize several recent reactions operating through routes of these types.

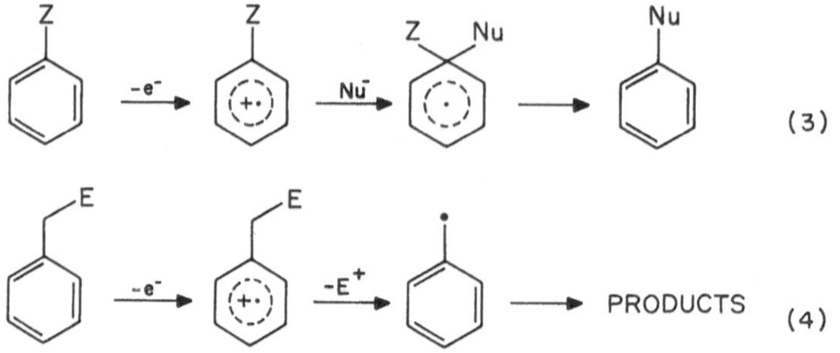

(3)

PRODUCTS (4)

Fig. 2.

2.1.1.1 *Nucleophilic Substitution Reactions*

Photocyanation of the arenes is probably the representative example of electron transfer initiated nucleophilic aromatic photosubstitution reactions where hydrogen serves as the group undergoing displacements [15]. In a recent study [16] this concept has been extended for the direct amination of polynuclear aromatic hydrocarbons (e.g. **1–3**) with ammonia or primary amines (**2**) via arene radical cation produced by irradiating arenes in the presence of 1,4-dicyanobenzene (DCNB). Another potentially useful application of this methodology was reported earlier [17] from the same group for the reduction of electron rich arenes to corresponding dihydroderivatives (e.g. **1–4**, Birch type reduction) using NaBH$_4$ as hydride donor to arene radical cation. An important aspect of this methodology is the selective reduction of electron-donating substituted rings of unsymmetrically substituted arenes in contrast to Birch-type procedures which favour the reduction of less electron rich aromatic rings.

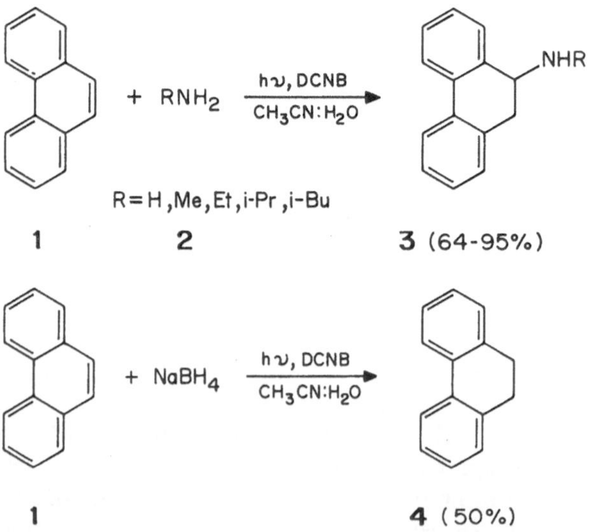

Scheme 1.

Investigations by our group have provided a novel application of PET generated arene radical cations for preparing variety of oxygen, nitrogen and carbocyclic aromatic compounds by intramolecular nucleophilic cyclisations. For illustration, coumarins (6) are synthesised [18] (70–90%) directly from the PET reactions of corresponding cinnamic acids (5). Ground state of 1,4-dicyano-naphthalene (DCN) is utilised as electron acceptor in this transformation. Photolysis in aerated solution led this reaction to occur in complete sensitized manner as shown in Scheme 2 (Eqs. 5–9). Several observations such as diffusion controlled fluorescence quenching of 5 with DCN and exergonic value for free energy change (ΔG_{et}) suggested [19] the electron transfer route for these processes. The regiospecificity of this cyclisation is in accord with the calculated electron densities (Huckle or MNDO programmes) at different carbons of the HOMO of arene radical cation [19]. Precocenes-I [20], a potent antijuvenile hormone compound and their several analogues, 2-alkylated dihydrobenzofurans [21], believed to possess antifungal and phytoalexin properties and carbocyclic benzannulated compounds [22] have been synthesised using this methodology.

The synthetic utility of these cyclisations are further extended by demonstrating [23] the efficient and regiospecific cyclisation of β-arylethylamines (7 and 8) to produce substituted dihydro indoles 9 and 10 (Scheme 3). A unique combination of two independently PET operating reactions discovered by our group have been reported [23] for realizing one-pot "wavelength switch" approach for benzopyrrolizidines (13) related to mitomycin skeleton (14) starting from 11. The strategy involves the construction of dihydroindole ring (12) from 11 via intramolecular cyclisation of arene radical cation by photolysing 11 in the presence of 9,10-dicyanoanthracene (DCA) at $> 300\,nm$ (light absorbed by 11 only) and followed by building the pyrrolizidine moiety of 13 via iminium cation intermediate [24].

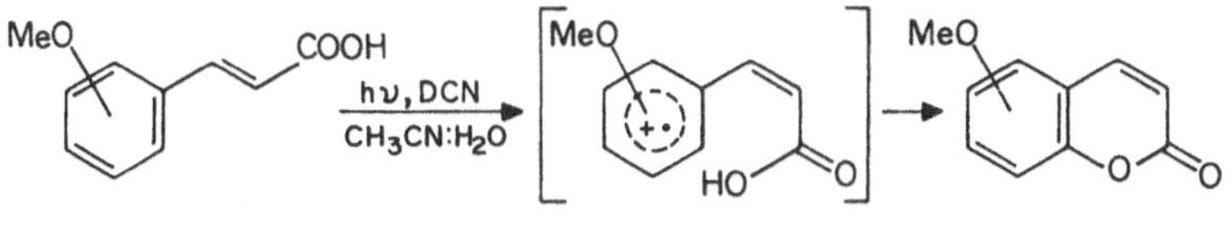

5		6 (70-90%)

$$5 + DCN \longrightarrow 5^{+\cdot} + DC\overset{-\cdot}{N} \quad (5)$$

$$DC\overset{-}{N} + O_2 \longrightarrow DCN + \overset{-\cdot}{O}_2 \quad (6)$$

$$2H_2O + \overset{-\cdot}{O}_2 \longrightarrow H_2O_2 + 2O\overset{-}{H} \quad (7)$$

$$3 H_2O_2 \longrightarrow 2H_2O + 2O_2 + 2\overset{+}{H} \quad (8)$$

$$5^{+\cdot} \longrightarrow 6 \quad (9)$$

Scheme 2.

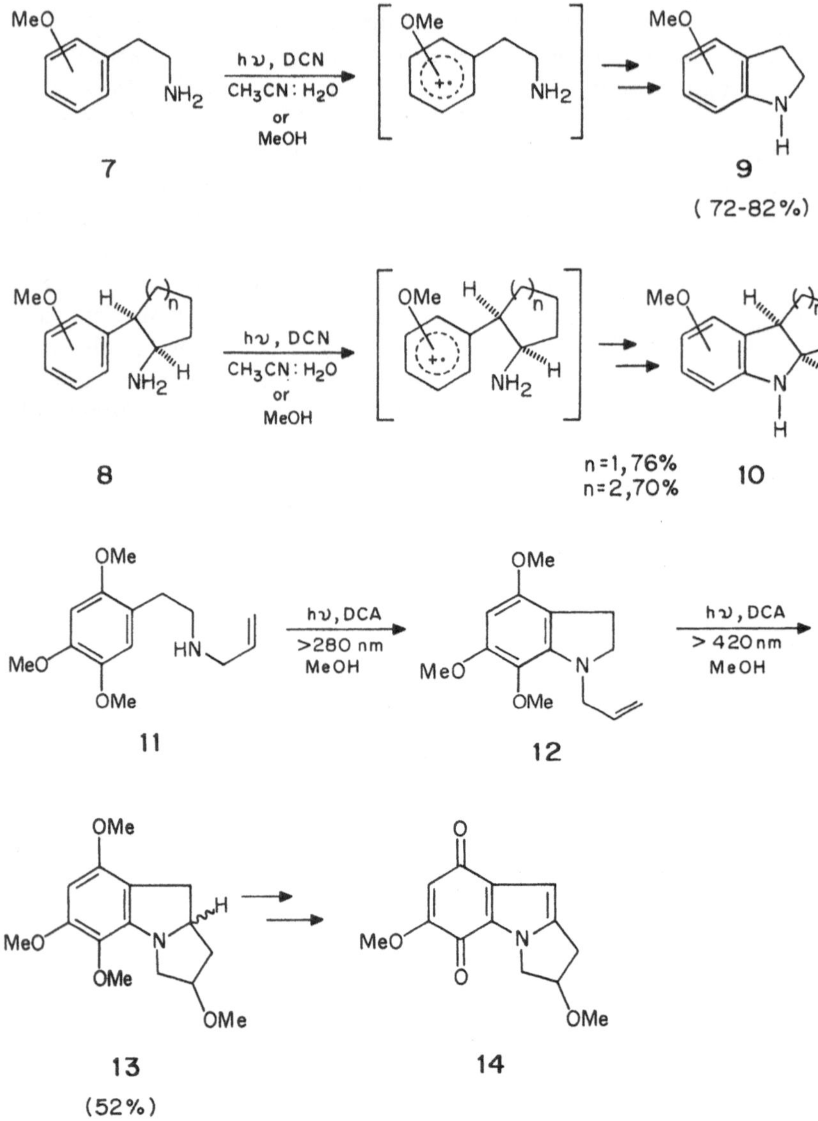

Scheme 3.

2.1.1.2 Cleavage Reaction at Benzylic Positions

The benzyl radicals generated by efficient deprotonation or electrofugal group loss from the benzylic position of arene radical cations (Eq. 4) have found interesting applications in organic synthesis [25]. Some of the examples pertaining to this class are exemplified in Sect. 2.5. A recent publication of Santamaria et al. [26] illustrates the use of PET generated benzylic radicals (via deprotonation step from arene radical cations) for selective and mild photo-oxidation of

$$ArCH_2R + {}^1DCA^* \longrightarrow ArCH_2R^{+\cdot} + DCA^{-\cdot} \quad (10)$$

$$DCA^{-\cdot} + MV^{++} \longrightarrow DCA + MV^{+\cdot} \quad (11)$$

$$MV^{+\cdot} + O_2 \longrightarrow MV^{++} + O_2^{-\cdot} \quad (12)$$

$$ArCH_2R^{+\cdot} + O_2^{-\cdot} \longrightarrow Ar\dot{C}HR + H\dot{O}_2 \quad (13)$$

$$Ar\dot{C}HR + H\dot{O}_2 \longrightarrow ArCH(R)-OOH \quad (14)$$

Fig. 3.

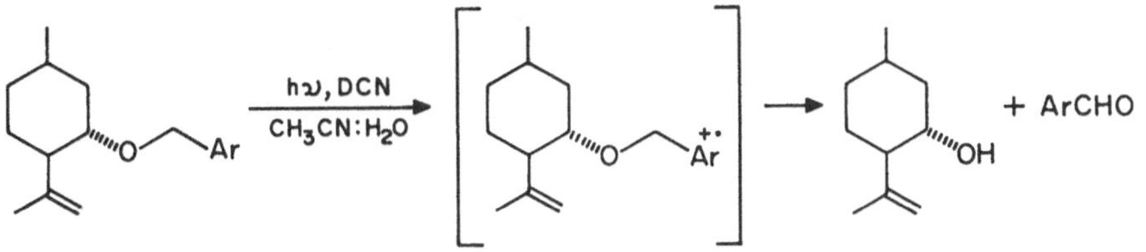

15

$Ar = \underline{p}\text{-MeO}-C_6H_4$

16
(75%)

Scheme 4.

benzylic groups to their corresponding hydroperoxides (57–100%) (Eqs. 10–14) using DCA as electron acceptor and methyl viologen (MV^{++}) as an electron relay. PET mediated substrate selective photo-oxidation of benzyl alcohol derivatives is also reported [27] using 10-methyl acridinium ion as an NAD^+ model. The benzyl radical formed by –C–C– bond dissociation reactions from PET generated arene radical cations have also found several application in organic synthesis [25]. One recent application of this cleavage may be depicted (Scheme 4) by citing our work [28] related to a novel debenzylation strategy for benzyl protected alcohols (e.g. **15–16**). An independent study by Nishida et al. [29] substantiated our photodebenzylation methodology.

2.1.2 Reactions with Arene Radical Anions

Aromatic halides and related compounds are good electron acceptors from a variety of donors such as amines, alkenes and arenes and the respective arene radical anions formed will undergo the nucleofugal group loss either from the arene rings or the benzylic position to give arene radicals (Eq. 15) or benzylic radical (Eq. 16) respectively. Aryl radicals thus formed may react with nucleophiles to give substitution ($S_{RN}1$) [30] products (Eqs. 17 and 18) or may abstract hydrogen from H-donors to give reduction products (Eq. 19) or can participate in intramolecular –C–C– bond formation reactions with π-systems.

$$Ar\overset{\underset{..}{}}{X} \xrightarrow{\;-\overset{-}{X}\;} \overset{.}{Ar} \;+\; \overset{-}{X} \qquad (15)$$

$$ArCH_2\overset{\underset{..}{}}{X} \xrightarrow{\;-\overset{-}{X}\;} Ar\overset{.}{C}H_2 \;+\; \overset{-}{X} \qquad (16)$$

$$\overset{.}{Ar} \;+\; Nu \longrightarrow Ar-\overset{\underset{..}{}}{Nu} \qquad (17)$$

$$Ar-\overset{\underset{..}{}}{Nu} \;+\; ArX \longrightarrow Ar-Nu \;+\; Ar\overset{\underset{..}{}}{X} \qquad (18)$$

$$\overset{.}{Ar} \;+\; \overset{.}{H} \longrightarrow ArH \qquad (19)$$

Fig. 4.

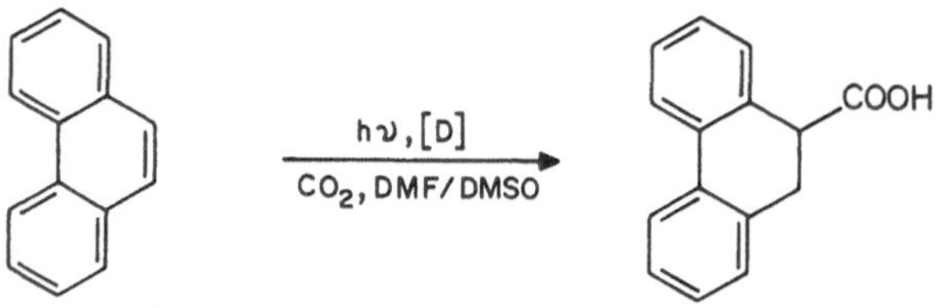

$$\xrightarrow[CO_2,\,DMF/DMSO]{h\nu,\,[D]}$$

1   17

[D] = $\underline{N}$, $\underline{N}'$ – DIMETHYLANILINE   (60%)

Scheme 5.

All these aspects of arene radical anion chemistry have been extensively studied and reviewed [10b, 11]. Dehalogenation of aryl halides by PET processes have been exploited for interesting transformations during organic synthesis [10b, 11, 31]. PET generated arene radical anions bearing no nucleofugal group have also been utilized for Birch type reductions of aromatic hydrocarbons [32]. Extension of the same concept has recently been reported by the Tazuke group [33] for the reductive photocarboxylation of aromatic hydrocarbons (e.g. 1–17) directly with CO_2 using N,N'-dimethyl aniline as electron donor (Scheme 5).

2.2 Reactions of Alkenes

Reactions of PET-generated alkene radical cations have been one of the important areas of research over the years and several reviews have been written on this subject [2, 5, 11]. A vivid summary of this topic has been provided by Mattay [10] recently. However, we will discuss here some representative examples of synthetic interest from olefin radical cations. The reactivity profiles of alkene radical cations may be illustrated on the lines of Mattay [10b] as shown below.

Since examples related to PET mediated isomerisations and rearrangements of alkenes have been intensively covered previously [5, 10, 34] and discussion pertaining to photooxygenation reactions have recently been reviewed [35] in

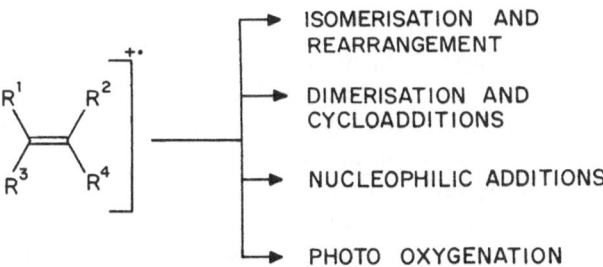

Fig. 5.

this series; we will restrict our discussion to dimerisations, cycloadditions and nucleophilic additions only.

2.2.1 Dimerisation and cycloadditions

The enhanced reaction rates and regioselectivities (head-to-head) in the alkene dimerisation (cyclobutane formation e.g. **18–19**) via radical cation catalysed reactions have led numerous studies in this direction [5, 10, 36–39]. The head-to-head stereochemistry of these dimerisations have been explained in terms of the addition of the radical cation to a neutral molecule giving the stabilised 1,4-radical cation. An interesting application of the cation-radical initiated (2 + 2)-cyclodimerisation strategy was reported by Mizuno et al. [40] for the synthesis of macrocyclic 2-m-dioxabicyclo (m-1, 2, 0) ring systems (**21**) from the PET reactions of **20** (Scheme 6).

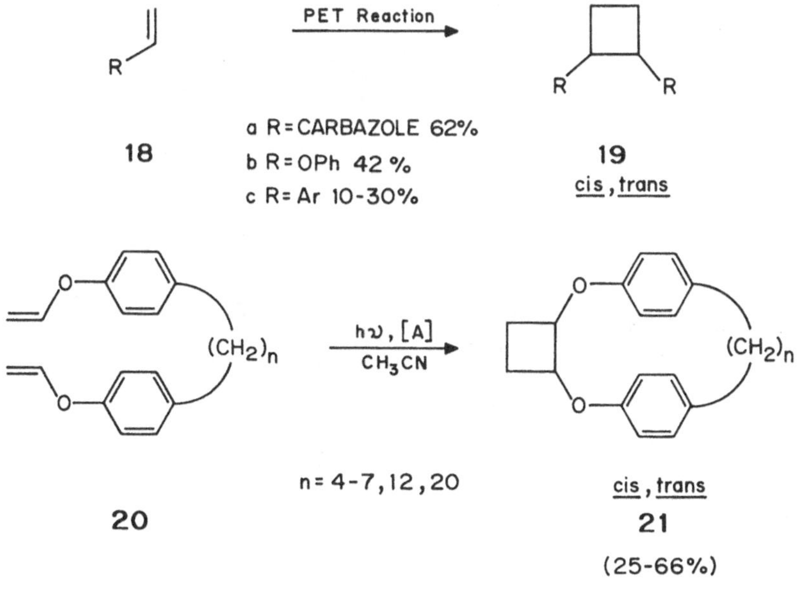

PET Reaction

a R = CARBAZOLE 62%
b R = OPh 42 %
c R = Ar 10-30%

18

19

cis, trans

$h\nu$, [A]
CH_3CN

$(CH_2)_n$

n = 4-7, 12, 20

20

21

cis, trans

(25-66%)

Scheme 6.

183

Soon the study related to the heterodimerisation followed [5, 10] and heterodimers such as **24** were prepared [41] from the PET reactions of two electron rich alkenes (e.g. **22** and **23**) in moderate yield (50%, Scheme 7). The yields of **24** may be significantly improved by minimising the competing dimerisations of starting alkenes by selecting appropriate reaction conditions [40]. Acyclic additions (alkylations) rather than cycloaddition have been reported [42] in the PET reactions of furans (**25**) with alkenes **26**. Nevertheless, the key step here also involved the nucleophilic attack of furans on the alkene cation radical similar to the dimerisation mechanism [43] but the tendency of regenerating the aromatic ring systems resulted in acyclic additions. For more examples please see the work of Arnold [44], Pac [45] and Mizuno [46].

In addition to the (2 + 2)-cyclodimerisations, the study of cation-radical catalysed Diels-Alder cycloadditions (4 + 2-cyclodimerisations) has provided a new and significant adjuvant to conventional Diels-Alder synthetic methodology [5, 10, 36] especially where both the dienes and dienophiles are electron-rich. PET generated cation-radical catalysed Diels-Alder cycloadditions [5, 10, 34, 47–49] utilizing the "special salt effect" have been extensively studied [5, 10, 34, 38, 48]. Extensive study of 1,3-cyclohexadiene (**28**) cyclodimerisation from several groups [5, 10, 34, 47–49] has culminated in a deep mechanistic and synthetic understanding of this methodology and synthetically good yields of both *endo* and *exo* cycloadducts (**29** and **30**) have been obtained. Bauld's [36b] "Hole" catalysed approach for the same has often complemented the PET methodology. Generally the *endo* isomer predominates in these cycloadditions but the exact *endo/exo* ratio depends on the electronic properties of reactants and reaction conditions [34]. Cation-radical catalysed mixed (4 + 2)-cycloaddition between **28** and electron rich dienophiles such as **31** have also

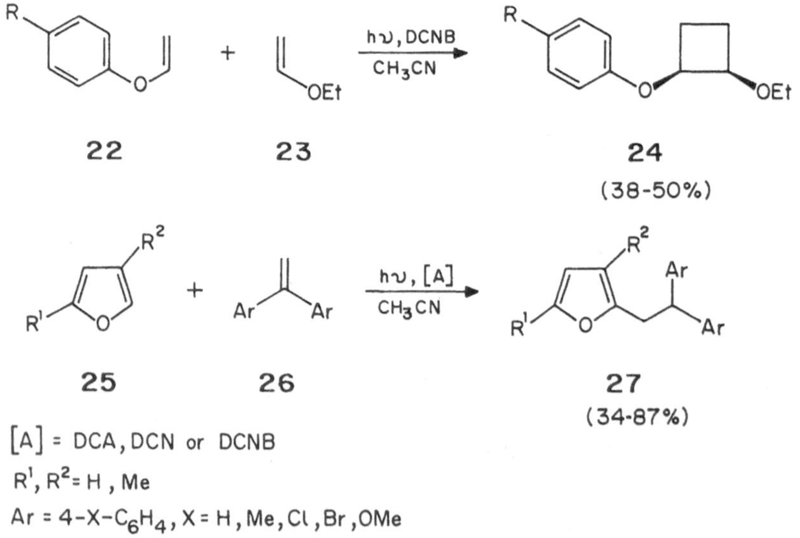

Scheme 7.

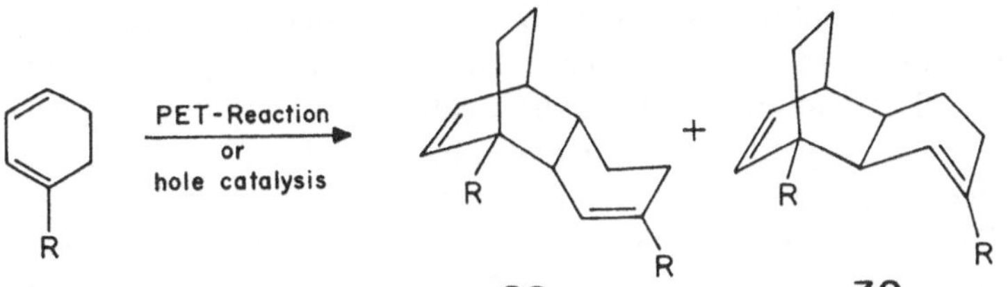

29-30	R	Yield(%)
a	H	70-75
b	OAc	56-83
c	OMe	73-94
d	Me	75

Scheme 8.

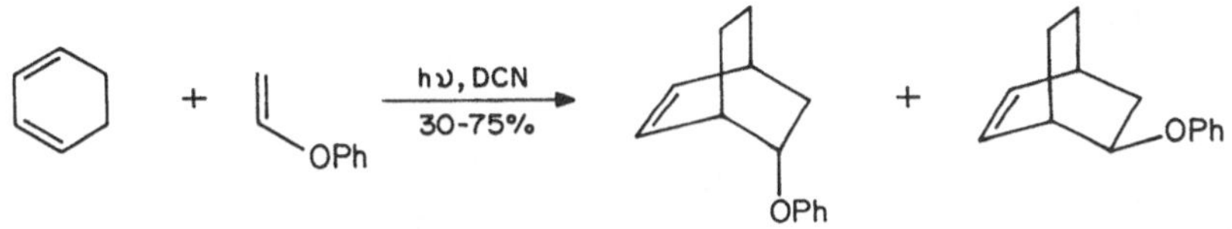

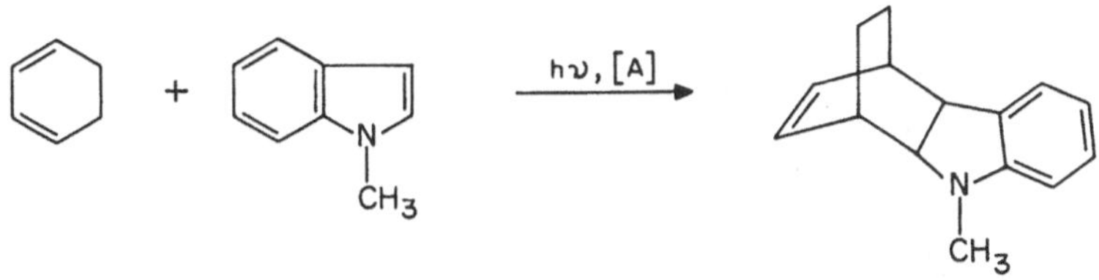

$$[A] =$$

38
(60%)

39
(-)-β-SELENINE

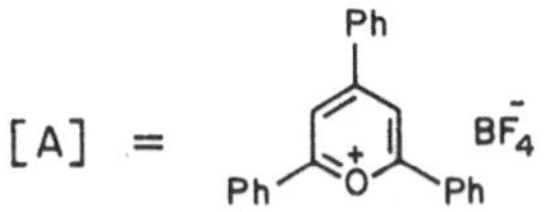

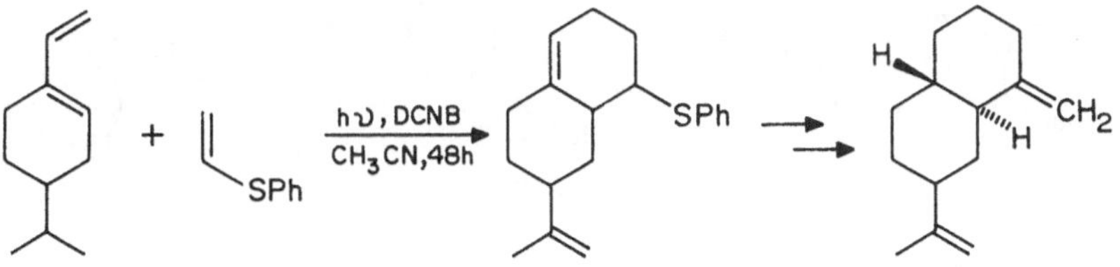

Scheme 9.

been studied (Scheme 8) in detail [34, 49]. Formation of both homo- and hetero-cycloadducts are observed but considerable selectivity in mixed cycloadditions may be achieved by PET reactions [34] than hole catalysed reactions [36b]. In a recent report [50], (4 + 2)-cycloaddition between indole (**34**) and **28** have been reported to produce adduct **35** in 70% yield. The effectiveness of the selectivity in PET initiated Diels-Alder reaction methodology has been demonstrated [51] by carrying out formal total synthesis of (−)-β-Selenine (**39**), a sesquiterpene natural product using the Diels-Alder product **38** as precursor as shown in (Scheme 9). Bauld's [36b] group has also studied the influence of the dienophile substitution on the periselectivity of cyclobutane formation during the Diels-Alder reaction. Cycloaddition reaction of diene **40** with electron rich olefin (e.g. **41**) has been reported [36b] to give mainly cyclobutane adduct **43**. Nevertheless, this periselectivity of cyclobutane formation has been suggested as being useful for an efficient indirect approach to formal Diels-Alder adducts (e.g. **40–42**). This strategy has been recently extended [52] by the same group for the Diels-Alder cycloadduct **46** from the PET reaction of diene **28** with enamide **44**, (Scheme 10).

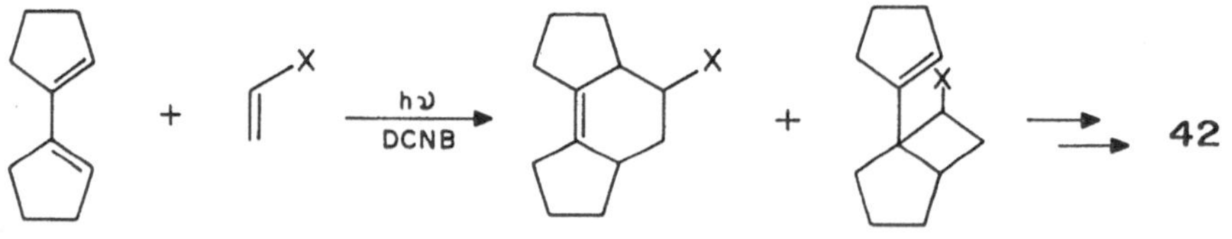

40	**41**	**42**	**43**
	X	42:43	
	NMeAc	0:100	
	OR	2-18:82-98	
	SPh	31:69	

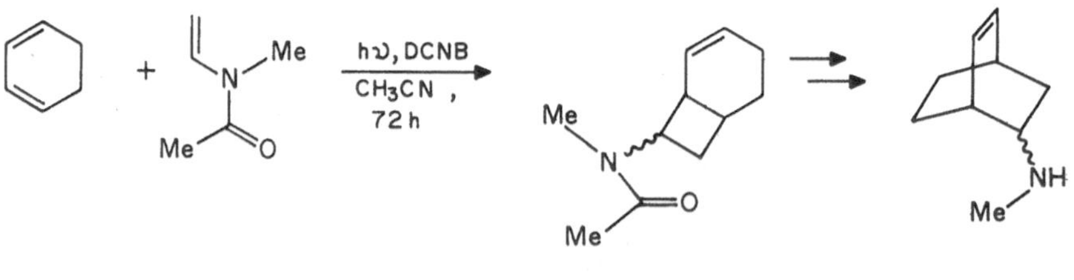

28	**44**	**45** (41%) anti:syn = 1·3:1	**46**

Scheme 10.

It may be appropriate to mention here (though it does not belong to our classification of radical ion pair reactions), the variant of photoinduced Diels-Alder reaction practised by Schuster's group [53] which has been termed the "Triplex Diels-Alder Reaction". According to this hypothesis, the exciplex formed by the excited singlet state of an electron deficient arene and a dienophile (e.g. 47) is trapped by a diene (e.g. 28) forming a triplex to give cycloadduct (48 and 49) [53e]. In this approach the stereoselectivity for the mixed cycloadduct is remarkably enhanced over the radical cation catalysed reaction and the configuration of alkene is retained in the cycloadduct. Recently the same group has reported [54] intramolecular cycloaddition of phenyl substituted alkenes to substituted 1,3-cyclohexadiene (50) to give cycloadduct 51 in 88% yield. A remarkable extension of the concept has been described [55] for the formation

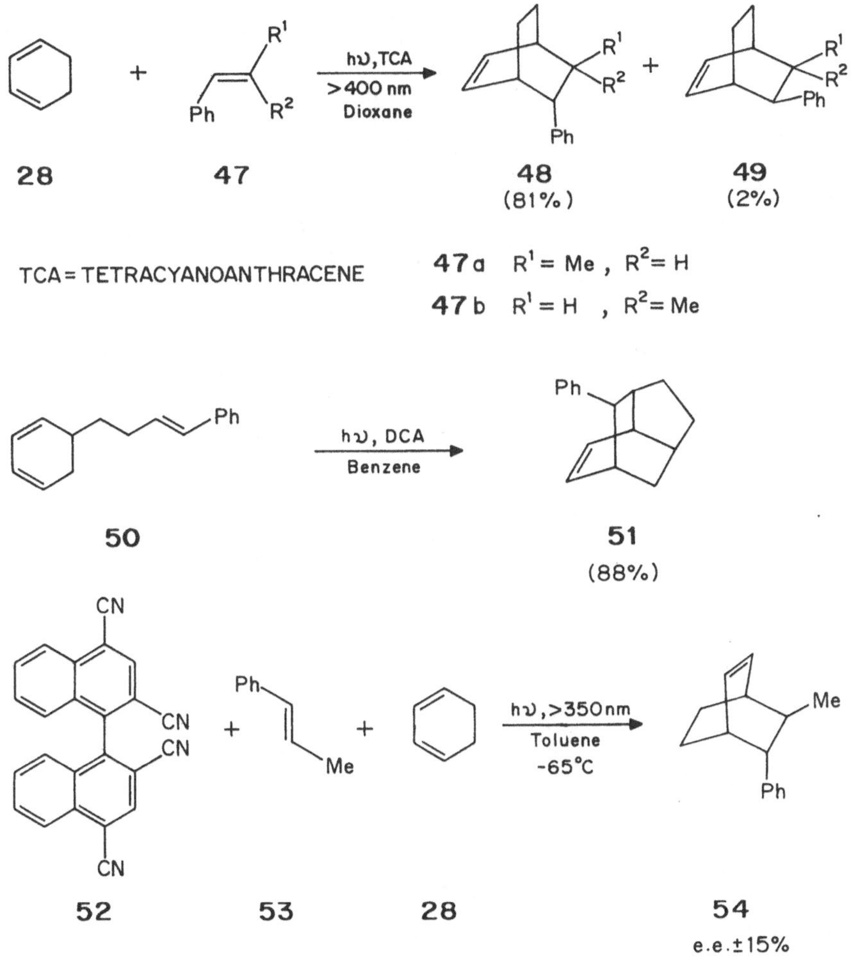

TCA = TETRACYANOANTHRACENE

47a R^1 = Me , R^2 = H

47b R^1 = H , R^2 = Me

Scheme 11.

of diastereomeric exciplex from chiral arene (52) and a prochiral dienophile 53 with diene 28 to achieve enantioselectivity in Diels-Alder adduct 54, though the induction observed was poor (e.e. 15%). Interested readers are advised to consult the work of Jones [56], Farid [57], Mizuno [58] and Mattay [59] on the involvement of ternary interactions in various photoreactions.

2.2.2 Nucleophilic Additions

Nucleophilic addition reaction is one of the most common reaction pathway available to organic cation radicals. Therefore, as expected alkene radical cations are also attacked by a variety of nucleophiles to give anti-Markonikov addition products. In this context the pioneering work of Arnold [60, 61] may be cited here by illustrating (Scheme 12) the addition of alcohol or cyanide ion

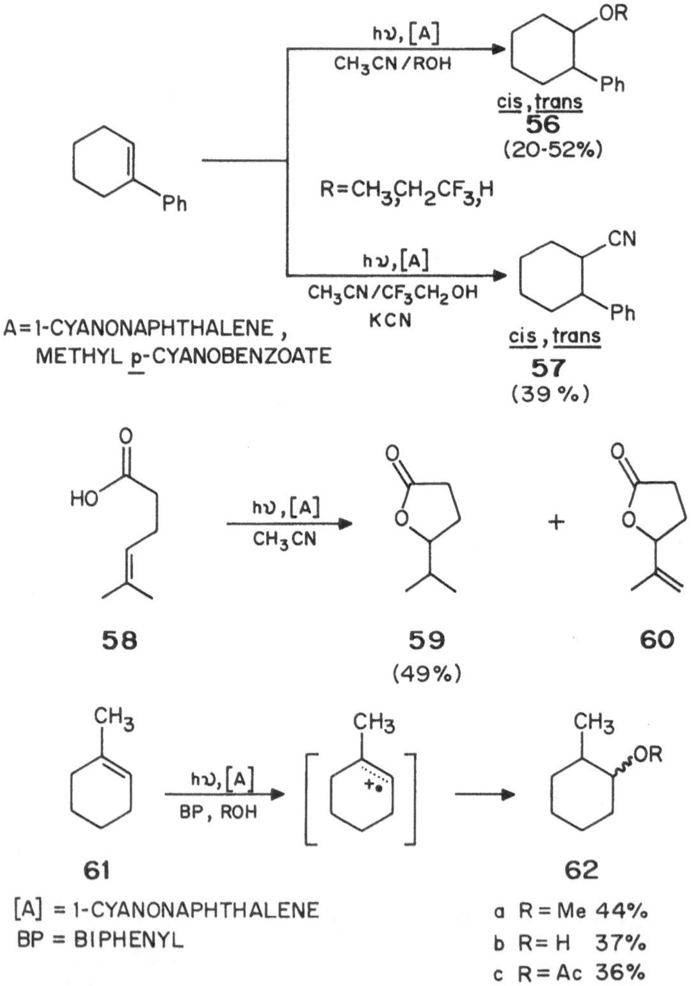

Scheme 12.

188

(e.g. **55–56** and **55–57**) to conjugated alkene **55**. Gassman et al. [62], have extended this discovery for the intramolecular addition of carboxylic acid to alkene radical cation for synthesising lactones **59** and **60** from photocyclisation of **58**. Improvement in the cyclisation yield is further suggested by using sterically hindered sensitizers [63].

An indirect PET methodology known as "redox photosensitization" has been developed by Pac [45] and Tazuke [64] for achieving higher yields of nucleophile addition product to alkene cation radicals. One recent example of this approach may be mentioned by illustrating anti-Markonikov alcohol addition (e.g. **61–62**) to non conjugated olefin **61** using biphenyl as cosensitizer [65]. More examples on this topic can be found in Farid [5] and Mariano's [11] reviews.

2.3 Reactions of Strained Ring Systems

PET reactions with strained ring compounds generally bring about their ionization followed by ring opening leading to isomerisation, rearrangement, nucleophilic addition and cycloaddition reactions [5, 10, 11]. Since the chemistry related to strained ring hydrocarbons has recently been reviewed [66], we will focus our attention mainly on the recent examples.

An interesting application of PET mediated bond cleavage reaction from azirine **63** has been reported by Mattay et al. [67] for synthesizing N-substituted imidazoles (**65**) via the (3 + 2) cycloaddition reaction of resultant 2-azaallenyl radical cations with imines **64**. Synthesis of pyrrolophane 3,4-dimethyl ester (**68**) has been reported recently [68] by the ring opening of **66** followed by intermolecular cycloaddition with dimethyl acetylene dicarboxylate (**67**) as shown in Scheme 13.

Substituted cyclopropanes [69] and oxiranes [70] have been shown to undergo efficient *cis,trans*-isomerisation by electron transfer sensitized photoreactions. Normally cyclopropane radical cations undergo ring-opening followed by rearrangement or nucleophilic additions depending upon the solvents used [71, 72]. Rao and Hixon [73] rationalised the PET addition of methanol to cyclopropanes by postulating the nucleophile capture of ring-closed cyclopropane radical cation, however, a recent study has described [74] this addition as involving a complete –C–C– bond dissociation step. In a recent report, an interesting application of the PET reaction of cyclopropanes has been described [75] for preparing 3,5-diaryl-2-isoxazolines (**70**) in good yield (91%) by direct NO insertion into the 1,2-diaryl cyclopropane (**69**) radical cation. Furthermore, the synthetic potential of these methodologies has been demonstrated [76] for the preparation of the cycloaddition product **72** from the PET reaction of 1,1,2-triarylcyclopropane (**71**) and vinyl ether. Gassman [77] has reported a novel ring expansion methodology for the ring opening of cyclopropanes (e.g. **75–76**) as shown in Scheme 14 Whiting et al. [78] have extended this methodology to other derivatives as well.

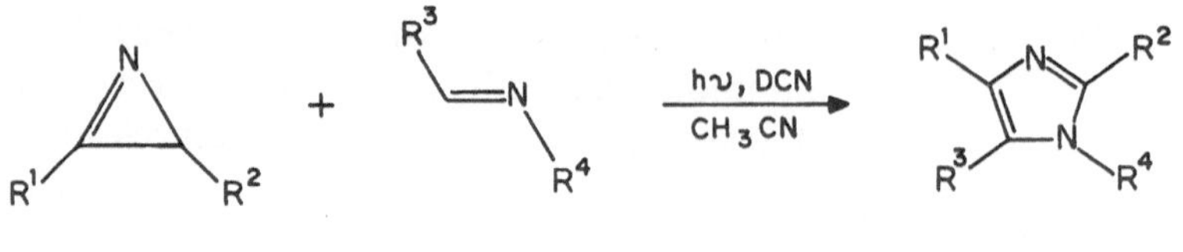

63		64		65 (%)
R^1	R^2	R^3	R^4	
Ph	Ph	Ph	n–Pr	87
Ph	H	Ph	n–Pr	12
n–Bu	H	Ph	n–Pr	3
Ph	Ph	n–Pr	n–Pr	25
Ph	H	n–Pr	n–Pr	35
n–Bu	H	n–Pr	n–Pr	40

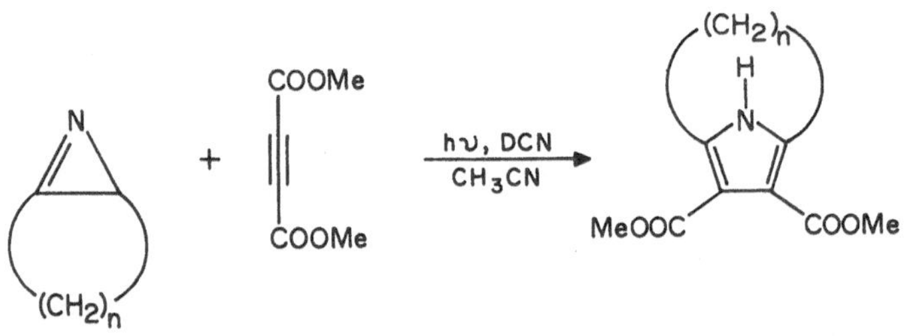

n = 5 (9%)

n = 6 (56%)

Scheme 13.

Similar to other strained ring compounds, cleavage reactions of cyclobutanes have also been described [79–81]. Recent examples of this topic and its implications may be found in the comprehensive review article by Mattay [10b].

2.4 Reactions of Amines

The studies related to the interactions of electronically excited arene molecules with tertiary amines have provided a basis for the present understanding of exciplexes and radical ion-pair phenomena [41, 82]. PET reactions of amines yield planar amine radical cations (Eq. 20) which are deprotonated to give α-amino radicals (Eq. 21) and usually cross-coupling (Eq. 22) between radical pairs of donor–acceptor terminates the photoreaction [32a, 83]. Mechanistic studies revealed contact ion pair (CIP) intermediate for these reactions [84, 85].

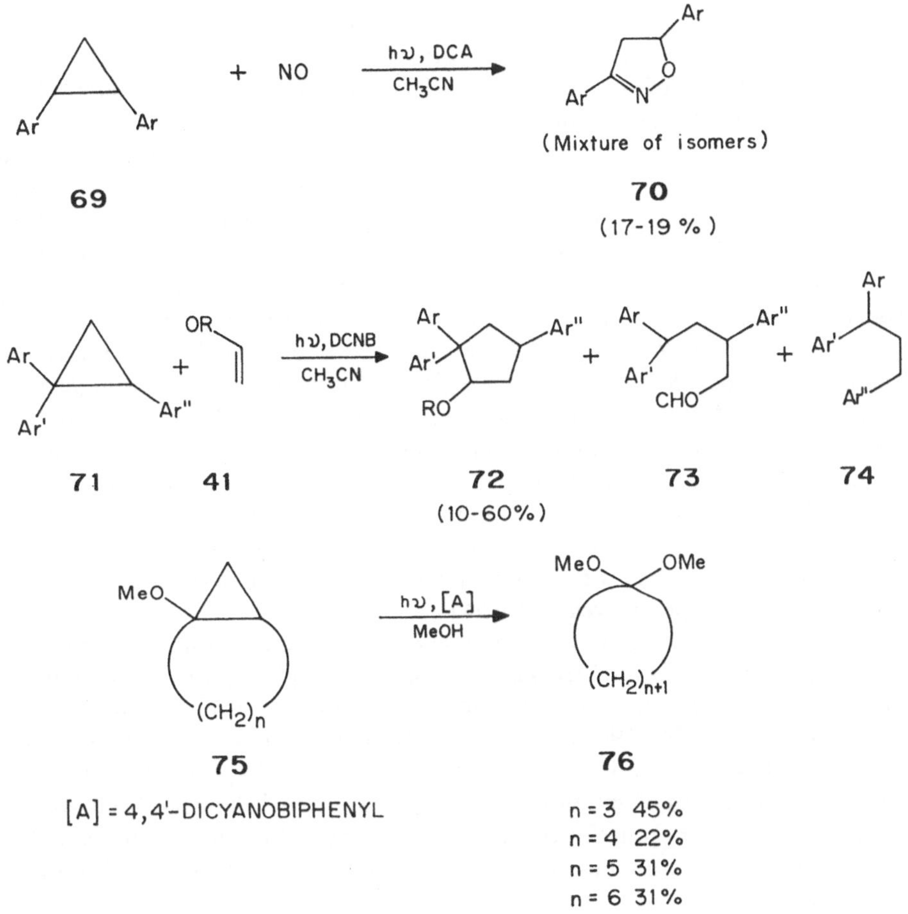

Scheme 14.

$$R_2N-CHR_2 + A^{*} \longrightarrow R_2N^{+\cdot}-CHR_2 + A^{-\cdot} \qquad (20)$$

$$R_2N-CHR_2 + A^{-\cdot} \longrightarrow R_2N-\dot{C}R_2 + A-H \qquad (21)$$

$$R_2N-\dot{C}R_2 + \dot{A}-H \longrightarrow R_2N-C-(A-H)-R_2 \qquad (22)$$

Fig. 6.

Ohashi et al. [86] in 1979 reported an unusual PET reaction of DCA in the presence of triethylamine in acetonitrile where the acetonitrile addition product to DCA (**78**) was observed and this unusual reaction was suggested as occurring via a longer-lived solvent separated ion pair (SSIP). This difference in the reactivity pattern encouraged us to explore "True sensitized" electron-proton-electron (E-P-E) transfer sequences [87] from the photoreactions of tertiary amine–DCN pairs (Eqs. 23–28) similar to the electrochemical reactions of amines [88]. The transformation as shown in Eq. 28 occurs due to reduced oxidation potentials of α-amino radicals [89]. Based upon the above concept we first reported [90] a clean and efficient transformation of N-hydroxyl amines (79) to nitrones 81. The (3 + 2)-cycloaddition of these nitrones gave a good yield of heterocyclic compound 82 as shown in Scheme 15. This E-P-E transfer sequence is further extended [24, 91] for generating regiospecific iminium cations. Iminium cations thus formed may be either trapped with internal nucleophiles to yield heterocyclic ring systems (e.g. 83–85) [24, 91] or hydrolysed to give secondary amines and aldehydes which have provided new and mild methodology for efficient N-debenzylations (e.g. 86–88) [92] and N-demethylation (e.g. 89–91) [93]. The regiospecificity of iminium cation generation have been demonstrated [24, 91, 94] to depend upon the α-deprotonation site of the corresponding amine radical cation which is subject to the kinetic acidity and stereoelectronic factor. The latter effect has been best demonstrated [91, 94] by generating regiospecific iminium cations and their in situ cyclisation from 92 to synthesize diastereoselective tetrahydro 1,3-oxazines (93) which has been utilized to prepare cis-α,α"-dialkylated piperidine and pyrrolidines 94 by a nucleophilic ring opening reaction [94].

Intermolecular trapping of PET generated iminium cations by Me_3SiCN have been recently reported by Santamaria et al. [95] for synthesizing α-amino nitriles (e.g. 95–96) in the alkaloid field and also for preparing 6-cyano 1,2,3,6-tetrahydropyridine (98) [96] from 1,2,3,6-tetrahydropyridine (97). Similar cyanation of catharanthine alkaloids has also been reported by Sundberg recently [97]. In situ trapping of iminium cation by allyltrimethylsilane has been utilised [98] recently by our group for the direct –C–C– bond formation reaction at the α-position of tertiary-amines. The concept involved the correlation of ion pair yield dependance on ΔG_{et} values.

An interesting application of PET-generated amine radical cations has been shown [87, 99] to produce α-amino radicals by efficient desilylation of α-methyl silylamine 99. High yields of both pyrrolidines and piperidines (101) are

$$R_2N\text{--}CHR_2 + {}^1\overset{*}{DCN} \longrightarrow R_2\overset{+\cdot}{N}\text{--}CHR_2 + \overset{-\cdot}{DCN} \quad (23)$$

$$R_2\overset{+\cdot}{N}\text{--}CHR_2 \longrightarrow R_2N\text{--}\overset{\cdot}{C}R_2 + \overset{+}{H} \quad (24)$$

$$\overset{-\cdot}{DCN} + O_2 \longrightarrow DCN + \overset{-\cdot}{O_2} \quad (25)$$

$$2H_2O + \overset{-}{O_2} \longrightarrow H_2O_2 + 2\overset{-}{OH} \quad (26)$$

$$3H_2O_2 \longrightarrow H_2O + 2O_2 + 2\overset{+}{H} \quad (27)$$

$$R_2N\text{--}\overset{\cdot}{C}R_2 + DCN \longrightarrow R_2\overset{+}{N}=CR_2 + \overset{-\cdot}{DCN} \quad (28)$$

Fig. 7.

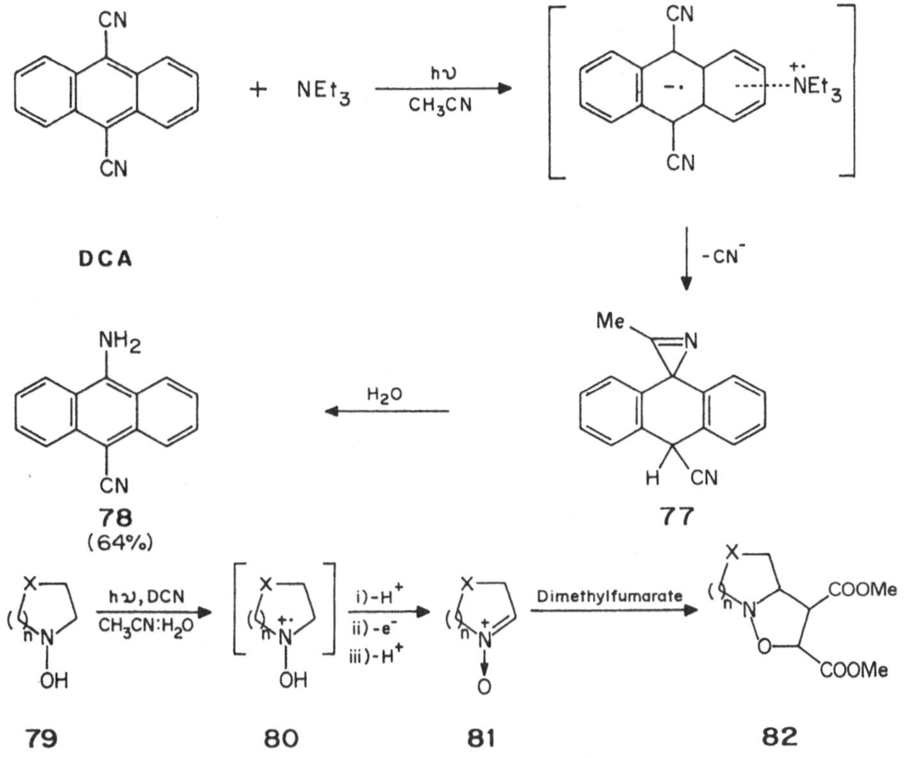

	n	X	82 (Yield %)
a	1	CH_2	76
b	2	CH_2	75
c	3	CH_2	78
d	4	O	60

Scheme 15.

obtained by the intramolecular cyclisation to π-bonds (Scheme 18). To investigate the stereochemical aspects of these cyclisations, cyclisation of **102** was studied and found to be non-stereoselective (a 1:1 mixture of **103** and **104** was formed) unlike the 3-substituted carbon centered radical carbocyclisations [100]. A plausible explanation for this difference has been advanced by considering the lower energy barrier between the two possible "chair" conformations of the transition state due to lone-pair flipping on the nitrogen [101]. Synthesis of diastereoselective 1-azabicyclo[*m.n.o*] alkanes (**106** and **107**) has been described [102] from the PET reaction of **105**. The stereochemistry of these cyclisations have been found to depend on the size of the new ring formed e.g. 1,5-*cis* and 1,6-*trans*. Mariano et al. [103] have recently reported an identical

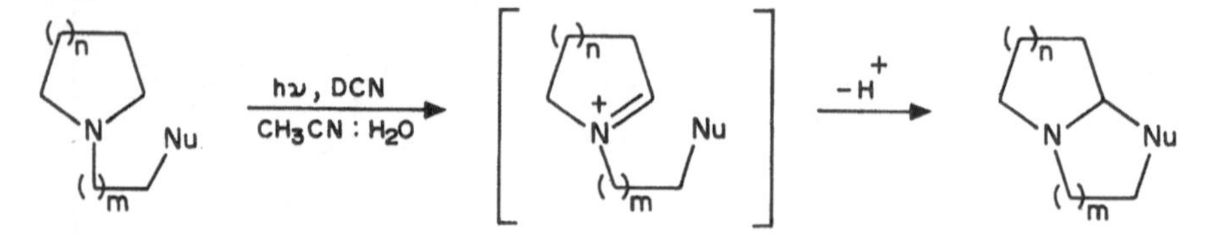

	n	Nu	(YIELD %)
a	1	CH$_2$	75
b	2	O	85
c	2	CH(COOEt)$_2$	76

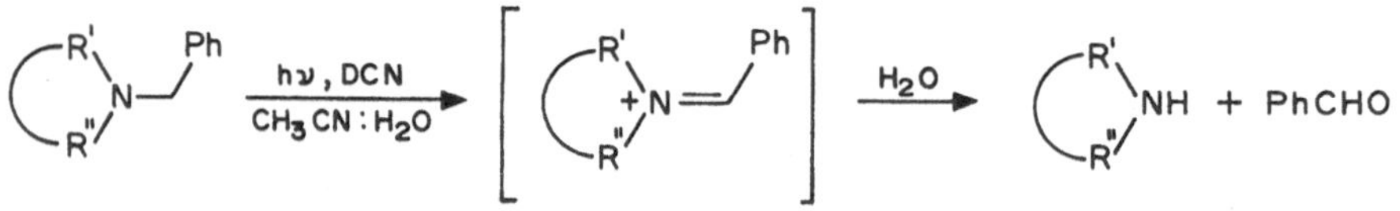

88
(75–90%)

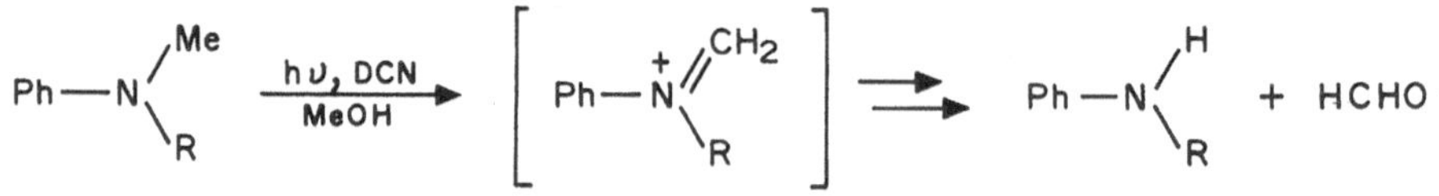

89
R = Me, Et, Bu

90

91
(72–82%)

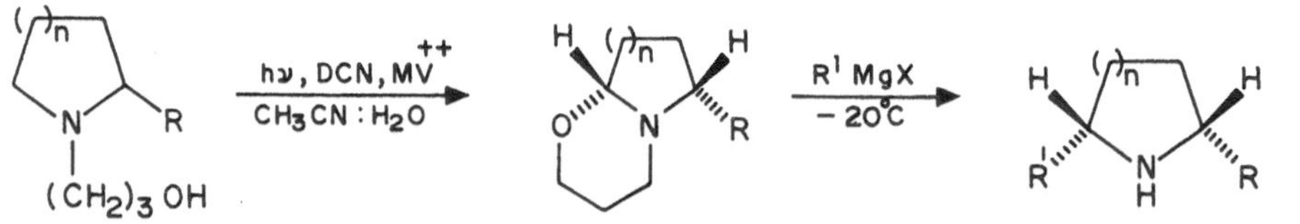

92

93

94

a n = 1 , R = n-Bu 88 %
b n = 2 , R = n-Bu 90 %
c n = 2 , R = Me   92 %

Scheme 16.

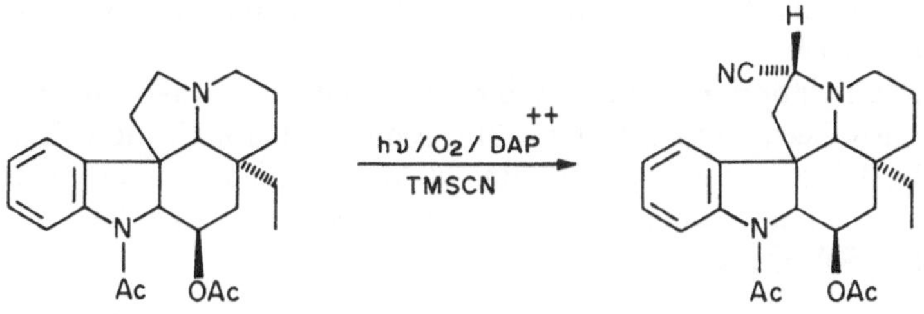

95 → **96** (88 %)

$h\nu$ / O$_2$ / DAP^{++}
TMSCN

DAP = $\underline{N}$, $\underline{N}'$-DIMETHYL-2,7-DIAZA-
PYRENIUM BIS (TETRAFLUOROBORATE)

TMSCN = TRIMETHYL SILYL CYANIDE

97 → **98** (69-87%)

$h\nu$ / O$_2$ / DCA
TMSCN

Scheme 17.

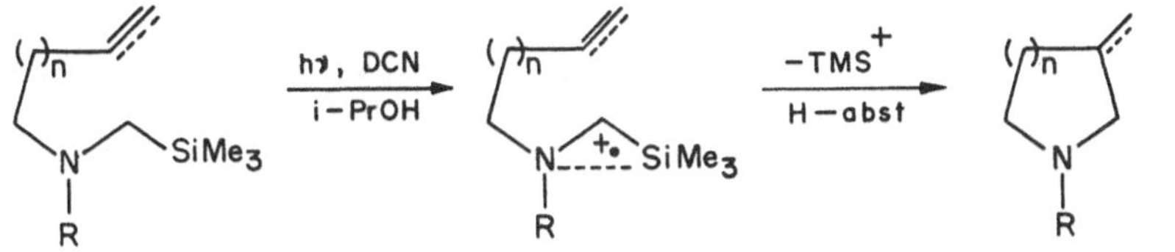

99 → **100** → **101** (70-78%)

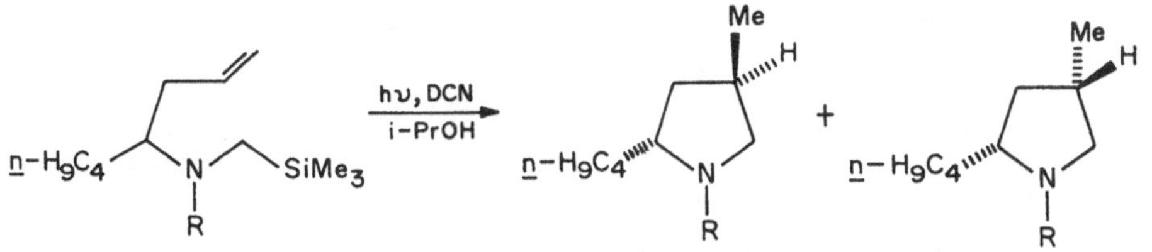

$h\nu$, DCN
i-PrOH

$-$TMS$^+$
H$-$abst

102 → **103** (1:1) **104**

$h\nu$, DCN
i-PrOH

Scheme 18.

195

approach for heterocyclisation reaction as exemplified from the transformations of **108–109** (Scheme 19)

The efficient desilylation from amine radical cation in media favouring SSIP formation has also been used [104] for the sequential double desilylation reaction of amine **110** to generate azomethine ylide **111** which upon cycloaddition with a different dipolarophile gives a stereoselective pyrrolidine ring system **112** as depicted in Scheme 20.

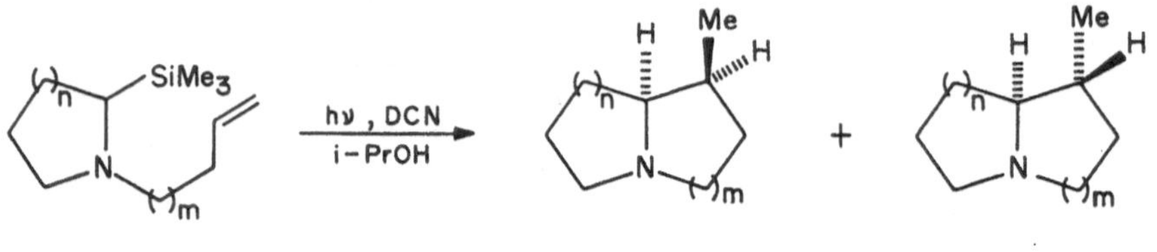

105 → **106** + **107**

$h\upsilon$, DCN / i-PrOH

			106 : 107	YIELD (%)
a	n = m = 1		97 : 3	90
b	n = m = 2		0 : 100	88
c	n = 1, m = 2		2 : 98	85
d	n = 2, m = 1		95 : 5	87

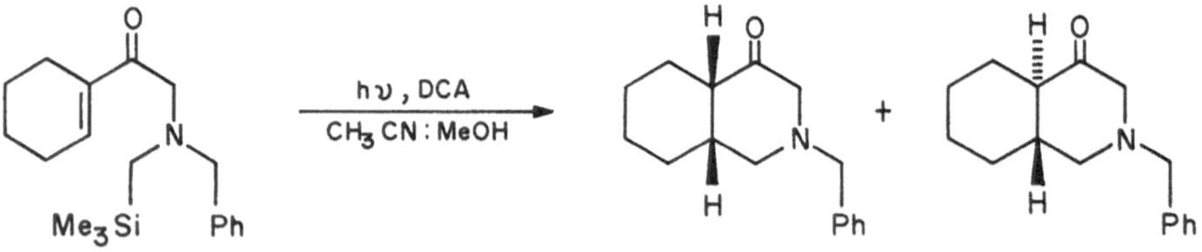

108 → **109 a** + **109 b**

$h\upsilon$, DCA / CH₃CN : MeOH

109 a (67%)   **109 b** (13%)

Scheme 19.

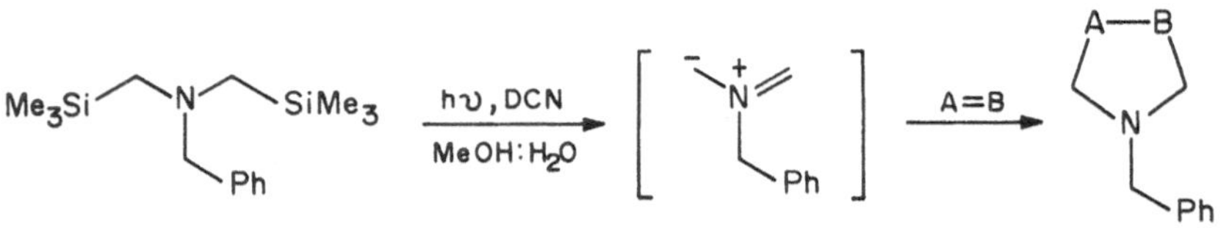

110 → **111** → **112**

$h\upsilon$, DCN / MeOH : H₂O   A=B

112 (55-83%)

A=B : BENZOPHENONE, ETHYL ACRYLATE,
METHYL CINNAMATE, PHENYL VINYL –
SULFONE, ACETYLENE DICARBOXYLATE ETC.

Scheme 20.

2.5 PET Dissociation Reactions

Perhaps one of the most general pathways available to the radical ions is their fragmentation to ions and neutral radicals which often serve as key reactive intermediates (Eqs. 29, 30) in various synthetic reactions. Cleavage reactions from both radical cations (Eq. 29) as well as from radical anions (Eq. 30) have been utilized for synthetic purposes [105]. Although this subject has been discussed in various forms in other sections, we will discuss this briefly here citing important examples of synthetic methods which proceed in a sensitized manner.

Due to sharp and selective reduction in bond dissociation energy of radical cations [106, 107], the cleavage of strong –C–C– or –C–heteroatom bonds has been observed during PET reactions. In this context, examples of –C–C– bond cleavage from β-phenylethers [108], bibenzyls and pinacols [25, 109, 110] may be mentioned. In an extensive study, Whitten et al. [111, 112] have reported clean –C–C– bond cleavage from the reaction of β-aminoalcohols and vinylogous amino alcohols.

Group 4A organometallics have been shown to be exceedingly electron rich [113] and upon ionisation both –M–C– and –M–M– bond dissociation reactions have been observed. One such example of heterolytic –C–M– bond cleavage has recently been reported [114] by our group for –C–Se– bond reactions and a novel methodology for deselenenylation from organoselenium compounds (e.g. 113–115) has been described. Similar cleavage reactions of benzylsilanes, germanes and stannanes radical cations have also been reported by Mizuno et al. [115] in alcoholic solvent and arylmethyl ethers are formed by trapping the benzyl cation. An important application of –Se–Se–bond cleavage has been reported [116] for in situ generation and reaction of electrophilic selenium species (PhSe$^+$) for a oxyselenenylation reaction (117–113) from the PET dissociation of diphenyldiselenide (116). In a significant application, –Se–Se– and –C–Se– bond dissociation strategy have been jointly utilised [114] for a one pot selenenylation and deselenenylation reaction (Scheme 21). Heterolytic –Si–Si–bond cleavage reaction has recently been reported [117] for the synthesis of cyclic silylether (119) from the PET reaction of 118. Selective β-desilylation from trimethyl silyl enol ether (120–121) involving the heterolytic cleavage of radical cations with the formation of α-oxo radicals and TMS$^+$ has been reported by Gassman et al. [118] using redox photosensitization reaction.

Analogous to the cleavage reactions of radical cations, a number of synthetically useful reactions from radical anions are also derived. Photo-decomposition of benzylesters and benzylsulfones are some of the simpler examples of this series [119]. Photoreduction of carboxylic esters (e.g. 122) to corresponding

$$D - \overset{+\cdot}{Y} \longrightarrow \overset{\cdot}{D} + \overset{+}{Y} \quad (29)$$

$$A - \overset{\cdot\cdot}{X} \longrightarrow \overset{\cdot}{A} + \overset{-}{X} \quad (30) \quad \textbf{Fig. 8.}$$

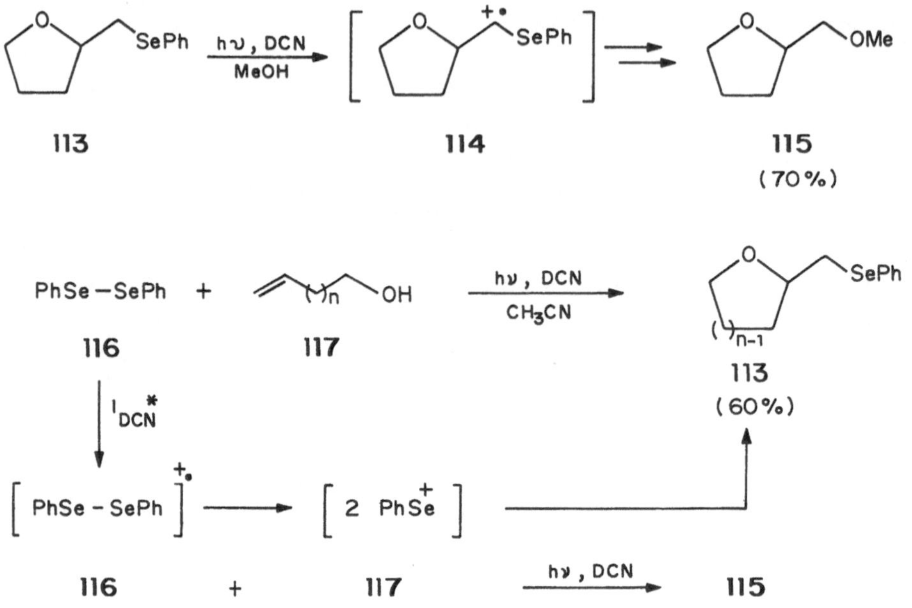

113

114

115
(70%)

116 + **117**

113
(60%)

116 + **117**

115

Scheme 21.

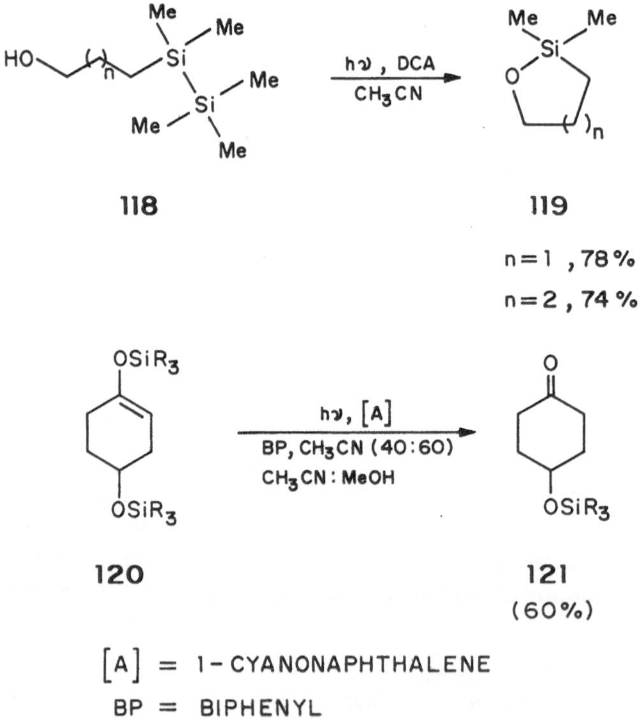

118

119

n = 1 , 78%

n = 2 , 74%

120

121
(60%)

[A] = 1-CYANONAPHTHALENE

BP = BIPHENYL

Scheme 22.

alkanes (123) in high yield (84%) is reported [120] by irradiating 122 in the presence of HMPT at 254 nm. The reaction is initiated by the transfer of an electron from HMPT to carboxylic ester (122) followed by fragmentation and H-abstraction. Based on a similar mechanistic paradigm, a practical method of photodecarboxylation (Scheme 23) is described [121] by the PET reaction of

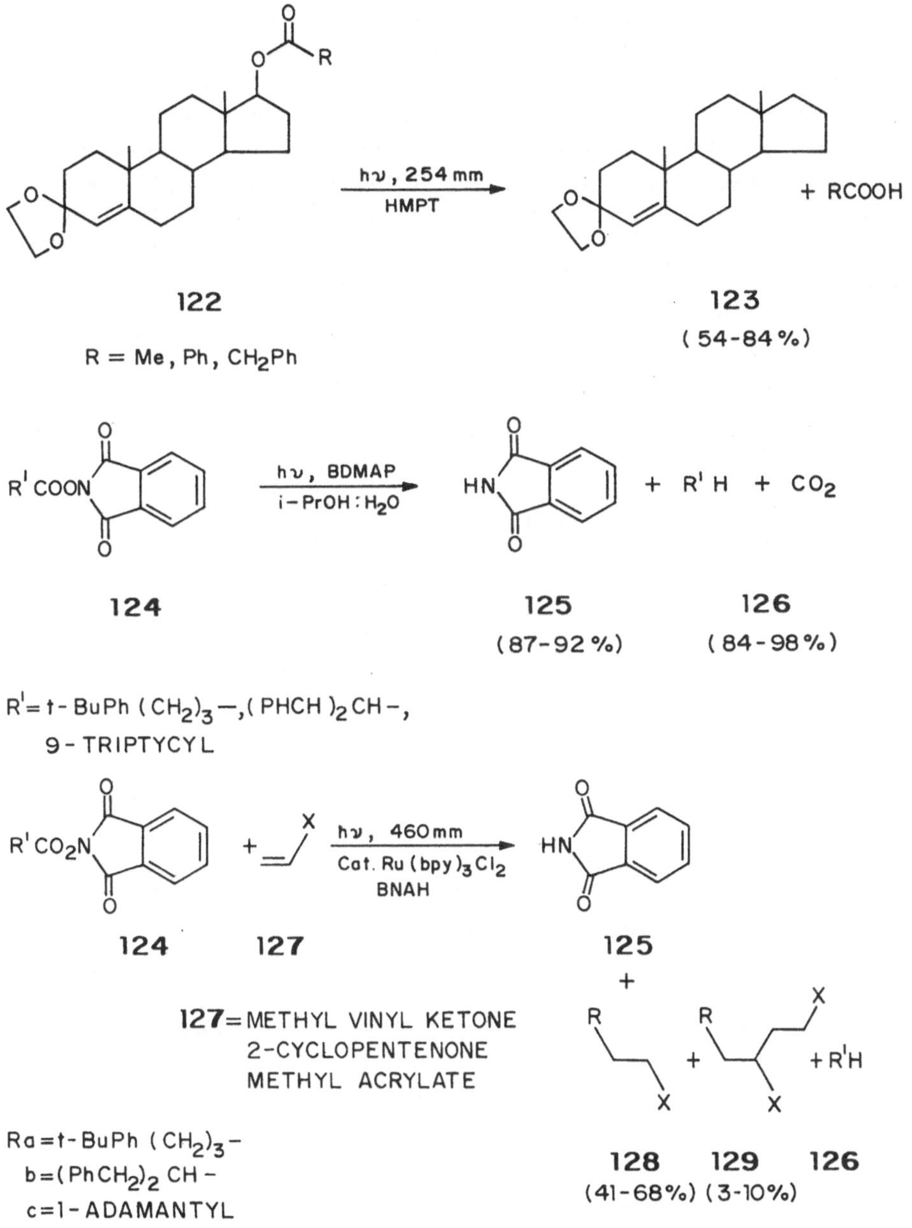

122   123

(54-84%)

R = Me, Ph, CH₂Ph

124   125   126

(87-92%)   (84-98%)

R' = t-BuPh (CH₂)₃—, (PHCH₂)₂CH—,
   9-TRIPTYCYL

124   127   125
                       +

127 = METHYL VINYL KETONE
        2-CYCLOPENTENONE
        METHYL ACRYLATE

Ra = t-BuPh (CH₂)₃—
b = (PhCH₂)₂CH—
c = 1-ADAMANTYL

128   129   126

(41-68%) (3-10%)

Scheme 23.

carboxylic acid derivatives N-acyloxyphthalimides (124) using 1,6 bis-(dimethylamino)pyrene (BDMAP) as the electron donor. This decarboxylation methodology is further extended [122] by the same author for generating carbon centered radical for utilisation in two radical type Michael addition reactions as shown in Scheme 23. The PET strategy employed redox combination of Ru(biPy)$_3$Cl$_2$ and BNAH in the aqueous solvent.

Yonemitsu et al. [123] have described a novel approach for detosylation of sulfonamides by a PET reaction using dimethoxybenzene as electron donor and a reductant (e.g. NaBH$_4$, ascorbic acid hydrazine etc.). A similar approach is reported [124] for the deprotection of aryl sulphonate esters from carbohydrate substrate (e.g. 130–131) using tertiary amines as electron donor. Improvement in these methodologies were subsequently reported by Nishida [125] and Art [126]. One of the most useful applications of heterolytic cleavage reaction from radical anions has been demonstrated by Saito et al. [127] for the selective deoxygenation of secondary alcohols (91%) in general and 2,3-deoxygenation of ribonucleoside (132–133) in particular.

The utility of a novel intra-ion-pair electron transfer cleavage reaction has been reported by Schuster et al. [128] for free radical polymerisation initiation

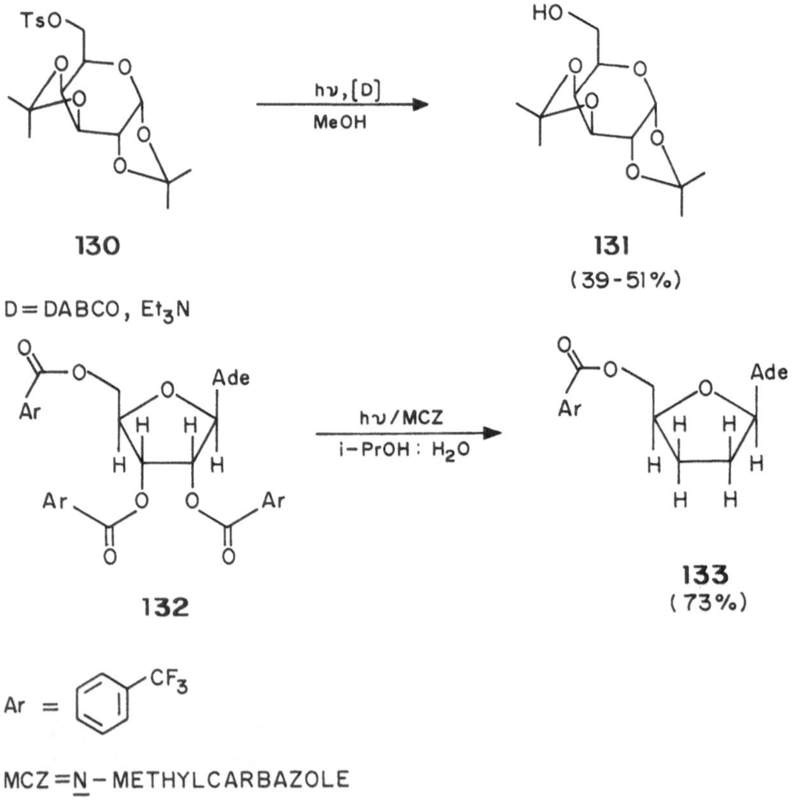

130

131
(39-51%)

D = DABCO, Et$_3$N

132

133
(73%)

Ar =

MCZ = N – METHYLCARBAZOLE

Scheme 24.

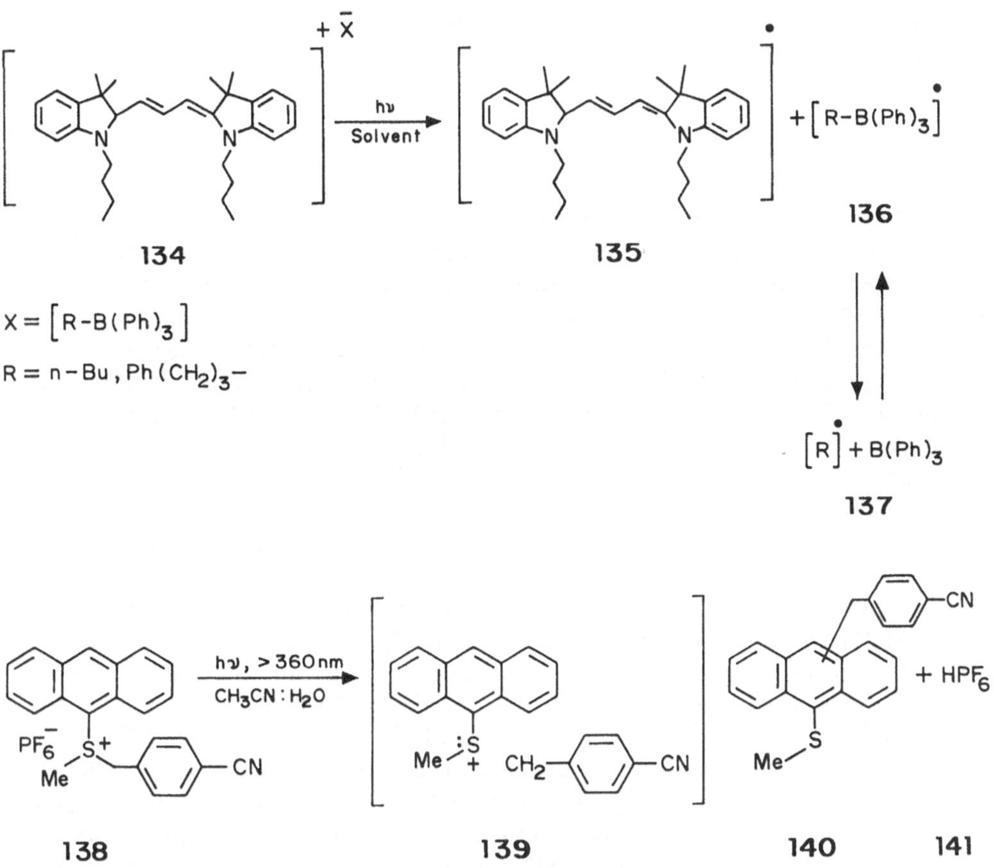

Scheme 25.

from triphenyl alkyl borate salts of cyanide dye (**134**) activated by visible light (Scheme 25). An intramolecular concerted bond cleavage/coupling process is described [129] using a phenyl anthracene sulfonium salt derivative (**138**) as a part of designing a novel "Photoacid" system (**141**). A systematic review on photoremovable protecting groups has been published recently by Binkeley et al. [130].

3 Donor–Acceptor Reactions

PET reactions where products arise by coupling of acceptor–donor pairs are summarised in this section. Usually these reactions do not involve direct coupling of radical ions, instead, recombination of radical pairs, formed by the proton or electrofugal group transfer/loss from one radical ion to another gives rise to the product.

3.1 Reactions of Excited Arene Acceptors

A considerable part of PET chemistry has focussed attention on the photo-additions of electron rich olefins to singlet excited cyanoarenes provided $\Delta G_{et} > 0$. Mechanistic details of these reactions have been reviewed by Arnold [131] and Lewis [85, 132]. However, the synthetic utility of these additions remained limited due to low yield and selectivity. Mizuno et al. [133] described the improvement in the yields of such additions and reported the synthesis of mono-allylated aromatic compounds 143 and 144 from the PET reactions of DCN with allyltrimethylsilanes (142) in CH_3CN. Efficient heterolytic cleavage of allyltrimethylsilane radical cation intermediates is one of the key steps in these photosubstitutions. The regioselectivity of these allylations has been demonstrated [134] to depend on the structure of the dicyanoaromatic compounds and reaction conditions. A novel correlation of ΔG_{et} with the regiospecificity of cyanoarene-olefin photoadditions has been discovered by Mattay et al. [135]. For the systems where ΔG_{et} is positive ($\Delta G_{et} >$ ca. 0.4 eV), *ortho-* and *meta-cycloaddition of alkenes* (146) to the arene ring of 145 occurs while in the case of alkenes such as tetramethylethylene (146) and ether 149 (where $0 < \Delta G_{et} < 0.4$ eV) only addition to the nitrile group is observed [136, 137] and azetines 147 and 2-azabutadienes 148 and 150 are produced [136, 137] in good yields. Identical correlation of ΔG_{et} have also been reported [138] from the photoreaction of α,α',α''-trifluoro toluene with electron-rich olefins. A remarkable application of PET additions [139] of olefin (152) to aromatic nitro compounds (151) have

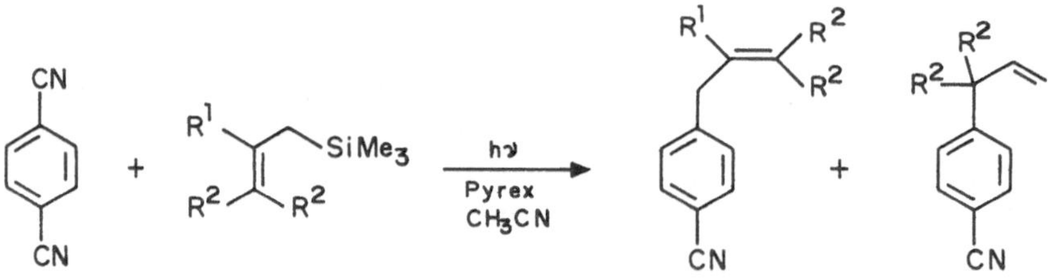

	142	143 (%)	144 (%)
a	$R^1 = R^2 = H$	66	–
b	$R^1 = Me$ $R^2 = H$	88	–
c	$R^1 = H$ $R^2 = Me$	36	55

Scheme 26.

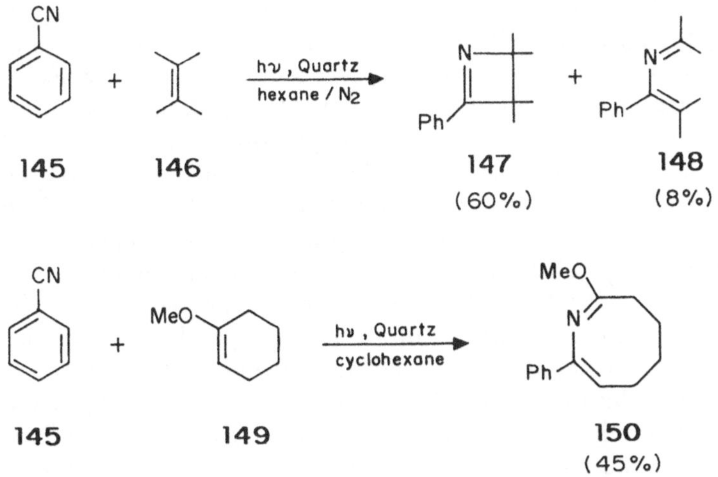

Scheme 27.

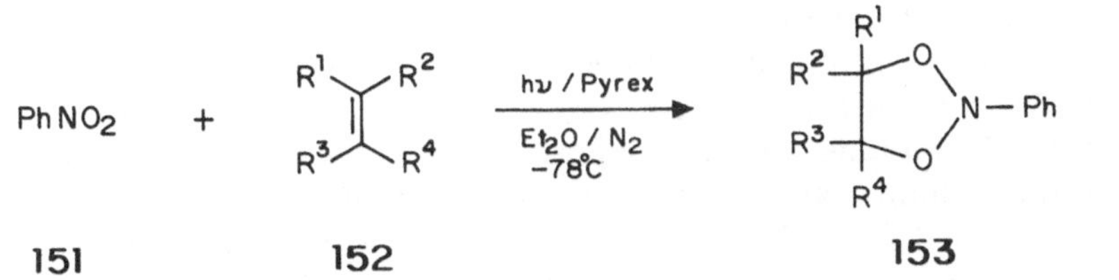

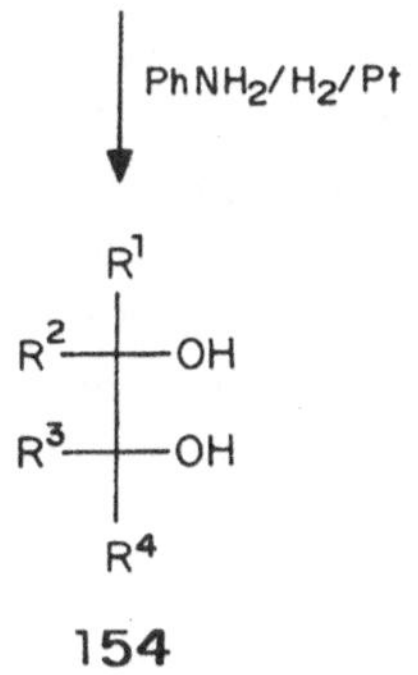

154	R^1	R^2	R^3	R^4	YIELD
a	H	H	$(CH_2)_4$	H	93 %
b	H	H	CH_3	CH_3	87 %
c	H	CH_3	CH_3	H	82 %
d	H	H	H	$-CH=CH_2$	89 %

Scheme 28.

been utilised [134] for the synthesis of cis-1,2-diols (**154**). Further examples of photoreactions of aromatic nitrocompounds of synthetic importance have been reported by Dopp et al. [140]. Photoadditions of alkylated benzenes to cyanoaromatics have been studied by several groups [25]. In these cases, cross coupling of the benzyl radicals followed by hydrogen cyanide elimination leads to the benzylated products [25]. The benzyl radicals are produced by the deprotonation of alkyl benzene radical cations. Besides, routes involving deprotonation for benzyl radicals, the PET additions of pinacols, pinacol ethers and bibenzyls to cyanoaromatics are reported to involve a –C–C– bond fragmentation step. Improvement in the yields of PET-mediated benzylation of cyanoaromatics has been reported [141] recently from the photoreaction of DCN and arylmethylsilanes, germanes and stannanes (155). The yields of benzylation products (156 and 157) are found to depend on the nature of M in 155 and is in the order of Si > Ge > Sn. Mariano and co-workers [142] have reported the addition of a trimethyl silyl substituted ethers, thio ethers and amines to singlet excited DCA and adducts 158, 159 and 160 respectively are formed (Scheme 30). Photoaddition of $(Me)_3SiN(Et)_2$ initially produces corresponding dihydroanthracene adduct 160 which spontaneously dehydrocyanates under the reaction conditions to give 161. This process was inefficient compared to the oxygen and sulphur analogues. A novel –Si–Si– bond cleavage reaction of hexamethylsilane (162) radical cations has been utilised [143] for photosilylation of cyanobenzene (e.g. 162–163). In a synthetically promising investigation, Schuster's (144) group has reported the photo-alkylation of dicyanoarenes (Scheme 31) using a conceptually new class of easily oxidisable reagents, alkyl triphenyl borate (165). Mechanistic studies supported the PET-initiated free radical processes for these reactions. Photoaddition of amines to the excited singlet state of aromatics is known [83, 85] to involve PET processes as key steps. Mechanistic and the regiospecificity of additions have been reported by Lewis and co-workers [85] from *trans*-stilbene amine pairs. An interesting application of these fundamental studies have emerged from the same group [145] synthesis of different ring systems (e.g. 169 and 170) by intramolecular amination of the *trans*-stilbene moiety (168) as shown in Scheme 32. It may be appropriate to mention here a few examples of the intramolecular coupling reaction between the acceptor–donor pair, though the PET is initiated by "redox photosensitization" using a light absorbing hole transfer agent. Photoadditions of DCNB to furans 171 have been reported [146] to produce adducts 172 and 173 by redox photosensitization using singlet excited states of various electron rich arenes such as phenanthrene and naphthalene (Scheme 33). The participation of

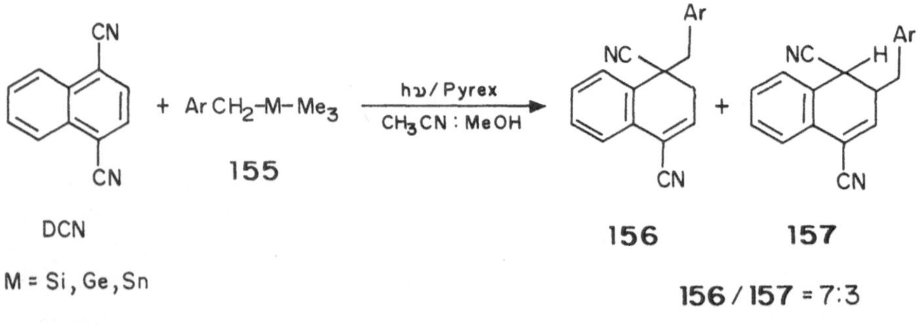

Scheme 29.

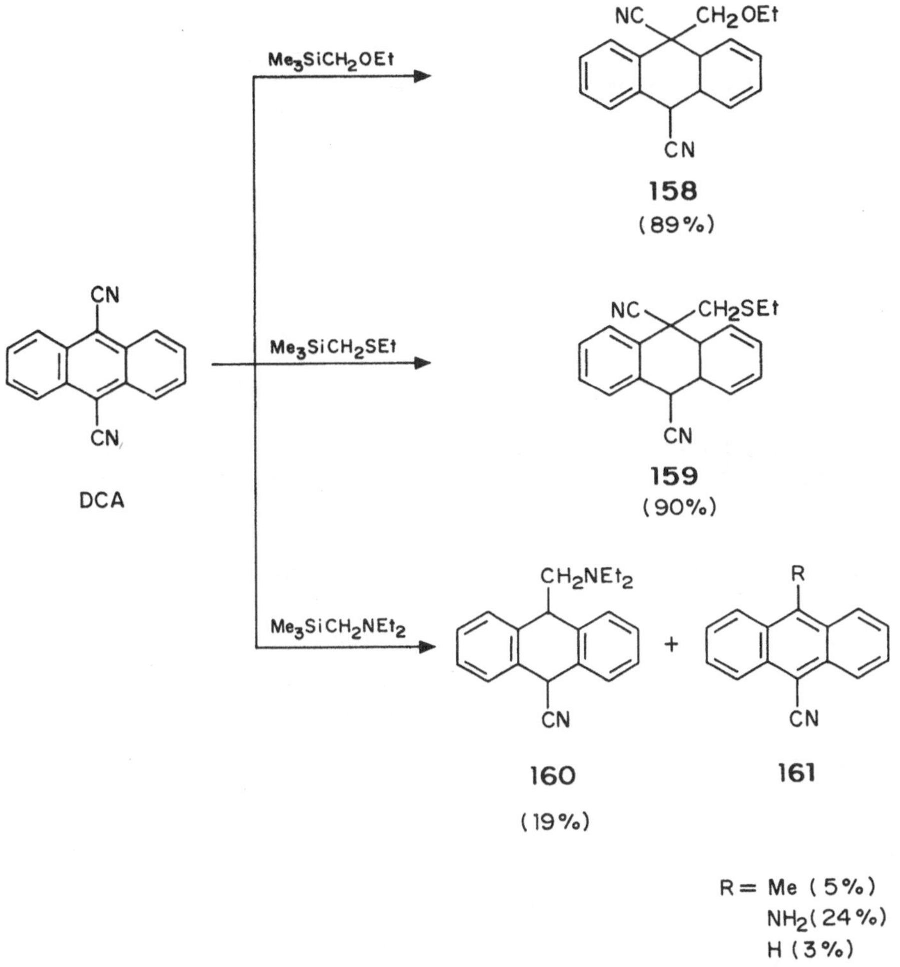

Scheme 30.

methanol in reactions indicated the involvement of out of cage pathways following the initial PET processes. Arnold [147] has recently reported a novel 1:1:1 additions of a nucleophile-olefin and arene from the reaction of DCNB with α-pinene (175) and β-pinene (176) in methanol. The products from α-pinene (177 and 178) are found to be racemic while products from β-pinene (179 and 181) have retained their optical activity (Scheme 34). High yields of DCNB photo-allylation is reported [148] from a phenanthrene sensitized photoreaction of group 4A organometallic allylated compounds. Photoadditions of benzotriazole (182) to aromatic hydrocarbons (183) by DCA photosensitization has been recently reported [149] to give 184 in 41–77% yield.

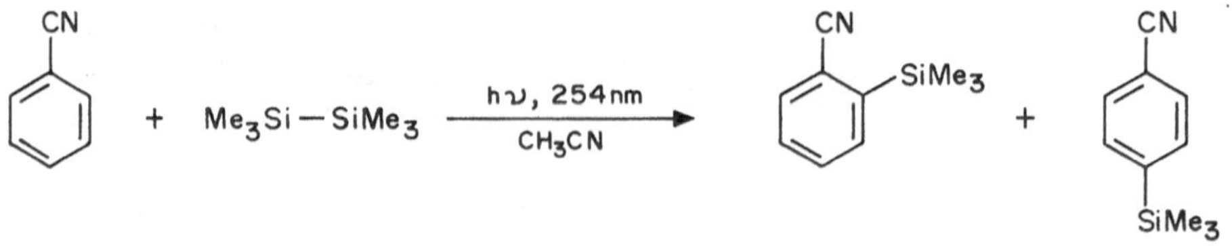

145   162   163   164

(6%)   (8%)

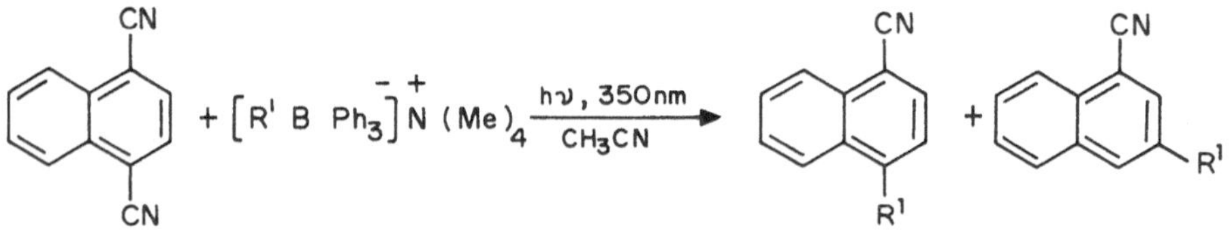

DCN   165   166   167

	165	166/167	YIELD (%)
a	$R^1 = PhCH_2$	0.6	95
b	$R^1 = Me$	1.4	100

Scheme 31.

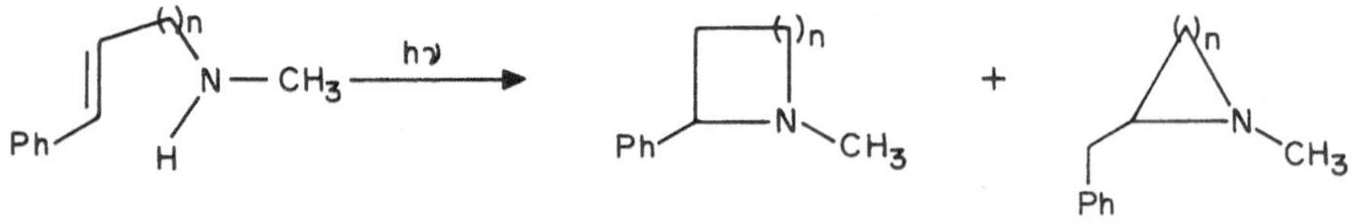

168   169   170

n	169/170	YIELD (%)
1	>20	15
2	14	63
3	2.4	57
4	0.15	82
5	>8	30

Scheme 32.

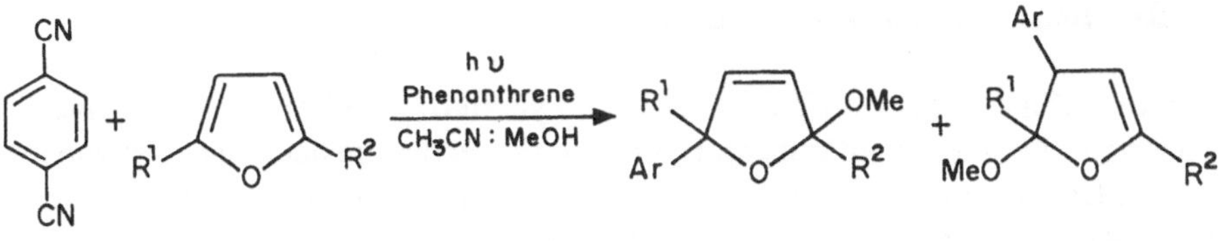

171 **172** **173**

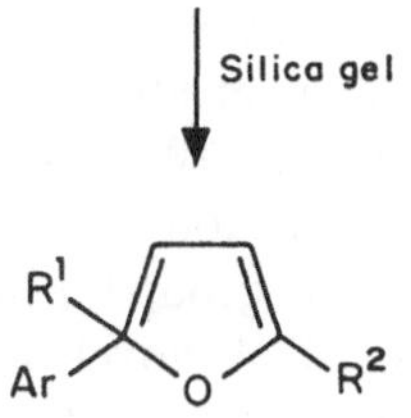

171	R^1	R^2	172	173
a	H	H	74 %	–
b	H	Me	57 %	11 %
c	Me	Me	8 %	31 %

174 R^1 = H

Scheme 33.

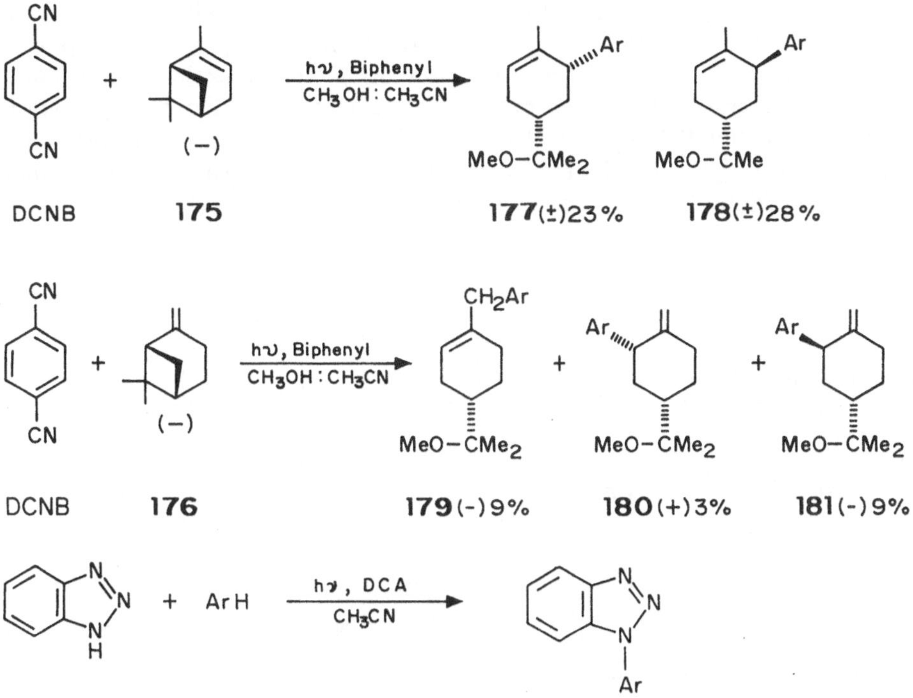

DCNB **175** **177**(±)23 % **178**(±)28 %

DCNB **176** **179**(−)9% **180**(+)3% **181**(−)9%

182 **183** **184** (39-77%)

Ar = ANISOLE, BIPHENYL, NAPHTHALENE

Scheme 34.

3.2 Reactions of Excited Arene Donors

PET reaction of electron rich arenes i.e. alkyl, alkoxy and amine derivatives of benzene and higher aromatic compounds with electron deficient species such as acrylonitrile, cyanoarenes and nitro arenes have generally led to the arene ring substitution provided ΔG_{et} is negative [25]. In a recent report, the role of ion pairs in the photoreaction of charge transfer complexes has been demonstrated by Kochi et al. [150] from the nitration of substituted anthracene 185 and tetranitromethane (186) by exciting within the charge transfer complex of 185 with 186. The formation of 187 has been suggested to arise from the cage-recombination of geminate ion pairs. A novel application of this reaction is reported [151] by the same group for the carboxylation of p-dimethoxy benzene (e.g. 188–190) Scheme 35. A novel and useful reaction of donor substituted arenes was described some time for the intramolecular cyclisation of chloroace-tamides to various electron-rich aromatic rings. Several biologically active compounds are synthesized utilizing this methodology and have been sum-marised in a comprehensive review by Sundberg [152]. However, the import-ance of this methodology may be highlighted by citing [153] an example of the synthesis of a precursor 192 of antileukemic compound cephalotaxine 193 from the photocyclisation of 191. In a recent work Moody et al. [154] have reported the synthesis of 7-substituted pyrrolobenzazocine (195), a skeleton of a biolo-gically active ergot alkaloid, from the photoreaction of dichloroacetyl trypto-phan methyl ester 194 in CH_3CN. (−) Indolactam-V (198), a tumour promoter has also been synthesized [155] from the same group using intramolecular photocyclisation of 196 as the key step.

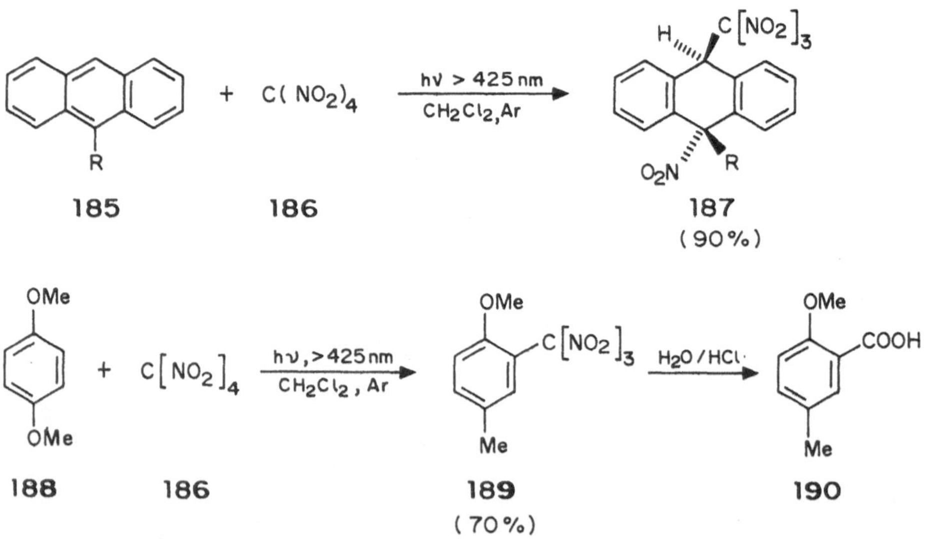

Scheme 35.

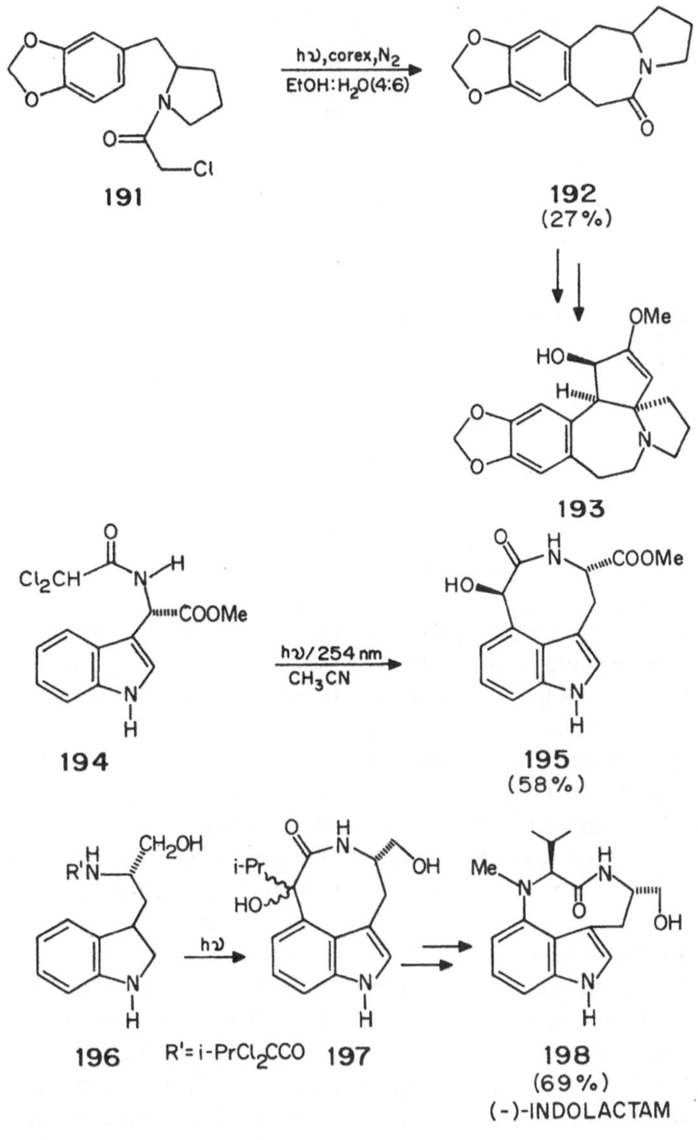

Scheme 36.

3.3 Reactions with Iminium Salts

Excited states of iminium ion functional groups have been shown to undergo one electron reduction readily from a number of unlikely one electron donors such as simple olefins, aromatic hydrocarbons, alcohols and ethers. PET reaction from these systems have led to the development of synthetically important carbon–carbon bond forming methodology. A number of studies

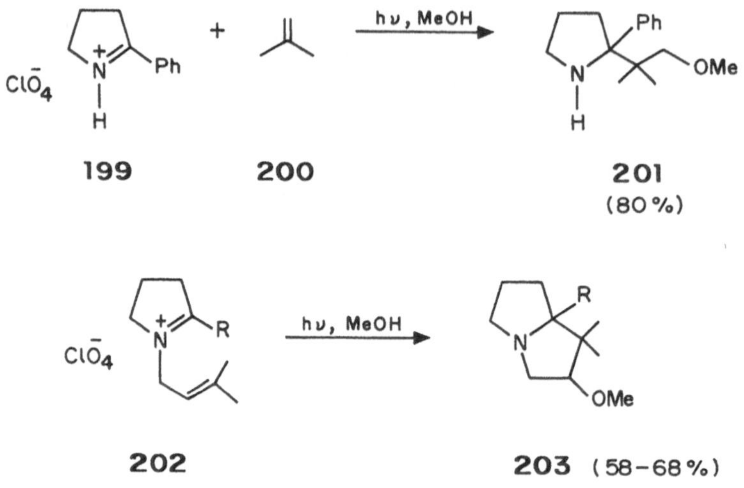

199　　　　　**200**　　　　　　　　　**201**
(80%)

202　　　　　　　　**203** (58–68%)

Scheme 37.

from Mariano's group [156] have reported the PET processes from compounds containing iminium salt groupings using different electron-donors. A general reaction type may be illustrated [157] by referring to the photoreactions of olefins **200** (π-oxidation potentials lower than ca. 2.6 eV) with excited states of iminiumsalts **199** which results in a novel –C–C– bond formation reaction and produces compound **201**. Interestingly the reaction with electron-poor olefins yield epimeric spirocyclic amine through alternate $\pi + \pi$ arene-olefin cycloaddition followed by ring expansion [156]. Intramolecular olefin-iminium salts photoadditions have been utilised [158] as a novel strategy for bi- or even poly-heterocyclic ring construction (e.g. **202–203**). Similar photoadditions are also described from allyl silanes [159], alcohols [157c], ethers [157c] arene donors [160].

Several applications of this methodology have appeared from Mariano's group, however, we will demonstrate the synthetic potential of this discovery by citing here the two most important examples. Synthesis of (+) Xylopinine (**205**), a protoberberine alkaloid, has been performed [161] from the photocyclisation of **204** in methanol. Similarly, a erythrane ring system (**207**) has been constructed [162] from the photoreactions of **206**. Some other interesting aspects of this chemistry have been related to a novel Pictet-Spengler [163] type cyclisation and the chemistry of allene cation radicals [164].

3.4 Reactions of Carbonyl Compounds

Carbonyl compounds undergo PET reaction with electron donating substrates viz., alkenes, amines and thio-compounds. Except in the photoreactions with electron poor alkenes, such as acrylonitrile [165, 166], carbonyl compounds usually act as an electron accepting partners. The chemical reaction pathways

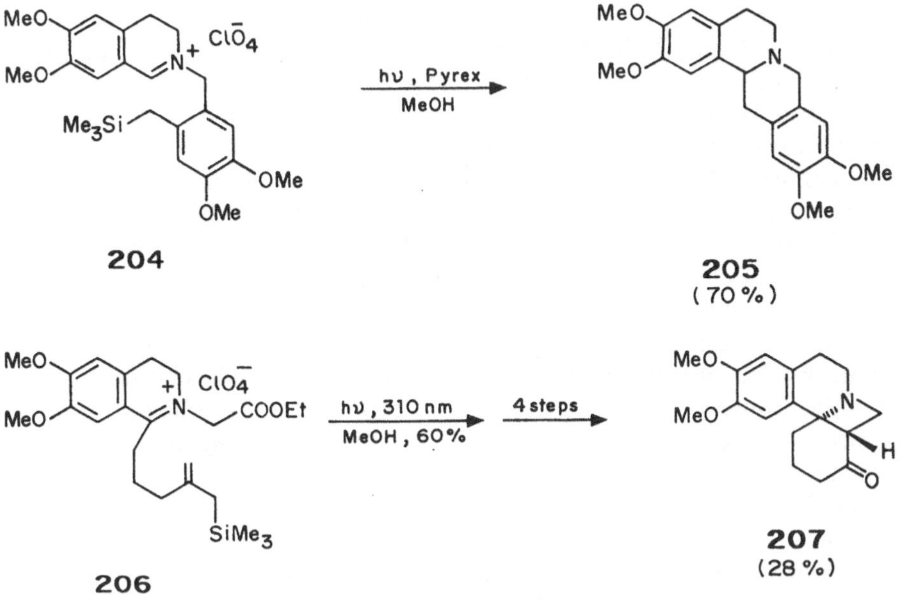

Scheme 38.

for these systems are controlled by the processes available to the radical ion intermediates.

PET reaction of carbonyl compounds with olefins form either oxetanes (Paterno-Buchi reaction, Eq. 31) by direct coupling or a radical pair reaction leading to coupling product or reduction. The carbonyl-olefin radical pairs are formed by proton transfer within their radical ion pairs (Eq. 32). Both these aspects of ketone-olefin photoreaction have been recently rationalized by Mattay et al. [167] from the photoreactions of 2,3-butanedione (**208**) with different olefins such as **209** and **210** as shown in Scheme 39. Photoprocesses of

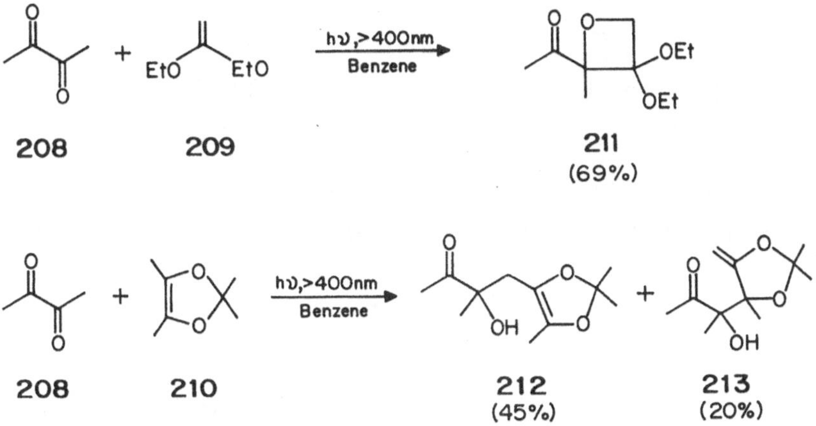

Scheme 39.

ketones with olefins with emphasis on the Paterno-Buchi reaction have been reviewed by Wagner [168] and more recently by Jones [166] and Carless [169].

In the PET reactions of ketone-amine pairs, proton transfer from the amine cation radicals to the ketyl radical anions is exceptionally efficient and radical pair formation (thus coupling or reduction) dominates other possible reaction modes (Eqs. 33, 34). In general coupling of the α-amino radical formed by proton transfer from initially formed amine radical cation ketyl radical anion terminates the reaction process. Early examples of ketone-amine photoreactions were reported by Cohen [170] and have been updated recently [10, 11]. Detailed mechanistic studies of these reactions have been discussed by Peters [171].

Another interesting reaction type detected from the photoreaction of tertiary amine–ketone system which have found use in alkaloids chemistry [172] involves sequential electron transfer and proton transfer followed by back electron transfer to yield iminium cations (Eq. 35) leading to dealkylation products upon aqueous workup. The regiocontrol in these dealkylations is dictated by preferential deprotonation at the less branched amine cation radical α-carbon.

Mariano's group [173, 174] in their interesting observations have reported the photo-additions of α-silylated tertiary amines **215** to excited triplet state of cyclohexenone (**214**) to produce adduct **216**. The addition which involves α-amino radical (Scheme 40) as intermediate is produced either by sequential PET desilylation or deprotonation of **215** depending upon the nature of solvent used. An alcoholic solvent favours desilylation while in acetonitrile deprotonation predominates. The application of these results has been reported recently [174] for the preparation of both fused (diastereomeric mixture **218** and **219**) and spiro-*N-N-hetero bicyclic* (**221**) systems (Scheme 41). Although the α-silylation of tertiary amines in the above study [173, 174] was designed to facilitate a –C–Si– bond rupture from the amine radical cations for generating α-amino radicals, recently [175], bridged bicyclic ring systems (e.g. **223**) have been constructed in synthetically acceptable yields (50%) by irradiating **222** in

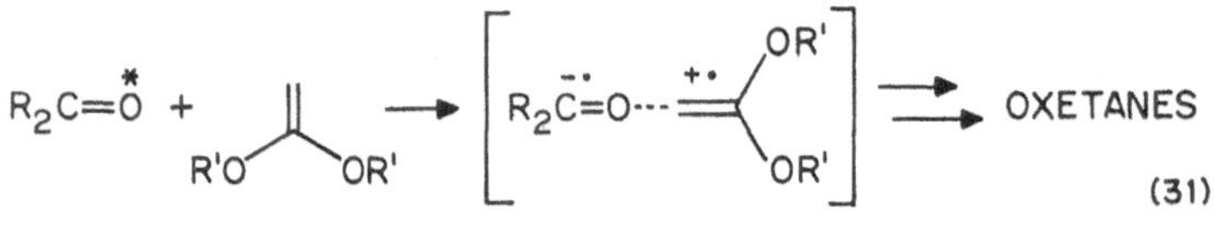

$$R_2C{=}\overset{*}{O} + \underset{R'O\quad OR'}{\overset{\|}{\bigg\|}} \longrightarrow \left[R_2\overset{-\bullet}{C}{=}O{\cdots}\overset{+\bullet}{=}\overset{OR'}{\underset{OR'}{\bigg\langle}} \right] \longrightarrow \text{OXETANES} \qquad (31)$$

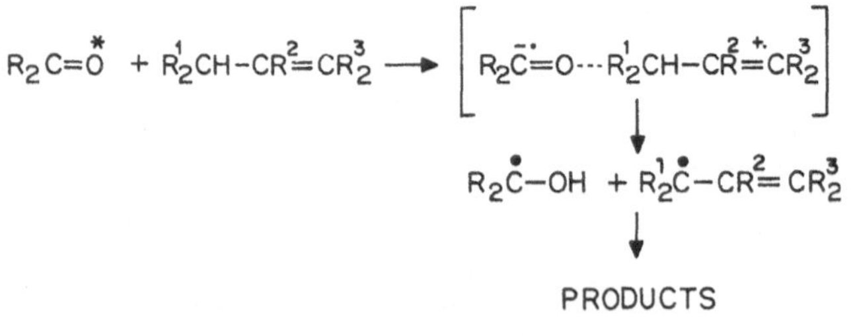

$$R_2C{=}\overset{*}{O} + \overset{1}{R_2}CH{-}\overset{2}{CR}{=}\overset{3}{CR_2} \longrightarrow \left[R_2\overset{-\bullet}{C}{=}O{\cdots}\overset{1}{R_2}CH{-}\overset{2}{CR}{=}\overset{3+\bullet}{CR_2} \right]$$

$$\downarrow$$

$$R_2\overset{\bullet}{C}{-}OH + \overset{1}{R_2}\overset{\bullet}{C}{-}\overset{2}{CR}{=}\overset{3}{CR_2}$$

$$\downarrow$$

$$\text{PRODUCTS} \qquad (32)$$

Fig. 9.

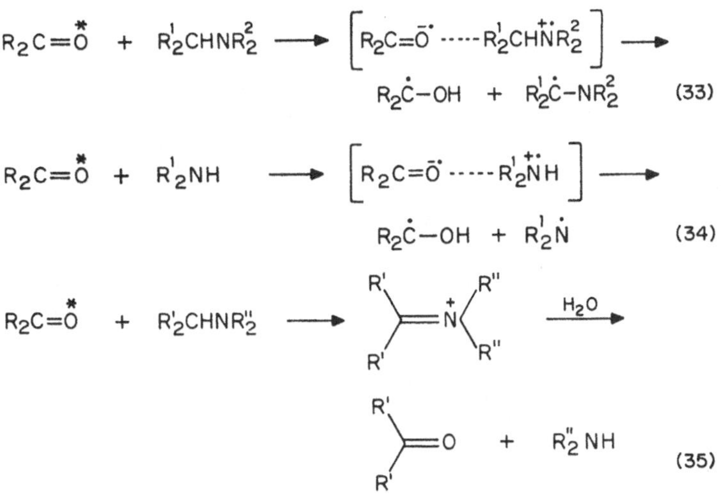

$$R_2C=\overset{*}{\overset{\shortparallel}{O}} \;+\; R'_2CHN\overset{2}{R_2} \longrightarrow \left[R_2C=\overset{\cdot}{\overset{-}{O}} \cdots\cdots R'_2CH\overset{+\cdot}{N}R^2_2 \right] \longrightarrow$$

$$R_2\overset{\cdot}{C}{-}OH \;+\; R'_2\overset{\cdot}{C}{-}NR^2_2 \qquad (33)$$

$$R_2C=\overset{*}{\overset{\shortparallel}{O}} \;+\; R'_2NH \longrightarrow \left[R_2C=\overset{\cdot}{\overset{-}{O}} \cdots\cdots R'_2\overset{+\cdot}{N}H \right] \longrightarrow$$

$$R_2\overset{\cdot}{C}{-}OH \;+\; R'_2\overset{\cdot}{N} \qquad (34)$$

$$R_2C=\overset{*}{\overset{\shortparallel}{O}} \;+\; R'_2CHNR''_2 \longrightarrow$$

(35)

Fig. 10.

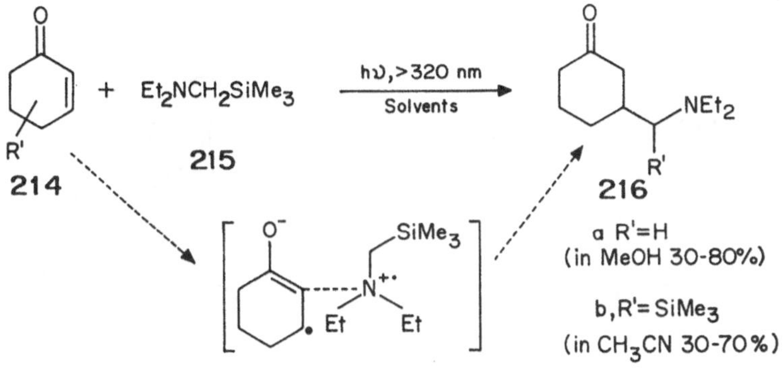

Scheme 40.

triton B. Conjugate addition of allylic group to β-enones has also been reported [176] recently via PET reactions of enones and allylic stannanes.

An interesting, although it does not belong to our present classification, PET initiated reductive cyclisation of δ–ε-unsaturated ketones (e.g. **224–225**) has been reported by Cossy and Portella [177] for synthesizing bicyclic pentanols (**225**) from the photoreaction of **224** in the presence of HMPT (neat) or triethylamine in acetonitrile. Synthesis of biologically active natural products such as (±) Hirsutine (**228**) [178], (±) Actinidine (**231**) [179] and an advanced precursor for C-5-Oxygenated Iridoids (**233**) [180] have been reported from the same group employing the above-mentioned methodology as key steps. Several other interesting reactions on the same subject have also been reported [181, 182].

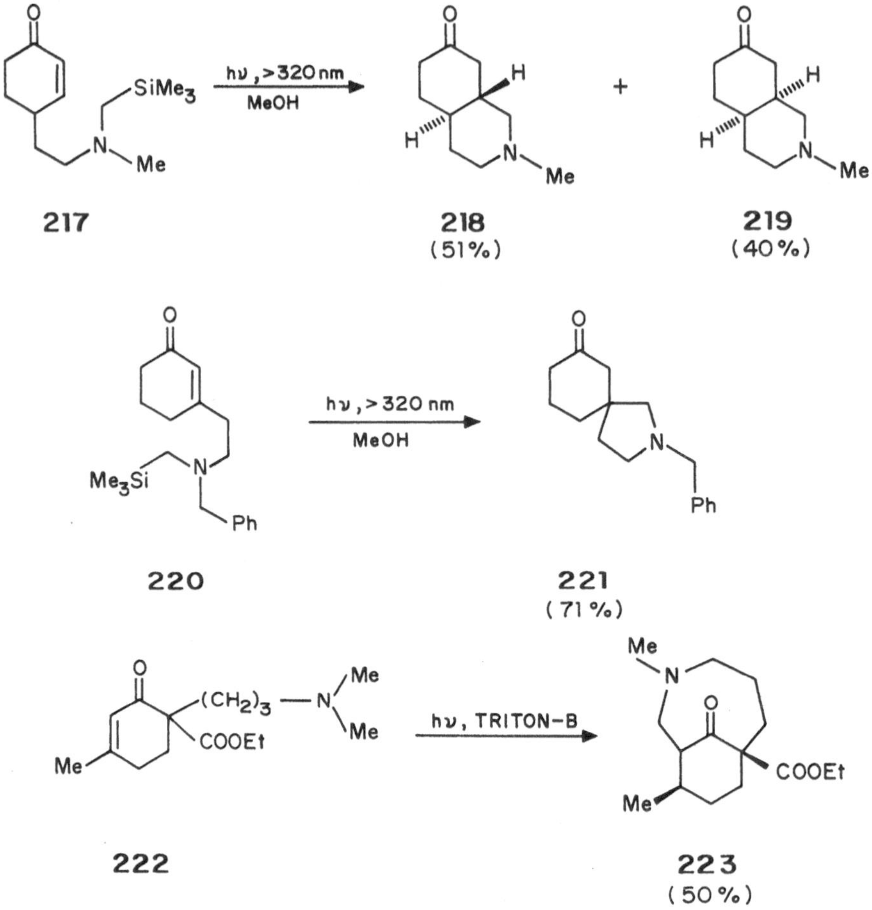

Scheme 41.

Unlike the photochemistry of ketones with amine donors, simple thio-ethers often do not lead to the reduction of ketones [183] due to the slower rate of proton transfer from the thioether cation radicals to the ketyl radical anions than back electron transfer. PET reactions of several other sulphur compounds with ketones have been summerised in Mariano's review [11].

3.5 Reactions of Imides

Cyclic carboxyimides, particularly phthalimides, have been shown to undergo PET reactions both inter- and intramolecularly with a variety of electron donating partners. Several examples of these reactions which have been put to good use in the synthesis of heterocyclic compounds are summarised in excellent reviews published by Mazzochi [184] and Coyle [185].

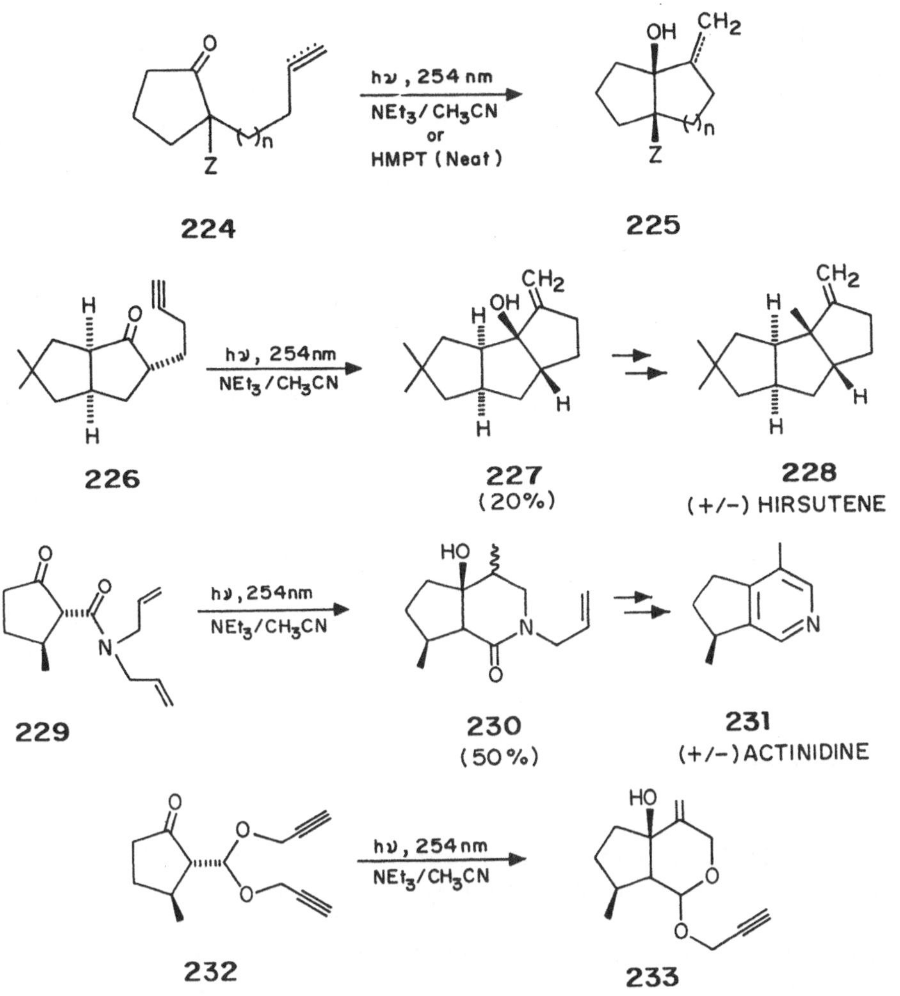

Scheme 42.

3.6 Reactions of Quinones

Similar to ketones, quinones have also been known for a long time to undergo PET processes with several donors such as aromatic hydrocarbons, olefins and amines etc. [10b, 11]. It is pertinent here to illustrate one of the synthetically important quinone–olefin [186] reactions which has been utilized for the preparation of Benz (a) anthracene 7,12-dione derivatives (**237**). In the present example, the excited state of naphthaquinone (**234**) reacts with ethene (**235**) to give **236**, provided the electron transfer is thermodynamically allowed (ca. $\Delta G_{et} < 0$). In the follow up processes, (**236**) is transformed to (**237**) in 30–50% yield. Another report from the same group has described [187] a novel method of photoallylation of naphthaquinone by allyl stannane.

215

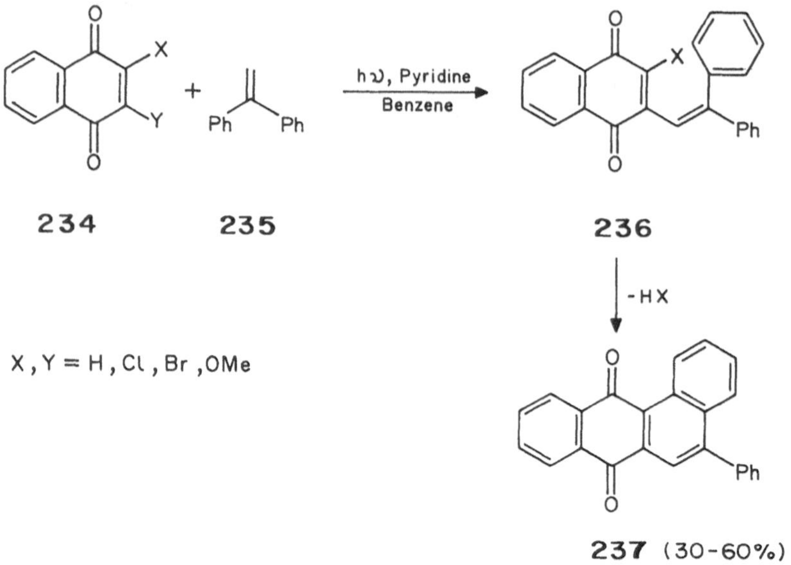

Ganesh Pandey

Scheme 43.

4 Concluding Remarks

We conclude this chapter on the note that the consequences of PET transformations from organic substrates may be characterised by evaluating the
reactivity profiles of initially formed ion radicals. The voluminous literature
accumulated on this subject in a relatively short period of time, has made it
possible to establish a general reactivity pattern of these reactive intermediates.
As a result, this has stimulated synthetic chemists to design newer synthetic
strategies. Though we are a long way from making predictions in such reactions,
the author is of the opinion that frequent use of these reactions for selective
transformations during complex molecule synthesis will prevail. It may be
essential to collect kinetic data for the chemical reactions of photogenerated
radical ions which may help to develop understanding of structure-activity
relationships, competing reaction pathways and relationships between photochemical and non-photochemical methods of generating radical ions. The study
of electron transfer processes on semiconductor surfaces and restricted environments will strongly influence further developments in this exciting area.

Acknowledgements. The author is indebted to all his colleagues whose names have figured in this
article for their dedication and intellectual contributions. I am specially thankful to Messrs M.
Karthikeyan and P. Yella Reddy for their help in preparing this review. I am grateful to CSIR, New
Delhi, for financing our research in this area over the past several years.

5 References

1. Eberson L (1987) In: Electron transfer reactions in organic chemistry. Springer, Berlin Heidelberg New York
2. Fox MA, Chanon M (1988) In: Photoinduced electron transfer reactions, part A–D. Elsevier, Amsterdam
3. Davidson RS (1983) Adv. Phys. Org. Chem. 19: 1; Lewis FD (1986) Adv. Photochem. 13: 165
4. Kavarnos GJ, Turro NJ, (1986) Chem. Rev. 86: 401
5. Mattes SL, Farid S (1983) In: Padwa A (ed) Organic photochemistry. Marcel Dekker, New York, vol 6, p. 223
6. Beens H, Weller A (1975) In: Birks JD (ed) Organic molecular photophysics. Wiley, London, vol 2, chapter 4; Mataga N, Ottolenghi M (1975) In: Foster R (ed) Molecular association. Academic, London, vol 2, chap 1
7. Rehm D, Weller A (1970) Isr. J. Chem. 8: 259
8. Gould IR, Ege D, Moser JE, Farid S (1990) J. Am. Chem. Soc. 112: 4290
9. Kellet MA, Whitten DG, Gould IR, Bergmark WR (1991) J. Am. Chem. Soc. 113: 358
10. a) Mattay J (1987) Angew. Chem. Int. Ed. Engl. 26: 825. b) Mattay J (1989) Synthesis 233
11. Mariano PS, Stavinoha JL (1984) In: Horspool WM (ed) Synthetic organic photochemistry. Plenum Press, New York, p 145
12. Fox MA (1986) Adv. photochem. Wiley, New York, vol 13, p 237
13. Becker HGO (1983) In: Einfuhrung in die Photochemie. George Thieme Verlag, Stutgart
14. Krogh E, Wan P (1990) Topics in Current Chemistry 156: 93
15. a) Mizuno K, Pac C and Sakurai H (1975) J. Chem. Soc. Chem. Commun. 553. b) Yasuda M, Pac C, Sakurai H (1981) J. Chem. Soc. Perkin Trans. 1746. c) Bunce NJ, Bergsma JP, Schmidt (1981) J. Chem. Soc. Perkin Trans. II 713
16. Yasuda M, Yamashita T, Shima K (1987) J. Org. Chem. 52: 753
17. Yasuda M, Pac C, Sakurai H (1981) J. Org. Chem. 46: 788
18. Pandey G, Krishna A, Rao JM (1986) Tetrahedron Lett. 27: 4075
19. Krishna A (1988) Ph.D. Thesis, Osmania University Hyderabad, India.
20. Pandey G, Krishna A (1988) J. Org. Chem. 53: 2364
21. Pandey G, Krishna A, Bhalerao UT (1989) Tetrahedron Lett. 30: 1867
22. Pandey G, Girija K, Karthikeyan M (1992) Unpublished results
23. Pandey G, Sridhar M, Bhalerao UT (1990) Tetrahedron Lett. 31: 5373
24. Pandey G, Kumaraswamy G (1988) Tetrahedron Lett. 29: 4153
25. Albini A, Sulpizio A (1988) Ref. 2 Part C, p. 88
26. Santamaria J, Jroundi R, Rigaudy J (1989) Tetrahedron Lett. 30: 4677
27. Fukuzumi S, Kuroda S, Tanaka T (1987) J. Chem. Soc. Chem. Commun. 120
28. Pandey G, Krishna A (1988) Synthetic Commun. 18: 2309
29. Nishida A, Oishi S, Yonemitsu O (1989) Chem. Pharm. Bull. 37: 2266
30. Rossi RA, Bunnett JF (1973) J. Org. Chem. 38: 1407
31. Lablache-Combier A, Ref. 2, part C, p. 134.
32. a) Barltrop JA (1973) Pure Appl. Chem. 33: 179. b) Rao VR, Ramakrishnan V (1971) J. Chem. Soc. Chem. Commun. 971
33. Tazuke S, Kazama S, Kitamura N (1986) J. Org. Chem. 51: 4548
34. Mattay J, Trampe G, Runsink J (1988) Chem. Ber. 121: 1991
35. Lopez L (1990) Topics in Current Chemistry 156: 117
36. a) Neunteufel RA, Arnold DR (1973) J. Am. Chem. Soc. 95: 4080. b) Bauld NL, Bellville DJ, Harirchian B, Lorenz KT, Pabon RA Jr., Reynolds DW, Wirth DD, Chiou HS, Marsh BK (1987) Acc. Chem. Res. 20: 37
37. a) Mattay J, Vondenhof M, Denig R (1989) Chem. Ber. 22: 951. b) Albini A, Spreti S (1986) J. Chem. Soc. Chem. Commun. 1426
38. Kojima M, Sakuragi H, Tokumaru K (1981) Tetrahedron Lett. 22: 2889
39. a) Al-Ekabi H, De Mayo P (1986) Tetrahedron 42: 6277. b) Al-Ekabi H, De Mayo P (1987) J. Org. Chem. 52: 4756
40. Mizuno K, Kagano H, Otsuji Y (1983) Tetrahedron Lett. 24: 3849
41. Mizuno K, Ueda H, Otsuji Y (1981) Chem. Lett. 1237
42. Mizuno K, Ishii M, Otsuji Y (1981) J. Am. Chem. Soc. 103: 5570

43. Heinrich N, Koch W, Morrow JC, Schwarz H (1988) J. Am. Chem. Soc. 110: 6332
44. a) Maroulis AJ, Arnold DR (1979) J. Chem. Soc. Chem. Commun. 351. b) Arnold DR, Borg RM, Albini A (1981) J. Chem. Soc. Chem. Commun. 138
45. Majima T, Pac C, Nakasone A, Sakurai H (1981) J. Am. Chem. Soc. 103: 4499
46. Mizuno K, Kaji R, Okada H, Otsuji Y (1978) J. Chem. Soc. Chem. Commun. 594
47. Mattay J, Gersdorf J, Mertes J (1985) J. Chem. Soc. Chem. Commun. 1088
48. a) Mlcoch J, Steckhan E (1985) Angew. Chem. Int. Ed. (Eng). 24: 412. b) Mlcoch J, Steckhan E (1987) Tetrahedron Lett. 28: 1081
49. Bellville DJ, Wirth DD, Bauld NL (1981) J. Am. Chem. Soc. 103: 718
50. Gieseler A, Steckhan E, Wiest O, Knoch F (1991) J. Org. Chem. 56: 1405
51. Harirchian B, Bauld NL (1989) J. Am. Chem. Soc. 111: 1826
52. Bauld NL, Harirchian B, Reynolds DW, White JC (1988) J. Am. Chem. Soc. 110: 8111
53. a) Calhoun GC, Schuster GB (1984) J. Am. Chem. Soc. 106: 6870. b) Calhoun GC, Schuster GB (1986) Tetrahedron Lett. 27: 911. c) Calhoun GC, Schuster GB (1986) J. Am. Chem. Soc. 108: 8021. d) Akbulut N, Schuster GB (1988) Tetrahedron Lett. 29: 5125. e) Hartsough D, Schuster GB (1989) J. Org. Chem. 54: 3. f) Akbulut N, Hartsough D, Kim JI, Schuster GB (1989) J. Org. Chem. 54: 2549
54. Wolfle I, Chan S, Schuster GB (1991) J. Org. Chem. 56: 7313
55. Kim J-I, Schuster GB (1990) J. Am. Chem. Soc. 112: 9635
56. Jones CR, Allman BJ, Morring A, Spahic B (1983) J. Am. Chem. Soc. 105: 652
57. Mattes SL, Farid S (1986) J. Am. Chem. Soc. 108: 7356
58. a) Mizuno K, Hashizume T, Otsuji Y (1983) J. Chem. Soc. Chem. Commun. 772. b) Mizuno K, Nakanishi K, Otsuji Y (1991) J. Chem. Soc. Chem. Commun. 90
59. Leismann H, Mattay J, Scharf HD (1984) J. Am. Chem. Soc. 106: 3985
60. a) Shigemitsu Y, Arnold DR (1975) J. Chem. Soc. Chem. Commun. 407. b) Maroulis AJ, Arnold DR (1979) Synthesis 819
61. Maroulis AJ, Shigemitsu Y, Arnold DR (1978) J. Am. Chem. Soc. 100: 535
62. Gassman PG, Bottorff KJ (1987) J. Am. Chem. Soc. 109: 7547
63. Gassman PG, De Silva SA (1991) J. Am. Chem. Soc. 113: 9870
64. Tazuke S, Kitamura N (1977) J. Chem. Soc. Chem. Commun. 515
65. a) Gassman PG, Bottorff KJ (1987) Tetrahedron Lett. 28: 5449. b) Gassman PG, Olson KD (1983) Tetrahedron Lett. 24: 19
66. Gassman PG (1988) Ref. 2 Part C: 70
67. Muller F, Mattay J (1991) Angew. Chem. Int. Ed. (Eng). 30: 1336
68. Muller F, Mattay J (1992) Angew. Chem. Int. Ed. (Eng). 31: 209
69. Wong PC, Arnold DR (1979) Tetrahedron Lett. 2101
70. Albini A, Arnold DR (1978) Can. J. Chem. 56: 2985
71. Roth HD (1987) Acc. Chem. Res. 20: 343
72. a) Arnold DR, Humphreys RWR (1979) J. Am. Chem. Soc. 101: 2743. b) Gassman PG, Olson KD, Walter L, Yamaguchi R (1981) J. Am. Chem. Soc. 103: 4977. c) Ipaktschi J (1970) Tetrahedron Lett. 3183
73. Rao VR, Hixon SS (1979) J. Am. Chem. Soc. 101: 6458
74. Dinnocenzo JP, Todd WP, Simpson TR, Gould IR (1990) J. Am. Chem. Soc. 112: 2462
75. Ichinose N, Mizuno K, Yoshida K, Otsuji Y (1988) Chem. Lett. 723
76. Tomioka H, Kobayashi D, Hashimoto A, Murata S (1989) Tetrahedron Lett. 30: 4685
77. Gassman PG, Burns SJ (1988) J. Org. Chem. 53: 5576
78. Clawson P, Lunn PM, Whiting DA (1984) J. Chem. Soc. Chem. Commun. 134
79. a) Kaupp G, Laarhoven WH (1976) Tetrahedron Lett. 941. b) Kadata S, Tsubono K, Makino K, Takeshita M, Kikuchi T (1987) Tetrahedron Lett. 28: 2857. c) Pac C (1986) Pure Appl. Chem. 58: 1249. d) Pac C, Fukunaga T, Go-An Y, Sakae T, Yanagida S (1987) Photochem. Photobiol. (A). 41: 37
80. Harm W (1980) In: Biological effects of ultraviolet radiation. Cambridge University Press, London; Beddard G (1982) In: Coyle JD, Hill RR, Roberts DR (ed) Light, chemical change and life. The Open University Press, Walton Hall, Milton Keynes, UK, p 178
81. Hartman RF, Camp JRV, Rose SD (1987) J. Org. Chem. 52: 2684
82. Masuhara H, Mataga M (1981) Acc. Chem. Res. 14: 312
83. a) Bellas M, Bryce-Smith DB, Clarke MT, Gilbert A, Klunkin G, Krestonosich S, Manning C, Wilson S (1977) J. Chem. Soc. Perkin Trans I. 2571. b) Yang NC, Shold DM, Kim B (1976) J. Am. Chem. Soc. 98: 6587

84. a) Lewis FD, Ho T-I, Simpson JT (1982) J. Am. Chem. Soc. 104: 1924. b) Hub W, Schneider S, Dorr F, Simpson JT, Oxman JD, Lewis FD (1982) J. Am. Chem. Soc. 104: 2044
85. Lewis FD (1986) Acc. Chem. Res. 19: 401; Lewis FD, Ho T-I, Simpson JT (1981) J. Org. Chem. 46: 1077
86. Ohashi M, Kudo H, Yamada S (1979) J. Am. Chem. Soc. 101: 2201
87. Pandey G (1992) Syn Lett. 546
88. Smith JRL, Masheder D (1976) J. Chem. Soc. Perkin Trans. II: 47 and references cited therein
89. Griller D, Lossing FP (1981) J. Am. Chem. Soc. 103: 1586
90. Pandey G, Kumaraswamy G, Krishna A (1987) Tetrahedron Lett. 28: 2649
91. Pandey G, Kumaraswamy G, Reddy PY (1992) Tetrahedron 48: 8295
92. Pandey G, Sudha Rani K (1988) Tetrahedron Lett. 29: 4157
93. Pandey G, Sudha Rani K, Bhalerao UT (1990) Tetrahedron Lett. 31: 1199
94. Pandey G, Reddy PY, Bhalerao UT (1991) Tetrahedron Lett. 32: 5147
95. Santamaria J, Kaddachi MT, Rigaudy J (1990) Tetrahedron Lett. 31: 4735
96. Santamaria J, Kaddachi MT(1991) Syn. Lett. 739
97. a) Sundberg RJ, Hunt PJ, Desos P, Gadamasetti KG (1991) J. Org. Chem. 56: 1689. b) Sundberg RJ, Desos P, Gadamasetti KG, Sabat M (1991) Tetrahedron Lett. 32: 3035
98. Pandey G, Sudha Rani K, Lakshmaiah G (1992) Tetrahedron Lett. 33: 5107
99. Pandey G, Kumaraswamy G, Bhalerao UT (1989) Tetrahedron Lett. 30: 6059
100. Beckwith ALJ (1981) Tetrahedron 37: 3073
101. Kumaraswamy G, (1991) Ph.D. Thesis, Osmania University, Hyderabad, 1991
102. Pandey G, Reddy GD (1992) Tetrahedron Lett. 33: 6533
103. Jeon YT, Lee C-P, Mariano PS (1991) J. Am. Chem. Soc. 113: 8847
104. Pandey G, Lakshmaiah G, Kumaraswamy G (1992) J. Chem. Soc. Chem. Commun. 1313
105. Saeva FD (1990) Topics in Current Chemistry 156: 59
106. a) Dinnocenzo JP, Farid S, Goodman JL, Gould IR, Todd WP, Mattes SL (1989) J. Am. Chem. Soc. 111: 8973. b) Wayner DDM, Griller D (1989) In: Liebman JF, Greenberg A (eds) From atoms to polymers. VCH Publishers, New York, chapter 3
107. Eaton DF (1981) J. Am. Chem. Soc. 103: 7235
108. Arnold DR, Maroulis AJ (1976) J. Am. Chem. Soc. 98: 5931
109. Sulpizio A, Albini A, d'Alessandro N, Fasani E, Pietra S (1989) J. Am. Chem. Soc. 111: 5773
110. Sankararaman S, Perrier S, Kochi JK (1989) J. Am. Chem. Soc. 111: 6448
111. Ci X, Whitten DG (1988) Ref. 2 Part C 553
112. a) Ci X, Kellett MA, Whitten DG (1991) J. Am. Chem. Soc. 113: 3893. b) Bergmark WR, Whitten DG (1990) J. Am. Chem. Soc. 112: 4042. c) Ci X, Whitten DG (1989) J. Am. Chem. Soc. 111: 3459. d) Ci X, Whitten DG (1987) J. Am. Chem. Soc. 109: 7215
113. a) Brown RS, Eaton DF, Hosomi A, Trayler TG, Wright JM (1974) J. Organomet. Chem. 66: 249. b) Hosomi A, Trayler TG (1975) J. Am. Chem. Soc. 97: 3682
114. Pandey G, Soma Sekhar BBV, Bhalerao UT (1990) J. Am. Chem. Soc. 112: 5650
115. Mizuno K, Yasueda M, Otsuji Y (1988) Chem. Lett. 229
116. a) Pandey G, Rao VJ, Bhalerao UT (1989) J. Chem. Soc. Chem. Commun. 416. b) Pandey G, Soma Sekhar BBV (1992) J. Org. Chem. 57: 4019
117. Nakadaira Y, Sekiguchi A, Funada Y, Sakurai H (1991) Chem. Lett. 327
118. Gassman PG, Bottorff KJ (1988) J. Org. Chem. 53: 1097
119. Kavarnos GJ, Turro NJ (1986) Chem. Rev. 86: 401
120. Portella C, Deshayes H, Pete JP, Scholler D (1984) Tetrahedron 40: 3635
121. Okada K, Okamoto K, Oda M (1988) J. Am. Chem. Soc. 110: 8736
122. Okada K, Okamoto K, Morita N, Oda M, Okubo K (1991) J. Am. Chem. Soc. 113: 9401
123. Hamada T, Nishida A, Yonemitsu O (1986) J. Am. Chem. Soc. 108: 140
124. Masnovi J, Koholic DJ, Berki RJ, Binkley RW (1987) J. Am. Chem. Soc. 109: 2851
125. Nishida A, Hamada T, Yonemitsu O (1988) J. Org. Chem. 53: 3386
126. Art JF, Kestemont JP, Soumillion JP (1991) Tetrahedron Lett. 32: 1425
127. Saito I, Ikehira H, Kasatani R, Watanabe M, Matsuura T (1986) J. Am. Chem. Soc. 108: 3115
128. Chatterjee S, Gottschalk P, Davis PD, Schuster GB (1988) J. Am. Chem. Soc. 110: 2326
129. Saeva FD, Breslin DT, Luss HR (1991) J. Am. Chem. Soc. 113: 5333
130. Binkley RW, Flechtner W (1984) Ref. 11, p 375
131. Arnold DR, Wong PC, Maroulis AJ, Cameron TS (1980) PureAppl. Chem. 52: 2609
132. Lewis FD, Petisce JR (1986) Tetrahedron 42: 6207
133. Mizuno K, Ikeda M, Otsuji Y (1985) Tetrahedron Lett. 26: 461

134. Mizuno K, Terasaka K, Ikeda M, Otsuji Y (1985) Tetrahedron Lett. 26: 5819
135. Mattay J (1985) Tetrahedron 41: 2393 and 2405
136. Cantrell TS (1977) J. Org. Chem. 42: 4238
137. Mattay J, Runsink J, Heckendorn R, Winkler T (1987) Tetrahedron 43: 5781
138. a) Mattay J, Runsink J, Rumbach T, Ly C, Gersdorf J (1985) J. Am. Chem. Soc. 107: 2557. b) Mattay J, Runsink J, Gersdorf J, Rumbach T, Ly C (1985) Helv. Chem. Acta. 69: 442
139. Charlton JL, Liao CC, De Mayo P (1971) J. Am. Chem. Soc. 93: 2463
140. a) Dopp D, Heufer J (1986) J. Photochem. 32: 243. b) Dopp D, Heufer J (1982) Tetrahedron Lett. 23: 1553. c) Dopp D, Muller D (1979) Recl. Trav. Chim. 98: 297
141. Mizuno K, Terasaka K, Yasueda M, Otsuji Y (1988) Chem. Lett. 145
142. Hasegawa E, Brumfield MA, Mariano PS, Yoon U-C (1988) J. Org. Chem. 53: 5435
143. Kyushin S, Ehara Y, Nakadaira Y. Ohashi M (1989) J. Chem. Chem. Commun. 279
144. Lan JY, Schuster GB (1985) J. Am. Chem. Soc. 107: 6710
145. a) Lewis FD, Reddy GD, Schneider S, cahr M (1989) J. Am. Chem. Soc. 111: 6465. b) Lewis FD, Reddy GD (1990) Tetrahedron Lett. 31: 5293
146. Pac C, Nakasone A, Sakurai H (1977) J. Am. Chem. Soc. 99: 5806
147. Arnold DR, Du X (1989) J. Am. Chem. Soc. 111: 7666
148. Mizuno K, Nakanishi K, Otsuji Y (1988) Chem. Lett. 1833
149. Xu J, He J, Qian Y (1991) J. Chem. Soc. Chem. Commun. 714
150. Masnovi JM, Kochi JK, Helinski EF, Rentzepis PM (1986) J. Am. Chem. Soc. 108: 1126
151. Sankararaman S, Kochi JK (1986) Recl. Trav. Chim. 105: 278
152. Sundberg RJ (1983) In: Padwa A (ed) Organic photochemistry. Marcel Dekker, New York, vol 6, p 121
153. Dolby LJ, Nelson SJ, Senkovich D (1972) J. Org. Chem. 37: 3691
154. Mascal M, Moody CJ (1988) J. Chem. Soc. Chem. Commun. 587
155. Mascal M, Moody CJ (1988) J. Chem. Soc. Chem. Commun. 589
156. a) Mariano PS (1983) Tetrahedron 39: 3845. b) Mariano PS (1983) Acc. Chem. Res. 16: 130
157. a) Mariano PS, Stavinoha JL, Pepe G, Meyer EF Jr (1978) J. Am. Chem. Soc. 100: 7114. b) Stavinoha JL, Mariano PS (1981) J. Am. Chem. Soc. 103: 3136. c) Mariano PS, Stavinoha J, Bay E (1981) Tetrahedron 37: 3385
158. a) Mariano PS, Stavinoha JL, Swanson R (1977) J. Am. Chem. Soc. 99: 6781. b) Stavinoha JL, Mariano PS, Leone-Bay A, Swanson R, Bracken C (1981) J. Am. Chem. Soc. 103: 3148.
159. a) Ohga K, Mariano PS (1982) J. Am. Chem. Soc. 104: 617. b) Ohga K, Yoon UC, Mariano PS (1984) J. Org. Chem. 49: 213
160. a) Borg RM, Heuckeroth RO, Lan AJY, Quillen SL, Mariano PS (1987) J. Am. Chem. Soc. 109: 2728. b) Lan AJY, Heuckeroth RO, Mariano PS (1987) J. Am. Chem. Soc. 109: 2738
161. Dai-Ho G, Mariano PS (1987) J. Org. Chem. 52: 704
162. Ahmed-Schofield R, Mariano PS (1987) J. Org. Chem. 52: 1478
163. Cho I-S, Mariano PS (1988) J. Org. Chem. 53: 1590
164. Haddaway K, Somekawa K, Fleming P, Tossel JA, Mariano PS (1987) J. Org. Chem. 52: 4239
165. Turro NJ (1978) In: Modern molecular photochemistry. Benjamin/Cummings. Menlo Park CA, chap. 11
166. Jones G (1981) vol 5: Ref. 5, p. 132
167. Mattay J, Gersdorf J, Buchkremer K (1987) Chem. Ber. 120: 307
168. Wagner PJ (1976) Topics in Current Chemistry 66: 1
169. Carless HAJ (1984) Ref. 11, p. 425
170. Cohen SG, Parola A, Parsons GH Jr. (1973) Chem. Rev. 73: 141
171. Simon JD, Peters KS (1983) J. Am. Chem. Soc. 105: 4875
172. Singh SP, Stenberg VI, Parmar SS (1980) Chem. Rev. 80: 269
173. Yoon UC, Kim JU, Hasegawa E, Mariano PS (1987) J. Am. Chem. Soc. 109: 4421
174. Xu W, Zhang X-M, Mariano PS (1991) J. Am. Chem. Soc. 113: 8863
175. Kraus GA, Chen L (1991) Tetrahedron Lett. 32: 7151
176. Takuwa A, Nishigachi Y, Iwamoto H (1991) Chem. Lett. 1013
177. a) Belotti D, Cossy J, Pete JP, Portella C (1986) J. Org. Chem. 51: 4196. b) Belloti D, Cossy J, Pete JP, Portella C (1985) Tetrahedron Lett. 26: 4591
178. Cossy J, Belotti D, Pete JP (1987) Tetrahedron Lett. 28: 4545
179. Cossy J, Belotti D (1988) Tetrahedron Lett. 29: 6113
180. Cossy J (1989) Tetrahedron Lett. 30: 4113
181. Cossy J, Leblanc C (1991) Tetrahedron Lett. 32: 3051

182. Cossy J, Aclinou P, Bellosta V, Furet N, Baranne-Lafont J, Sparfel F, Souchaud C (1991) Tetrahedron Lett. 32: 1315
183. a) Guttenplan JB, Cohen SG (1969) J. Chem. Soc. Chem. Commun. 247 and J. Org. Chem. 1973 38: 2001. b) Ando W, Suzuki J, Migita T (1971) Bull. Chem. Soc. Jpn. 44: 1987
184. Mazzochi PH (1981) vol 5: Ref. 5, p. 421
185. Coyle JD (1986) In: Photochemistry in organic synthesis. The Royal Society of Chemistry, London, p 301; Coyle JD (1984) Ref. 11, p. 259
186. Griffiths J, Hawkins C (1973) J. Chem. Soc. Chem. Commun. 111; Maruyama K, Otsuki T, Tai S (1985) J. Org. Chem. 50: 52
187. Maruyama K, Imahori H, Osuka A, Takuwa A, Tagawa H (1986) Chem. Lett. 1719

Fluorescent PET (Photoinduced Electron Transfer) Sensors

Richard A. Bissell, A. Prasanna de Silva, H.Q. Nimal Gunaratne,
P.L. Mark Lynch, Glenn E.M. Maguire, Colin P. McCoy and
K.R.A. Samankumara Sandanayake

School of Chemistry, Queen's University, Belfast BT9 5AG, Northern Ireland

Table of Contents

Topics in Current Chemistry, Vol. 168
© Springer-Verlag Berlin Hiedelberg 1993

Fluorescent PET (photoinduced electron transfer) sensors are considered to be those molecular systems where the binding of ions and other species leads to the perturbation of the competition between the de-excitation pathways of fluorescence and electron transfer. The early developments in this field are traced and the design logic of these sensors is detailed. A variety of examples drawn from different areas of chemistry are classified according to the 'fluorophore-spacer-receptor' format and their photophysical behaviour is rationalized in terms of fluorescent PET sensor principles. Cases are pointed out where such experimental data are unavailable but desirable. During these discussions, the relevance of twisted fluorophore-receptor systems and the contrast with integrated fluorophore-receptor systems is noted. The utility of the fluorescence 'on-off' phenomenon in these PET sensors for the area of molecular photoionic devices is pointed out.

1 Historical Beginnings

The collision of opposites is a fertile approach to new discoveries. The interaction of electron donors and acceptors in a photon field gave rise to not one but three important phenomena in photochemistry—charge transfer absorption [1], exciplexes [2] and photoinduced electron transfer [3]. First in Stuttgart, then in Göttingen, Germany, Weller established the groundwork concerning the last two phenomena [4]. In the present context, it is important that fluorescence was the first experimental key to these discoveries. Exciplex research began when a red-shifted emission devoid of vibrational fine structure appeared in the fluorescence spectra of aromatic hydrocarbons, e.g. (1) mixed with amines, e.g. (2) in relatively apolar solvents, at the expense of the single-component emission. The solvent-dependent fluorescence studies established that exciplexes were very dipolar species and that such charge separation maximized in solvents of high polarity. Research in light-driven electrochemistry has gone from strength to strength since then [5].

As the story unfolds, the next scene takes us to Australia, when Selinger in Canberra investigated (3) [6]. His previous association with Weller may have had a bearing on the selection of this structure. In fact, (3) was the subject of several studies with regard to intramolecular exciplex formation and/or electron transfer [7]. (3) and a related molecule (4) showed significant pH dependence of their fluorescence spectra in an aqueous micellar medium. Selinger's analysis hinged on the removal of exciplex and electron transfer phenomena upon protonation of the nitrogen centre of the amine group. The power of the phenomenon can be attributed to two features. First, the intramolecular nature of the interaction coupled with the favourable characteristics of the three-methylene chain spacer aided the strong quenching of the anthracenic fluore-scence. Second, a single protonation nullifies this quenching because of the identification of the essential role of the amine nitrogen centre. Selinger's indication that the pH-dependent excited state properties of (3) were controlled by its ground state pK_a value was to prove prophetic.

Next stop – Gunma, Japan. Shizuka made a detailed photophysical exam-ination of several phenyl alkylamines [8] (5). For our present purpose, his most important discovery was their pH-dependent fluorescence and the associated equivalence of excited and ground state pK_a values. Also, the controlling nature of the spacer upon the magnitude of the phenomenon was clarified. This meant that the large body of information concerning chain cyclization [9] could now be sourced for the efficient design of pH-sensitive fluorescent systems.

As with most discoveries, these ideas then passed through a lull before take-off. Then the Chicago-New York axis of Yang and Turro in the USA investi-gated the first example (6) in the host-guest mould [10]. Their focus was on the demonstration of exciplex emission of (6) in aqueous solution due to protection by an encapsulating β-cyclodextrin. Ever since Weller's experiments, exciplex emissions had been observed to fall off rapidly in intensity as solvent polarity was increased [4]. A formally related example due to Tazuke from 1982 must, however, be pointed out, where a hydrophobic polymer microdomain in mixed and neat aqueous solution allows exciplex emission from pyrene and dimethyl-aniline pendants [11]. The importance of (6) in the present context stems from the correlation of pH-dependent emissions from the naphthalene moiety and from the exciplex. Co-occurrence of externally switchable photophysics and supramolecular phenomena would later become common in the research literature [12].

Soon afterwards, Verhoewen in Amsterdam, The Netherlands, demonstrated the switching capabilities of (7) [13] – the beginnings of the possibilities in the molecular electronics area. Selective protonation neutralized a charge transfer emission process originating from an aliphatic amine and also involving a remote aromatic amine. The switching of optically observable phenomena involving several different molecular sites by means of an externally controllable ionic input was a special novelty of this work.

The parallel nucleation of supramolecular chemistry [14] in Wilmington and Los Angeles, U.S.A. and Strasbourg, France, in the laboratories of

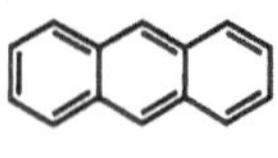

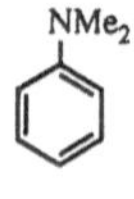

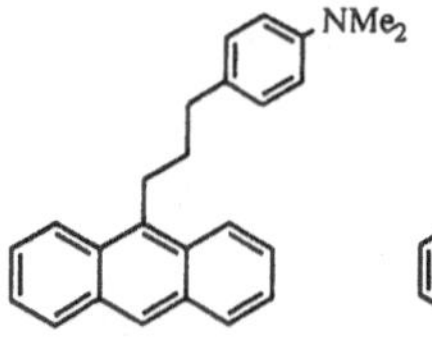

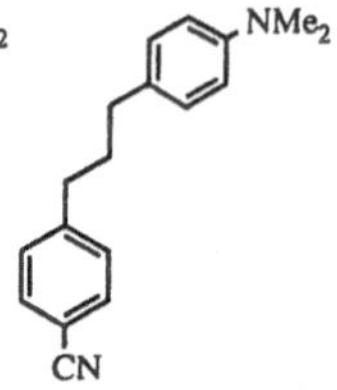

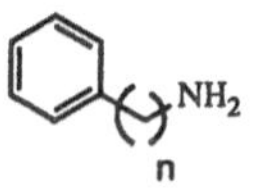

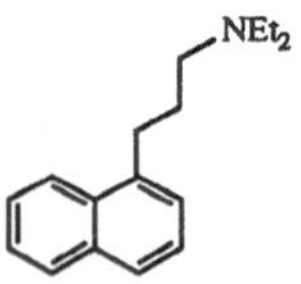

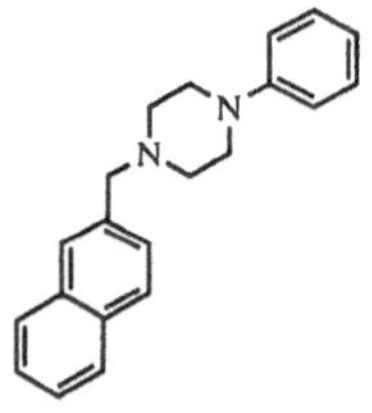

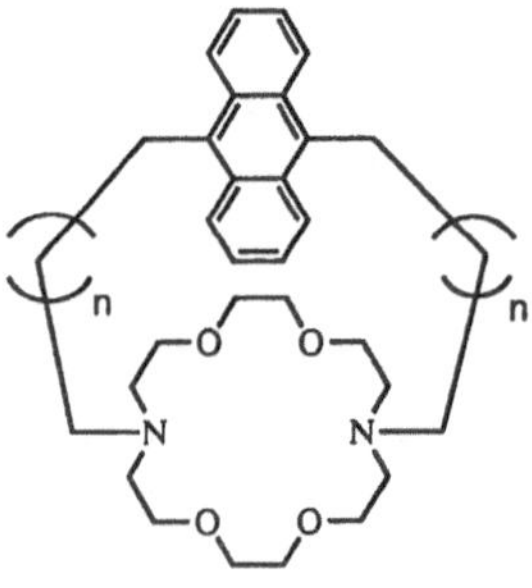

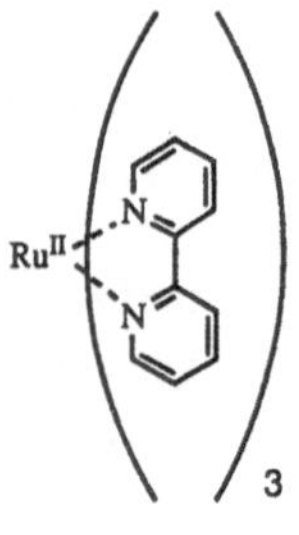

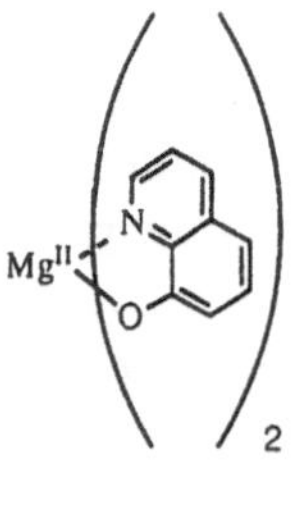

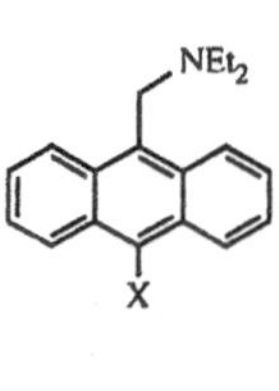

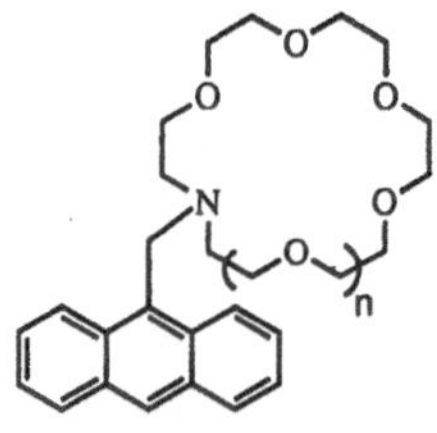

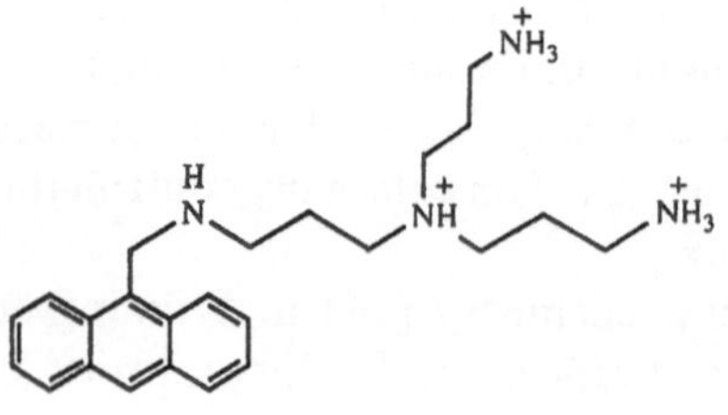

Pedersen, Cram and Lehn respectively gives us the other thread to the present story. The first example of fluorescence enhancement caused by a supramolecular interaction between a metal ion and a neutral receptor (**8**) was reported by the Bordeaux-Strasbourg collaboration of Bouas-Laurent, Desvergne and Lehn [15]. An older example of luminescence arising from a strong coordinate interaction between a neutral ligand and a metal ion is (**9**) [16]. Anionic ligands,

(1)

(2)

(3)

(4)

(5)

(6)

(7)

(8)

(9)

(10)

(11)a; X = Cl
b; X = CH_2NEt_2

(12), n=0,1

of course, have a long history of forming fluorescent adducts e.g. (10) with non-transition metal ions [17]. (9) and (10) notwithstanding, (8) remains the first reported molecule where the competition between photoinduced electron transfer and fluorescence was biased by an ion other than protons.

The generality of this design logic and its potential for sensing purposes began to emerge when two reports from Colombo, Sri Lanka demonstrated that protons [18] and alkali cations [19] could be sensitively monitored by the fluorescence output of simple and related molecules (11) and (12) respectively. Other examples quickly arrived, but non-metal based anions remained the missing link in the general scheme until 1988 when Czarnik at Columbus, U.S.A. brought phosphate and others into the fold with (13) [20].

2 Introductory Comments on the Design Logic

When compared with most of the material in this series, the theoretical view taken in this article would appear to be rather naive. However, the simple theory is sufficient to understand, and even predict, the guest responsive fluorescence behaviour of many sensor molecules as well as permitting considerable conceptual expansion. These points are examined separately in a companion review [21]. Those readers requiring more sophisticated and general treatments will find them in other chapters in this series [3g] and elsewhere [3f].

A guest responsive fluorescent sensor must necessarily possess a guest binding site (receptor) and a photon-interaction site (fluorophore) [22]. In the present design, these two sites are separately identifiable within a supramolecular structure because of an intervening spacer unit [21] (Fig. 1). The symbolism for the receptor module does not necessarily imply a macrocycle. It is taken to be a generic representation for binding sites of various forms. Thus single point binders [23], clefts [24], hydrogen bond donor/acceptor arrays [25] including nucleic acid strands [26], charge transfer donor/acceptors [27] and other acyclic receptors [28] are included in the symbol besides macrocycles. The design in Fig. 1 means that the separate sets of guest-binding and photophysical properties of the terminal modules serve as quantitative predictors for the sensory properties of the three-module super-molecule. Additivity of this kind while useful for predictive purposes must not apply to all interactions. Sensing implies that the act of guest binding at the receptor site is communicated across the spacer to cause a change of photo properties of the fluorophore site, i.e. the isolation of the spacer must break down with respect to at least one trans-spacer

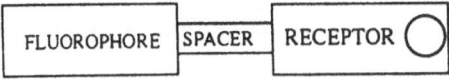

Fig. 1. The assembly of functional components in a fluorescent PET sensor

interaction. PET is the sufficiently long-range interaction [29] that leads to the sensory behaviour, by entering into competition with fluorescence for the deactivation of the excited fluorophore module. Let us close this short section by noting that the sensor action results from the guest complexation by the three-module supermolecule. Attention is therefore drawn to the two layers of supramolecular construction that continually pervade the thinking in this area.

3 An Illustration of Sensor Design and Evaluation

The logical approach can be best presented by following a single example from conception through to completion. The goal was to construct a fluorescent sensor molecule for sodium or potassium ions with maximum selectivity commensurate with convenient synthesis. In terms of a functional analysis, this means that the sensor would contain a selective receptor for alkali cations and a fluorophore with convenient excitation/emission wavelengths. An all-oxygen crown ether would be a suitable receptor with some degree of discrimination between the different alkali cations with nearly complete rejection of protons [30]. Previous experience with azacrown ether sensors had illustrated the overpowering extent of proton interference that is possible when Brønsted basic sites are available [19]. The alkali cation selectivity profile of the all-oxygen crown ethers could also be biased in a controlled manner by altering their ring size. However, another important consideration must be made at this stage for the receptor to be a functional module within the PET sensor supermolecule. That concerns the electrochemical activity. Aliphatic all-oxygen crown ethers would have prohibitively high oxidation- and even higher reduction-potentials. In order to bring the receptor electroactivity into an accessible range, we can employ benzo-annelated all-oxygen crown ethers, while preserving the positive features discussed above. 1,2-dimethoxybenzene would be an approximate electrochemical model for benzocrown ethers in the absence of such data on the latter. The model has an oxidation potential of $+ 1.45$ V (vs. SCE, acetonitrile) $(E_{ox.receptor})$ [31].

 Shifting our attention to the fluorophore, we consider anthracene derivatives because of our prior experience with them [18] and (more importantly) because of the wealth of data available on them in both the electrochemical and photophysical spheres. Even at an intuitive level, anthracenes are a good starting point because such rigid, well-delocalized systems are usually efficient fluorophores [32]. It is also important that the benzocrown ether receptor is transparent to the communication wavelengths of the fluorophore in excitation and in emission. Anthracene derivatives satisfy this criterion very well in this instance. Further progress can be made via three avenues and the most general path is explored first.

a) If PET between fluorophore and receptor modules is to control the sensory action of the supermolecule, the photo energy contained within the electronically excited fluorophore must be (in the first instance) thermodynamically capable of driving the electron transfer process [33]. By scanning tables of singlet energies (E_S) [34] and reduction potentials ($E_{red.fluorophore}$) [35], we can conclude that 9-cyanoanthracene (with values of 3.04 eV and -1.58 V respectively) can be profitably paired with benzocrown ethers. While the solvent conditions under which these redox potentials have been obtained will frequently differ from the eventual operating regime of the assembled sensor and can therefore jeopardize the validity of such designs, this protocol still remains a valuable starting point. The previous deduction can be schematized with a frontier orbital energy diagram for the PET pair (Fig. 2).

$$\Delta G_{ET} = -E_S - E_{red.fluorophore} + E_{ox.receptor} - e^2/\varepsilon r \tag{1}$$

The Weller Equation, Eq. (1) [4, 33] allows the quantitation of the thermodynamic driving force for PET (ΔG_{ET}) as -2.53 kcal mol^{-1}, given that the attractive potential between the radical ion pair $-e^2/\varepsilon r$ is approximately -0.1 eV under the operating conditions. Such a small exergonic driving force is preferred (over a large exergonicity) since it can be easily switched over to an endergonic condition by a suitable charge perturbation. Such a perturbation occurs when, driven by mass action, an alkali cation lodges in the crown ether cavity. This should cause an increase in the oxidation potential of the (cation-bound) benzo crown ether. Related evidence is available from photochemical experiments [36]. Direct electrochemical evidence is obviously preferable and a growing body of voltammetric data on alkali cation-perturbed macrocyclic compounds support the above assumption [37]. However, the magnitude of oxidation potential increase in simple benzocrown ethers sadly remain unquantitated. Nevertheless, a schematic frontier orbital energy diagram for a cation-bound situation can be constructed (Fig. 3). The ΔG_{ET} value for the present condition is expected to be substantially endergonic. The cation-free (Fig. 2) and cation-bound (Fig. 3) situations can be qualitatively different in that a visually powerful stimulus, i.e. fluorescence is maintained in an 'off' and 'on' condition respectively. Herein lies the strength of the fluoresent PET sensor

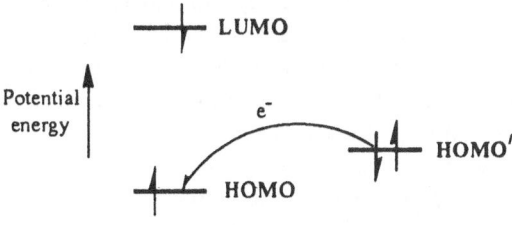

Fig. 2. Frontier orbital energy diagram for a fluorophore–receptor pair where PET is feasible

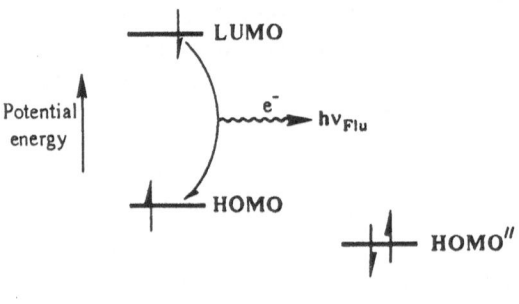

lowest excited singlet   cation- bound
state of fluorophore   receptor

Fig. 3. Frontier orbital energy diagram for the fluorophore–receptor pair when the receptor is occupied by a cation

logic. For the logic to succeed, however, the fluorescence rate constant must be orders of magnitude slower or faster than the PET rate constant when alkali cation – free or – bound respectively. Before we leave avenue a), we must emphasise that it embodies the heart of the fluorescent PET sensor logic and, hence, it is indispensable for the thoughtful sensor designer. Also, as indicated above, this avenue is quite general because it employs established theory and basic molecular data which are commonly available.

b) The second avenue heads to the library. Here we draw directly on any existing knowledge with much less deductive reasoning than in a). The question we ask ourselves is the following: does 1,2-dimethoxy benzene efficiently quench the fluorescence of a suitable structure? If the answer is favourable, it is also worth inquiring for any evidence of an electron transfer mechanism for the quenching act. Indeed, the work of Foote [38] suggests that 9-cyano anthracene is such a structure. During such a search it would be natural to come across related PET pairs such as 9-cyano anthracene/2-methoxy naphthalene [39] and acridine orange/1,2-methoxy benzene [40]. In the mind of a receptive designer, the latter pair could spark off an approach to an alkali cation sensor with communication wavelengths well into the visible region.

c) The final avenue takes us into do-it-yourself territory. The most convincing selection of fluorophore might come from this personal approach, simply because we can *see* the result of fluorophore-receptor interaction. A simple laboratory experiment of adding (commercially available) dibenzo-18-crown-6-ether into a dilute solution of 9-cyano anthracene under ultraviolet illumination will display the fluorescence quenching phenomenon. Even 1,2-dimethoxy benzene would give this result in no uncertain terms. However, the former experiment can be followed up by adding anhydrous potassium acetate to produce a visibly noticeable increase of sky blue fluorescence. When we first observed this double result in benzene/2-propanol, this solvent composition being necessary to maintain the organic and inorganic species in solution, it was clear that an operational fluorescent sensor for alkali cations was only a synthesis away.

Sequential examination of approaches a), b) and c) in chronological order is the most rational and leads us to consider the spacer module in the sensory supermolecule. The choice of the optimal spacer structure is governed by the following features: (i) favourable electron transfer kinetics – short spacers would be preferred [29], though longer structures with rigid σ bond frameworks with W-arrangements also permit fast rates [29c, 41], (ii) Synthetic accessibility – methylene spacers with aromatic termini imply covalent component assembly via a reactive benzylic functionality, (iii) modular behaviour within the super-molecule – single methylene spacers would prevent the folding of the sensor into configurations containing direct interactions between fluorophore and receptor [8, 9, 42]. So the target sensor supermolecule is (14) [43]. Synthesis of (14) is aided by Seshadri's previous preparation of (15) [44]. Similar preparation of (16), followed by Vilsmeir-Hach formylation [45], oxime formation and dehy-dration [46] gives (14). The recent availability of (17) [47] may open up single-step routes to (14) from the corresponding benzocrown ether. The significance here is that the 9-cyano anthracen-10-yl methyl unit may be grafted onto various receptors containing electron rich aromatic functionalities.

Now comes the moment of truth. Does the sensory supermolecule (14) show a 'switching on' of fluorescence upon addition of anhydrous sodium acetate? Yes, it does. Figure 4 shows a family of fluoresence emission spectra obtained with a 10^{-5}M solution of (14) in methanol and with 360 nm excitation, while progressively increasing salt concentration. The spectra are very similar to that obtained for 9-cyano,10-methyl anthracene in terms of wavelengths and shape. It is also notable that the electronic absorption spectrum of (14) is virtually unaltered by salt addition. The maximum sodium ion-induced fluorescence enhancement factor is 15 and large enough to justify the use of the term 'switching on'. Other alkali cations produce much weaker effects as seen in the

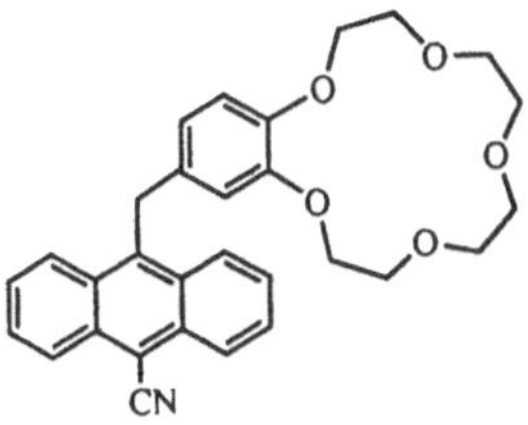

(14)

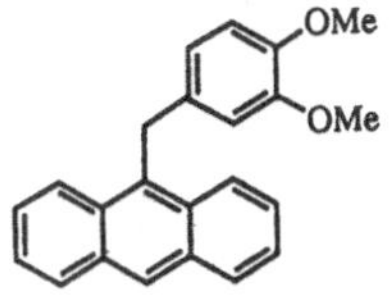

(15)

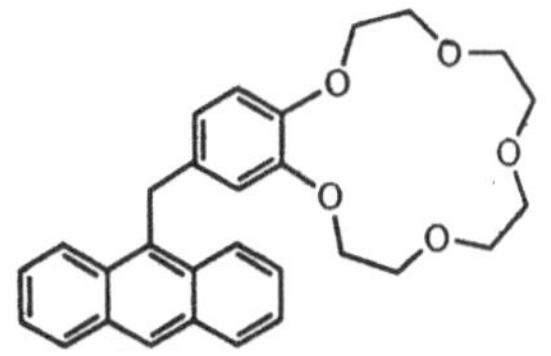

(16) 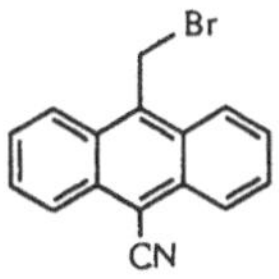   (17)

collected fluorescence intensity (I_F)–pM_{total} profiles (Fig. 5), and protons have a still weaker influence. The rising parts of these profiles can be fitted to Eq. (2) where (L_{total}) is the sensor concentration.

$$\log[(I_F - I_{Fmin})/(I_{Fmax} - I_F)]$$
$$= \log[(M_{total}^+) - (L_{total})(I_F - I_{Fmin})/(I_{Fmax} - I_F)] + \log\beta \qquad (2)$$

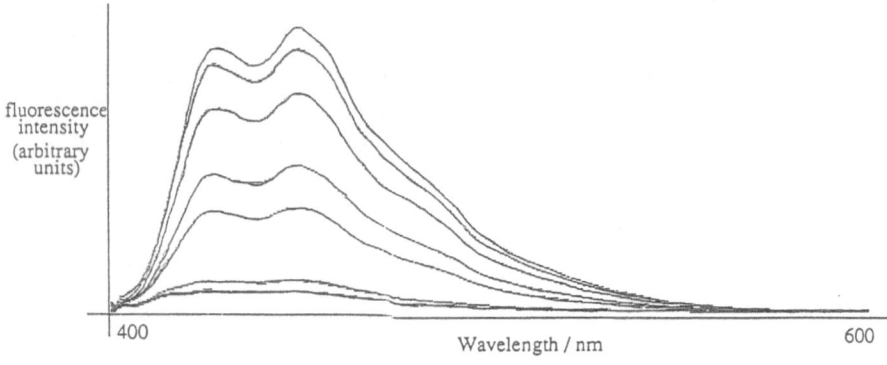

Fig. 4. Family of fluorescence emission spectra of 10^{-5} M **(14)** in methanol with 360 nm excitation. pNa_{total} values, in order of decreasing fluorescence intensity, are 1.6, 2.3, 2.6, 3.0, 3.3, 4.3 and ∞. These are obtained with anhydrous sodium acetate

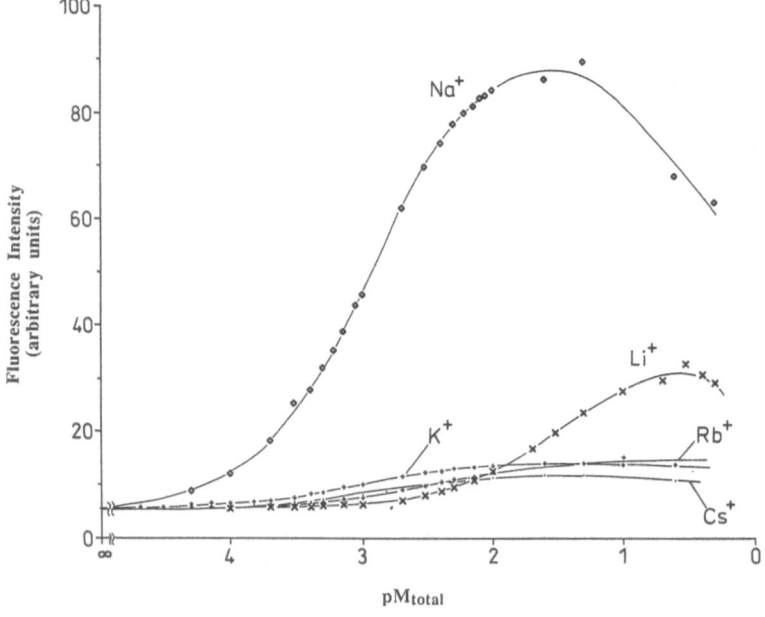

Fig. 5. Family of fluorescence intensity– pM_{total} profiles for 10^{-5} M **(14)** in methanol with 360 nm excitation upon contact with various anhydrous alkali metal acetates

Importantly, the goodness-of-fit parameters are satisfactory and the coefficient of the term involving (M^{+}_{total}) is near unity in all cases. Also, it must be noted that the stability constants for the alkali cation–benzocrown ether interaction available in the literature [30, 48] agree closely with the values extracted from Fig. 5 for the alkali cation–(14) association. All of this very gratifyingly support the expectations of the fluorescent PET sensor design logic in terms of a supermolecule with modular behaviour [43].

Not all the experimental results are so easily accounted for. Firstly, it is interesting that the fluorescence enhancement factor is most favourable by far for sodium ions, even though the log β values do not suggest any substantial selectivity over potassium ions. This means of course that the sensory selectivity is jointly determined by a binding selectivity and a fluorescence enhancement selectivity. The former component is quantitatively predictable in favourable cases like (14) but the latter component can only be qualitatively rationalized at present in terms of higher charge density of sodium over potassium ions. In general, the distinction between nesting and perching configurations of complexation (and variants thereof) can also have a bearing on this rationalization. The fluorescence enhancement factor due to lithium is smaller in spite of the high ionic charge density because the poor fit in the benzo-15-crown-5 ether cavity drives the lithium ion away from the benzoannelated oxygen atoms towards the harder aliphatic oxygen centres. Thus, the effective increase of oxidation potential of the 1,2-dioxybenzene system is attenuated (see next paragraph, for anion-pairing complications).

Secondly, the fluorescence quantum yield of sodium ion-perturbed (14) is only 0.019 which is well below unity, i.e. the 'switching on' is nowhere near ideal. In other words, the PET process has not been sufficiently suppressed by sodium binding. One of the reasons lies in the fact that three of the five ligating oxygen centres make no direct contribution to the PET active electrophore and, indeed, contribute negatively by dispersing some of the ionic charge. Another charge dispersal mechanism makes up the other reason. Axial anion pairing to the macrocycle-bound cation becomes significant at high salt concentrations in a solvent of moderate dielectric constant such as methanol. Evidence pointing in this direction can be found in Fig. 5 where the fluorescence intensities, instead of reaching a plateau at high salt levels, pass through a maximum and fall off as anion-pairing takes hold.

Finally, the sensory behaviour of (18) is not as good as that of its smaller ring cousin (14). On the positive side, however, the cation-binding constants remain quantitatively predictable and fluorescence enhancement selectivity peaks at potassium ion. The smaller fluorescence enhancement factor of 3 for (18) with potassium ion can be attributed to lower intrinsic charge density and more charge dispersal to aliphatic oxygen centres than those discussed in the previous paragraph.

This story has a happy ending, however. Taking advantage of the retardation of PET in neutral systems in solvents of lower polarity, we were able to tune the sensory performance of both (14) and (18) to very respectable levels in terms

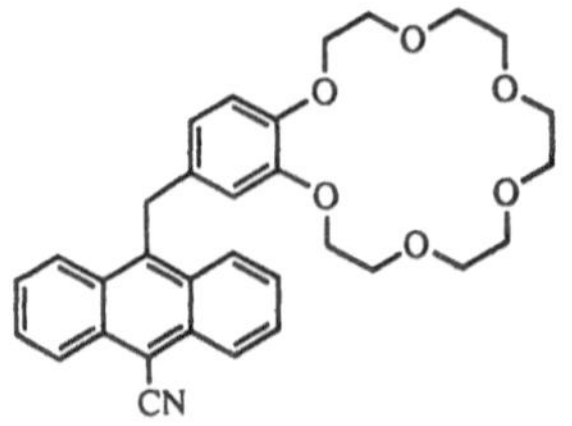

(18)

of both the fluorescence enhancement factors and absolute fluorescence quantum yields [49]. We are looking forward to monitoring membrane-bounded alkali cations with (14) and (18) without any significant interference from co-existing protons.

4 The Possibilities for Modification of the Sensor Response Range

4.1 Expansion of the Response Range

One of the most powerful features of fluorescent PET sensors is that their emission can be switched on or off by the appropriate guest. This leads, on one hand, to efficient sensing of ions within a moderate concentration range and on the other hand to ion controlled optical switches. Commonly, ion binding equilibria lead to comparable concentrations of bound and free sensor molecules, only over 2 pM units [50]. This range corresponds to the ion-bound/free sensor ratio going from 10:1 to 1:10. Thus, the fluorescence intensity is a sensitive and single valued function of the ion concentration only between these limits. This is an analogue device situation and is commonly exploited during the fluorescent monitoring of ion concentration. The limitation of this sensing dynamic range to 2 pM units has been recognized for nearly a century in the related case of absorption indicators [51] and is commonly overcome by employing different sensors for different pM ranges. The combination of several such sensors can result in systems which can handle wide pM excursions. Universal pH indicators are qualitative or semiquantitative examples [52]. Some of the more recent approaches for fluorescent cases involve multiple fibre optic systems [53] or hybrid sensor molecules with multiple ion-reception sites [54]. Even a single sensor compound can give rise to an extended range when anchored on a fibre optic tip [55]. Quasilinear fluorescent intensity–pH profiles with a wide dynamic range can be simply developed with fluorescent PET sensors because of their modular behaviour [56]. A family of PET sensors with identical fluorophore modules, but chosen different ion receptor modules, will give identical I_F–pH profiles for each member, except that they will be centred at their respective pK_a values. It is a straightforward computational matter to optimize the linearity of the overall response function by choosing the appropriate pK_a values. Such systems are molecular mimics of the glass pH electrode behaviour. The facts that the glass pH electrode is a ubiquitous device in a variety of scientific fields and that its miniaturization has limits imposed by its macromolecular ionic membrane nature, create new application areas for its molecular counterparts.

4.2 Contraction of the Response Range

A virtual, but non-trivial, contraction suitable for some situations can be achieved by applying large pM excursions across and beyond the analogue range. Thus, the experimental regime would only involve rapid jumps between regions A and E in Fig. 6. Under this restricted, but realistic, regime we have essentially a two-state system with digital action, as far as fluorescence output is concerned. Thus, external control of the analogue – digital duality of sensors is possible. The system is switched between its two states by external large-magnitude modulation of the local ion concentration. Such modulations can be achieved by ion binding/release from switchable receptors. Photoswitchable ion-receptors or one-way versions thereof are vibrant research areas [57]. Another approach is via application of membrane or electrode potentials. The membrane bounded ion concentrations can be altered by several pM units by altering/removing the interfacial charges [58]. It must be noted that this virtual contraction will only apply if we maintain control of the ion concentration as a large-amplitude input variable.

We have also observed another situation in which the response range is contracted. Sensors such as (19) which carry multiple charges in media of low dielectric constant/polarity such as $CHCl_3/MeOH$ mixtures display such a

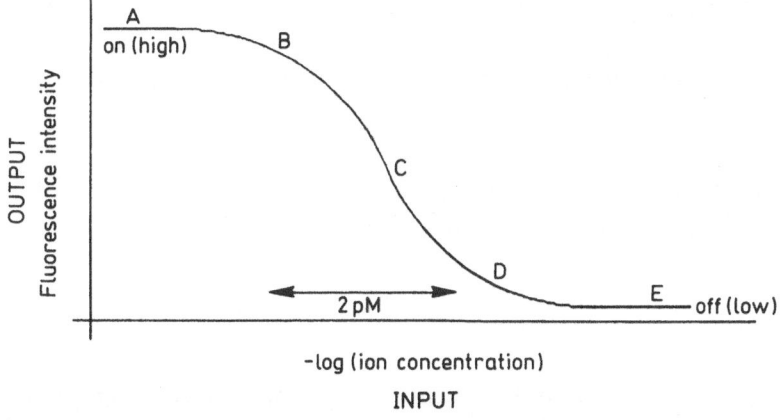

Fig. 6. The interpretation of a fluorescence intensity–pM profile of a fluorescent PET sensor as an input/output characteristic of a molecular photoionic device

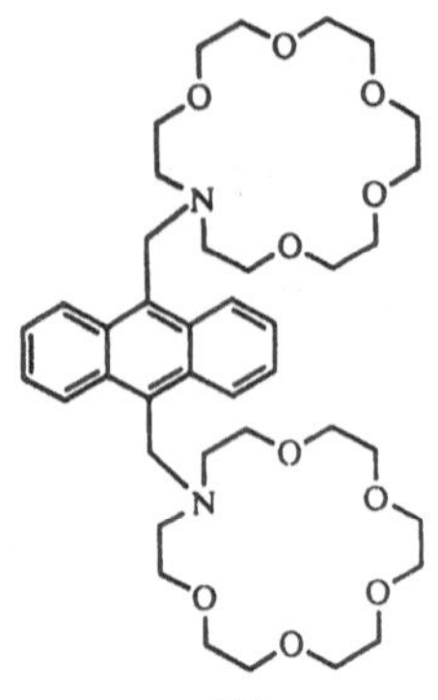

(19)

contraction [59]. These observations are interpreted as arising from the high electric field strengths developed in such situations. The non-unity activity coefficients arising from such interactions are well known in the Debye-Hückel theory of electrolyte solutions. The conventional sensing range of 2 pM units, as pointed out above, assumes simple ion-binding equilibria and unit activity coefficients. So the removal of the latter assumption can naturally lead to a new sensing range.

Another appealing approach has been recently outlined by Gabor and Walt [60] for conventional fluorescent pH sensors which should be equally valid for PET systems. In this method an absorption pH sensor with a similar pK_a value to that of the fluorescent sensor is employed to attenuate the fluorescent signal in a pH dependent manner so as to cause a considerable sharpening of the gradient of the fluorescent intensity–pH profile. If we think laterally, the use of Q-switch dyes to sharpen temporal light pulses is not dissimilar in principle [61].

Even more drastic modifications of the response range can be imagined and, when realized, should lead to sensors with specialized applications and molecular photoionic devices with a variety of input/output characteristics. The latter point takes on special significance when we remember that the availability of the corresponding devices in vacuum- and solid state-electronics led the surge in electronics technology which touches our lives today in so many ways.

5 A Survey of the Available Fluorescent PET Sensors and Potentially Relevant Systems

5.1 Systems Responsive to Protons

The majority of the fluorescent PET sensors for protons possess an amine moiety as the receptor module. One reason for this prevalence is that a wealth of established physicochemical data on proton binding are available for a variety of amines [23]. A further reason lies in the simplicity of the synthetic routes by which such sensor molecules can be accessed. Amino alkyl aromatic hydrocarbons are the simplest of these from a design viewpoint because the required redox potentials are easily obtainable. In contrast, carboxylate, phenolate and pyridine moieties are poorly represented as yet, even though phenolic compounds make up a substantial fraction of the older class of fluorescent pH sensors and indicators based on photoinduced proton transfer. Therefore, cases composed of receptors other than amines are deferred until Sect. 5.1.6.

5.1.1 Explicitly Designed Cases and Those Which Explicitly Consider PET in Competition with Fluorescence

The pH dependent photophysics of phenylalkylamines (5) [8] illustrates the features that simple molecules can provide during mechanistic investigations in a new field. The fluorescence quantum yields (ϕ_F) and lifetimes (τ_F) of (5, n = 1, 2 and 3) when protonated are similar to those of the parent fluorophore, toluene. However, differences appear as the pH is increased. The ϕ_F values of the proton-free (5) (ϕ_{min}) allow the ranking n = 2 > n = 3 > n = 1, which is also supported by the lifetime data. The unique importance of folded conformations in (5, n = 3) rationalizes this deviation from the simple expectation of decreasing PET rate with increasing spacer length.

Another collection of lifetime and quantum yield data are available for a variety of amino alkyl naphthalimide derivatives (20) with and without protonation of the side chain [62]. While complications are caused by the use of aqueous acid in polar organic solvents due to the solvent perturbation of the dipolar excited states of these fluorophores, this data set provides additional information regarding structural influences on PET rates. For instance, the addition of an extra methylene spacer to (20a) reduces the proton induced fluorescence enhancement (FE) from 2.1 to 1.3 due to the retardation of PET in the proton-free state. Due to the moderate FE values encountered in this family, the fluorescence emission at intermediate pH values decays in a biexponential manner. While steady state ϕ_F measurements are sufficient to establish sensory behaviour, associated τ_F determinations enrich such experiments by revealing individual rate constants. PET rate constants of $10^{11}\,s^{-1}$ are found in the aminomethyl anthracene sensors (21) in their proton-free state [63]. The use of two or more amine units in e.g. (21b) enhances the PET rate via statistical and related effects and gives rise to large FE values. However, this situation also leads to the sensing action being displaced to lower pH, due to the necessary but difficult perprotonation of all amine moieties [63, 64]. Therefore, the multiplicity of receptor units is an additional variable that is available to the sensor designer.

The use of heterocyclic units as fluorophores introduces electrostatic interactions arising from their dipolar excited states with internal charge transfer (ICT) character. The other partner in such interactions would be the proton-bound receptor which is held a short distance away by the spacer module. Such

(20)a; R$_1$ = R$_2$ = Me, X = 3-NH$_2$
b; R$_1$, R$_2$ = (CH$_2$)$_4$, X = 4-OMe

(21)a; X = H
b; X = CH$_2$N((CH$_2$)$_2$OH)$_2$

(22)

electrostatic interactions are held responsible for the proton-induced bath-ochromic shift in the electronic absorption spectra of the 7-methoxy coumarin derivatives such as (22) which, as expected, show strong proton induced FE values of > 30 [65]. Thus, electrostatic forces and PET processes are sufficiently long-range to overcome the isolation role of the spacer module. Nevertheless, the equality of ground- and excited-state acidity constants found with aromatic hydrocarbon [8, 18] fluorophores is still maintained. This can be rationalized as being caused by the kinetic limitation to the attainment of a new protonation equilibrium during the short lifetime of the excited singlet state of the fluoro-phore which is a short distance away in spite of the transient electrostatic interaction. Thus the excited state acidity constant is an apparent one with the excited state measurement only able to reflect the equlibrium in the ground state.

(23), due to Grigg and Norbert [66], expands the potential of fluorescent PET sensors by introducing the distinctive characteristics of metal-containing systems into an area monopolized so far by all-organic fluorophores. Besides its intrinsic interest as a photofunctional porphyrin, (23) possesses the additional special feature of the proton-free quantum yields being controlled by the nature of the amine substituent. The thermodynamic driving force for PET (ΔG_{ET}) is found to be linearly related to these ϕ_{Fmin} values. While the application of classical Marcus theory [67] and PET sensor design principles suggests a more complex relation, the experimental results could represent a linear segment of the general curve. (24) [68] is the follow-up to (23) and its relatively long-lived luminescence can be potentially employed in a time-resolved mode to reject matrix fluorescence. Application of Eq. (1) to (24) reveals that the PET processes are expected to be substantially exergonic only when aromatic amines are employed as the receptor module. The experimental results largely bear out this expectation. Interestingly, members of series (24) possessing aliphatic amines display a proton-induced luminescence quenching mechanism.

(23)   (24)

The experimental results discussed thus far can be summarized as follows. Fluorescent PET sensors for protons display a working range of about 2 pH units as seen in conventional fluorescent sensors based on photoinduced proton transfer [69]. However, the two classes do differ in that PET sensors have indistinguishable proton binding constants in ground- and excited-state experiments. The proton induced fluorescence enhancements decrease as longer spacers and amine groups of poorer electron donicity are employed. The Weller equation, Eq. (1) serves as an important design aid when the appropriate redox potentials are available.

The implicit assumption in the fluorescent PET sensor design is that the major competitors for deactivation of the singlet excited state are electron transfer and fluorescence. While the occurrence of electron transfer has yet to be experimentally demonstrated for many sensors, this assumption appears well justified under the usual sensing conditions of polar solvents and fluorophores of high quantum yield. Indeed, this contributes to the logical and experimental simplicity of PET sensory action. Evidence for the latter is that for aminoalkyl aromatic hydrocarbons, of all the electronic spectroscopic parameters characterizing the transition between ground and lowest singlet excited states only one, the fluorescence quantum yield, is proton-controlled [22]. However, there are refreshing situations when one must look beyond the confines of the above assumption. (6) in aqueous β-cyclodextrin solution [10] displays exciplex emission as an additional phenomenon, which relies on the amine moiety remaining in the proton-free state. Since this situation is opposite to the PET sensor action expected of the locally excited naphthalene emission, the overall experimental outcome is a dual fluorescence where one component switches off as the other switches on upon pH variation. Competing exciplex emission involving three sites is seen in (7) in 2-propanol whereas excision of the remote aromatic amine results in a simpler donor–acceptor exciplex emission at shorter wavelengths. Fluorescence–pH titration of such excised versions of (7) reveal reduced basicity values as compared to the parent amine. This effect has received confirmation [18] and can be ascribed to steric inhibition of solvation of the protonated amine by the fluorophore as well as its weak electron withdrawing effect across the methylene spacer [63]. A virtual excision of the central aliphatic amine moiety is efficiently achieved by selective protonation by a weak acid. Under these conditions a long range electron transfer can be inferred to occur across the conformationally semi-rigid skeleton of (7).

The modular construction of fluorescent PET sensors implies that encapsulation of the fluorophore unit need not interfere with the sensory behaviour provided that the receptor unit maintains (at least some) contact with its original environment. Transparent molecular encapsulants in the form of cyclodextrins have been employed to enhance fluorophore quantum yields by impeding dynamic quenchers [70] or causing fluorophore deaggregation [71]. However, the most dramatic results are obtained when organic phosphors are substituted for the fluorophore unit, due to the much longer lifetime of phosphor excited states [72]. The PET sensor principle is thus extensible to the phosphorescence phenomenon [73]. Besides its intrinsic interest, including the successful competition of electron transfer with intersystem crossing or phosphorescence, there is a practical ramification. Phosphorescent sensing can take advantage of its longer (millisecond) timeframe to allow interference-free measurements in intrinsically fluorescent environments. Aminomethyl bromonaphthalene derivative (25) in β-cyclodextrin illustrates these ideas as applied to proton sensing [73]. Even though examples of this class are still rare, differences in behaviour from their fluorescent counterparts are already emerging. Due to the millisecond lifetime of the phosphorescent excited states, any resulting electrostatic interaction has ample opportunity to perturb protonation–deprotonation equilibria. In the present case, the proton binding constant changes from 7.9 to 6.9 upon moving from the ground- to the lowest triplet-excited state.

(25)

5.1.2 Cases Developed for Other Species but Which Can Also Respond to Protons

The electron rich nature of nitrogen centres leads to their frequent use in PET type sensor designs for various species. While these designs are frequently successful in this respect (Sects. 5.2–5.4), a strongly sensitive response to protons is an undesirable, but natural, accompaniment. In many instances protons elicit the largest sensory response of any ion due to the creation of N–H covalencies. For example, the azacrown ether derivative (12, n = 0) displays a fluorescence enhancement factor of 6 with sodium ions, but the best performance is with protons which yield an FE value of 36 [19]. Several other sensors can be added to this example and, interestingly, all of these cases are also unified by the presence of anthracene fluorophores. This set consists of nitrogen-bridgehead cryptands [15, 74] designed for alkali cations, azacrown ethers for alkane diammonium ions [59], ammonium ion arrays for phosphate ions [20], ethylene diamine derivatives for zinc chloride [64] and cyclams for zinc and cadmium ions [75, 76]. The fluorescence of the potassium ion sensory coumarocryptand (26) [77] can also be expected to respond strongly to protons.

Some receptors designed for physiological applications have their nitrogen centres conjugated directly with aromatic units in order to minimise proton binding. Tsien's calcium ion receptor (27a) [78] and the alkali cation binding benzoannelated cryptands (28a) [79] and (28b) [80] are of this type. When these are incorporated as receptor modules in PET sensors the proton response is not exceptional as seen in FE values of 11 and 21 with rubidium ions and protons, respectively, in the case of sensor (28c) [80]. The calcium ion sensor (27b) [81] is remarkable in that the calcium-induced FE of 92 is to be compared with FE values of ca. 1 and 24 upon binding one and two protons, respectively. This rare situation is brought on by the presence of several proton binding sites at various distances from the fluorophore module. Further protonation eventually re-establishes the status quo, but only just. The FE value following triprotonation at the rather extreme condition of pH 1 is only 120.

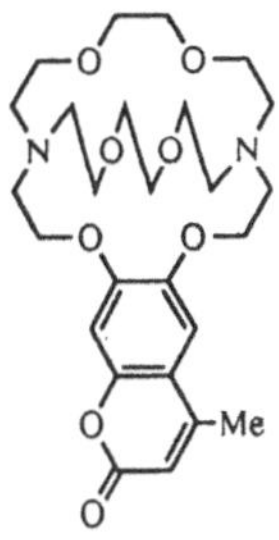

(26)

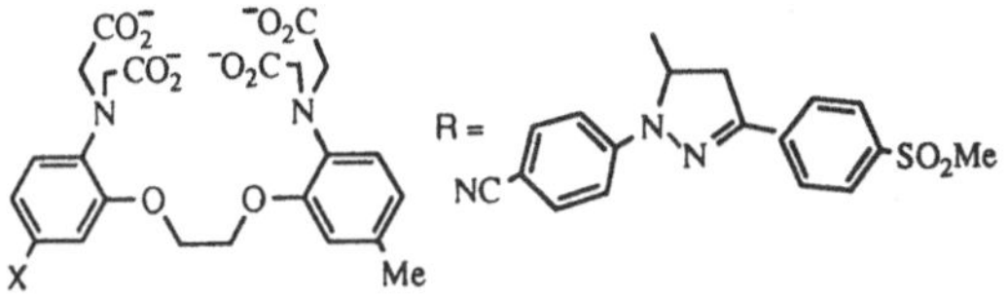

(27)a; X = H
    b; X = R

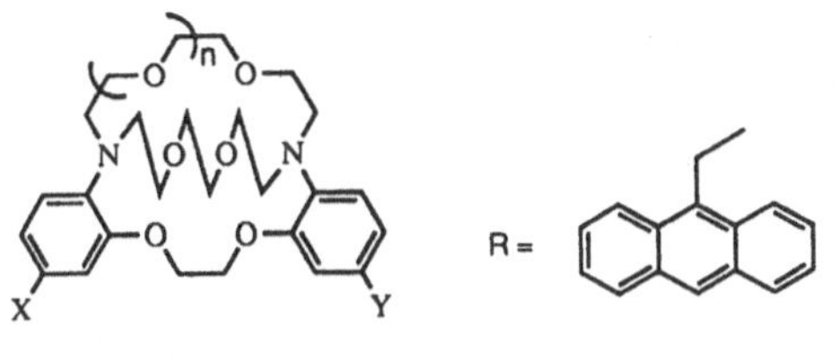

(28)a; n = 0, X = Y = H
    b; n = 1, X = H, Y = Me
    c; n = 1, X = R, Y = Me

Several rather basic azamacrocycles act as receptor components within 'lumo/fluorophore-spacer-receptor' systems (29) [82], (30) [83] and (31) [84], which are at least structurally related to PET sensors. However, the lack of any substantial luminescence/fluorescence quenching in the cation-free state suggests that the thermodynamic criteria for PET (Eq. (1)) are not met in some of these cases. The low luminescence quantum yields seen in some variants of (31) [84] are due to the intrusion of metal centered lowest excited states. Related cases, with or without macrocyclic units, (32) [85] and (33) [86] carry the

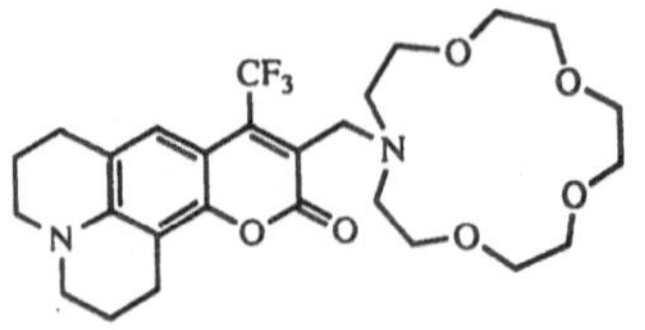

(29)

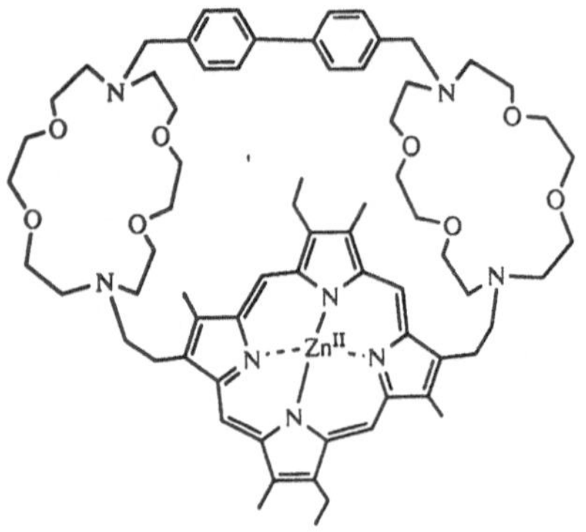

(30)

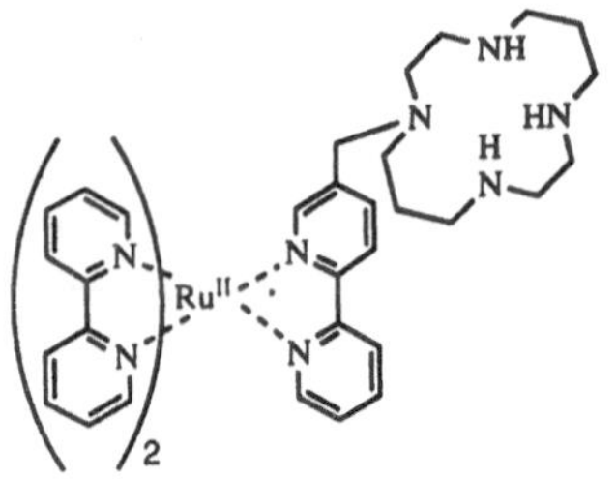

(31)

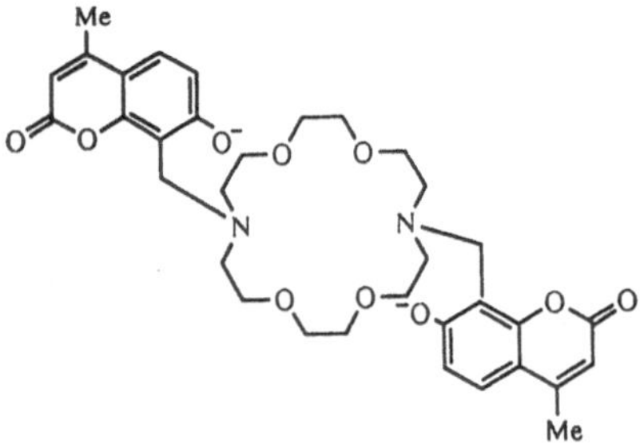

(32)

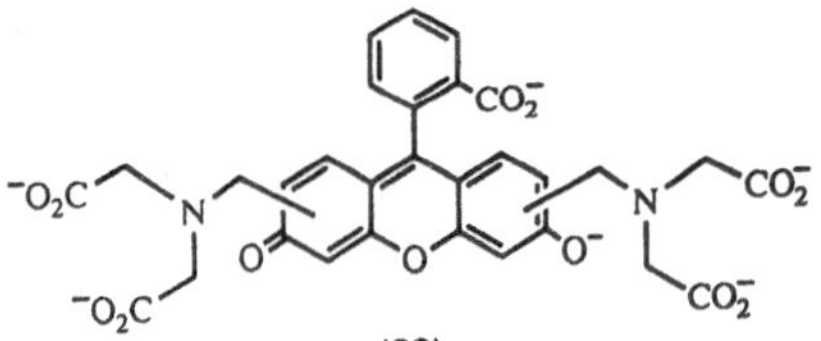

(33)

complication of protonatable anionic fluorophores, though (33) is exceptional in showing PET type sensing action. Nevertheless, PET type processes can be induced in some of these systems by redox-active cations and will be examined in Sect. 5.4.

Aminoalkyl anthracene derivatives have already featured prominently in this discussion with regard to their proton responsive fluorescence. Some of these in their protonated form also appear in a fluorescent signalling role in association with macromolecular species which is however unconnected with PET action. These mono- or oligocations bind to polyanions such as DNA and heparin [87, 88]. On one hand, this can lead to excited state association between two sensor cations residing on the same polyanionic strand. Acridine orange with its cationic π-system also exhibits a related behaviour which has been exploited by life scientists [89]. The excimer-like emission arising therefrom will disappear if the polyanion is depolymerized, e.g. heparin with heparinase, thus opening up a route to fluorimetric assays of hydrolytic enzymes. On the other hand, electronic energy transfer from the nucleic acid bases to the intercalated π-system of the sensor cation can be demonstrated by examining the excitation spectrum of its emission [88].

5.1.3 Cases Where Proton Responsive Fluorescence is Examined Empirically

pH is a common independent variable which is considered during optimization procedures for fluorescence analysis of compounds, especially in the pharmaceuticals area. Anaesthetic (34) has been the subject of two investigations a decade apart [90, 91]. Both studies further strengthen the deduction in Sect. 5.1.1 about the equality of proton binding constants measured in the excited- and ground state in 'fluorophore-spacer-receptor' systems. The later work includes confirmatory data from time-resolved experiments [91] and also shows the ease of interpretation of the full data set in the pH range where PET sensor action can now be inferred to prevail. The typical complexities (or richness, depending upon one's viewpoint) are found in the pH range where photo-induced proton transfer [69] is expected to be found. Due to the presence of Brønsted basic sites within and outside the fluorophore, (34) presents an unusual opportunity for the quantitative comparison of the two proton sensing approaches within the one sensor. This opportunity has been well exploited experimentally in the present instance [90, 91]. It is worth pointing out that this same opportunity seems yet unexploited to the same extent in several other cases, e.g. (35) [92], (36) [93] and (37) [94]. The familiarity with the usage of quinine (35) in sulphuric acid as a fluorescence standard must not blind us to the fact that the weaker fluorescence transition at more alkaline pH values corresponds to PET sensor type action involving an extrafluorophoric amine receptor. Unfortunately (36) has not been studied in the pH range where the protonation

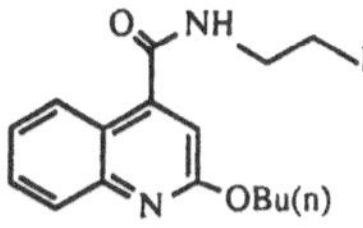

(34) 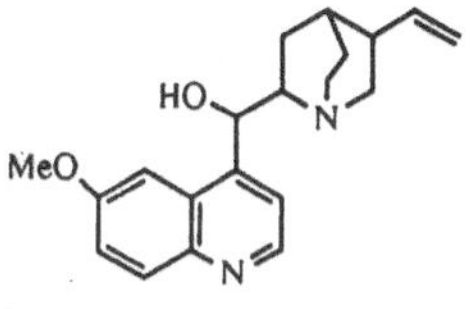 (35) 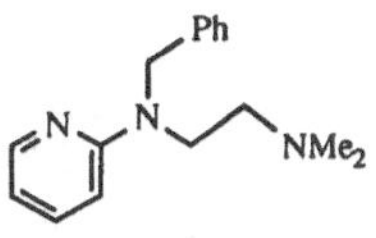 (36)

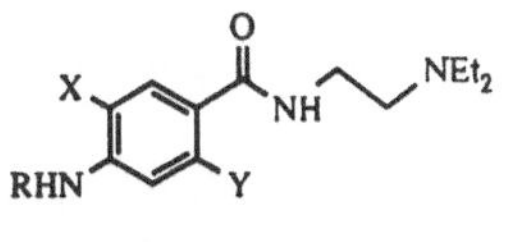

(37)a; X = Cl, Y = OMe, R = H
b; X = Y = R = H
c; X = Cl, Y = OMe, R = MeCO

status of the amine side chain may be altered [93]. In contrast, extensive pH and structural variations on (37) show PET sensor type action only when relatively electron poor fluorophores are employed, in keeping with the expectations outlined in Sect. 3. The pH dependent fluorescence of (33) [86] (Sect. 5.1.2) justifies its inclusion in this section as well.

5.1.4 Cases Which Mention Proton Responsive Fluorescence in Passing

Photophysicists have often examined aminoalkyl aromatics during investigations of exciplex and related phenomena. So it is natural that one can locate examples where the effect of protonation is considered briefly as an adjunct to the main thrust of the research. Usually, protonation is used as a convenient means of disabling the lone electron pair of the amino group from participating in the exciplex interaction as discussed previously for (6) in Sect. 5.1.1. (38) [95] considers exciplex-related charge transfer state formation across methylene spacers. (39) [96] and (40) [97] bring in the additional phenomenon of excimer emission due to their homobifluorophoric nature. (40) [97] and its mono-fluorophoric counterpart (41) [97] are remarkable in being the first cases in which we can find essentially complete 'on-off' fluorescence switching with protons. (42) [98] is a later case where conformational restrictions preserve exciplex emission under a variety of conditions. The exception, of course, is provided by protonation. (40) [97] also employs the amine in a different role as a handle to allow perturbation of the spacer characteristics during intramolecular excimer formation. Protonation is considered to provide dipole induction in the fluorophore pair.

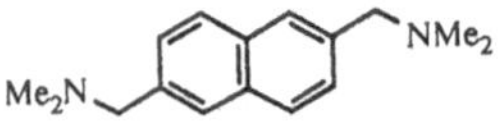

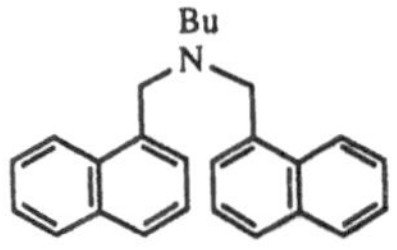

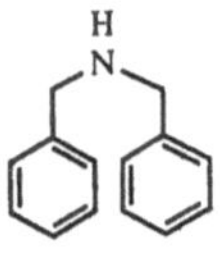

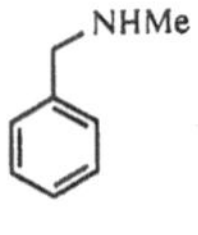

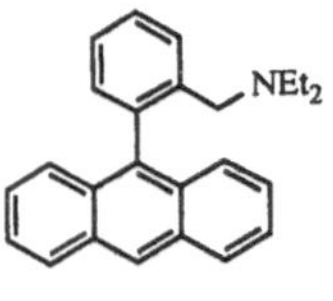

5.1.5 Cases Where Proton Responsive Fluorescence is Not Reported

As mentioned in the previous section, photophysical/chemical investigations are a natural resource regarding aminoalkyl aromatics. In some of these instances, protonation studies have not been on the experimental agenda. The time seems ripe now for the evaluation of these cases for fluorescent PET sensing action toward protons, especially where the thermodynamic criteria in terms of Eq. (1) are favourable. A variety of aminoalkyl naphthalenes and anthracenes have been available from the work of Davidson and Mataga for the past two decades [99]. Chandross is another early contributor to this pool [100]. The reviews quoted earlier [3] contain more examples which have accumulated since then.

A separate vein of examples worthy of investigation can be found in a surprising location – pharmaceutical chemistry. In fact, this field abounds with aminoalkyl aromatic systems of varied structure. At least one reason for this wealth is that biogenic amines are of this structural type and serve as a continuing inspiration for the drug discovery effort. An early example is (43), first described in 1951 as a prospective antimalarial [101]. (43) has been found in our laboratory to display pH sensitive fluorescence [102] as expected of a PET sensor. The spacer of three carbon atoms limits the PET rate in the proton-free form which leads to a modest but useful proton-induced fluorescence enhancement. The fluorescence properties of another antimalarial (44) with structural features required of a PET sensor have been recently reported in a different context [103] and deserves sensory evaluation.

(43)

(44)

5.1.6 Cases Composed of Proton Receptors Other Than Amines

The basis for the use of carboxylates as proton receptors within fluorescent PET systems lay in the fact that the carboxylic acid group was more electron withdrawing than the carboxylate moiety when attached to benzene [104]. Benzoic acid and its esters can be reduced at accessible potentials [105]. Indeed, benzoic acid has been implicated in electron transfer quenching of fluorophores for thirty years [106], even though hydrogen bonding can also play its part [107]. Some early examples of fluorophores linked via spacers to (aliphatic) carboxylic acids do show clear pH dependence of fluorescence with quenching when the carboxylate group is protonated. The geometric preferences seen with (45) [108] and relatives suggest that the proximity of the carboxylic acid to the indole fluorophore allows hydrogen bonding to open a quenching channel. On the other hand, it is difficult to picture a similar quenching action in (46) [109] which also displays a sigmoidal pH dependence of fluorescence intensity. In contrast (47) [110] could be rationally designed as a PET sensor by considering available redox potential data for 1,3-diaryl Δ^2-pyrazolines [111] and for benzoic acid [105] for use in Eq. (1). It is notable that (47) is a 'fluorophore-spacer-aromatic carboxylate' system. The confidence in the design could be

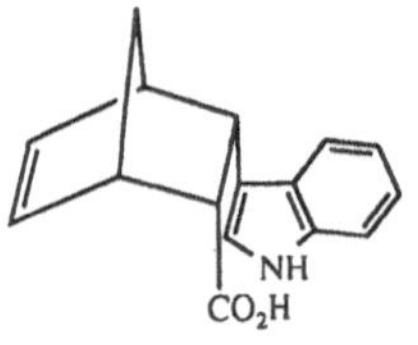

(45) 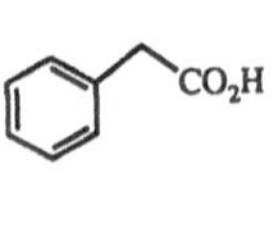 (46)

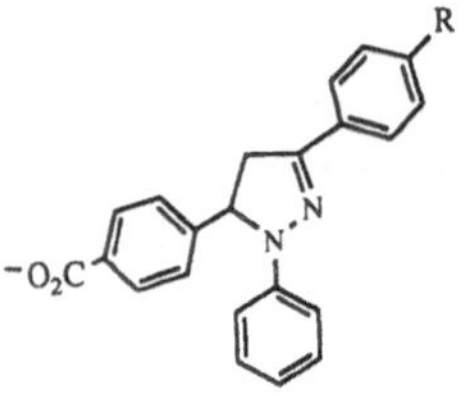

(47) 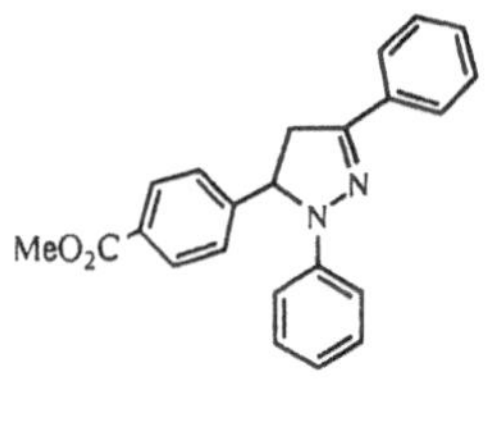 (48) 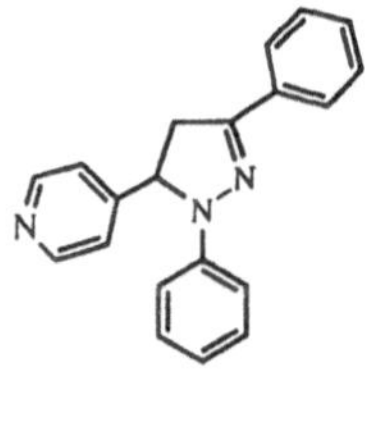 (49)

increased by noting the intramolecularly quenched fluorescence of the benzoic acid ester (48) [112] since, for the present purposes, esters are derivatives of carboxylic acids which are locked against prototropy. The overall sensory behaviour of (47) and relatives are similar to those PET sensors with amine receptors with the notable difference that the proton-dependence of fluorescence is opposite. From a molecular photoionic device viewpoint, this means that a YES or NO response in photon output is possible to a proton input signal with e.g. (22) or (47), respectively.

The Brønsted acid/base pair pyridinium/pyridine is unusual because redox potentials are available for both partners in an accessible range [35]. This allows detailed anticipation of the sensory behaviour of systems containing pyridine receptors. (49), whose fluorescence properties have received some attention before [113], is a member of a family of PET sensors for protons whose excitation and emission wavelengths spread over a substantial fraction of the accessible optical spectrum without compromising sensory action [114].

The aromatic nature of the two receptors examined as PET sensor components in this section is expected to limit the convenient synthetic accessibilities of sensor designs. Nevertheless, it must be hoped that synthetic ingenuity will permit expansion of this subset which, by its very presence, has offered valuable evidence of the generality of the fluorescent PET sensor concept.

5.2 Systems Responsive to Alkali Cations

The macrocyclic revolution in metal ion coordination chemistry [115] soon had repercussions on the design of fluorescent sensors for alkali cations. Before the advent of PET sensors, the first examinations of crown ether receptors with adjacent π-electron systems such as naphtho- and benzo crown ethers (50) [116] and (51) [117, 118] showed small but significant alkali cation induced fluorescence and phosphorescence modulations. The latter point was subsequently addressed in another time-resolved study on (52) [119]. Two important results to emerge from these studies were the significance of spin-orbit coupling effects with the heavier alkali cations and coordination-induced ligand rigidification. The latter effect was of particular importance in bifluorophoric systems such as (51) [117]. Further development of crown ethers with integrated fluorophores with more pronounced ICT character in their excited states such as (53) [120] and (54) [121] uncovered a further general result. Cation induced spectral wavelength shifts in either direction [120, 121] are a natural consequence of ionic charge coupling with excited state dipoles. The parallel phenomenon is a

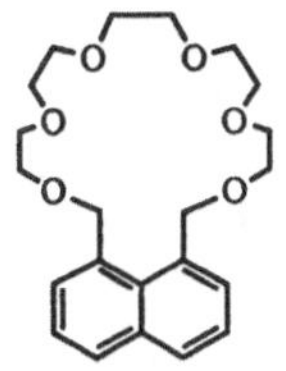

(50)

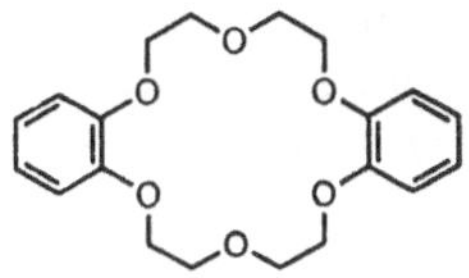

(51)

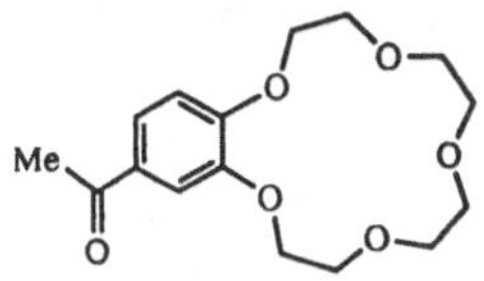

(52)

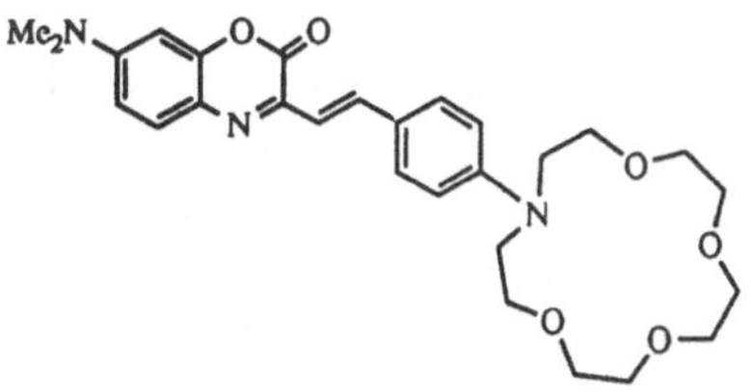

(53)

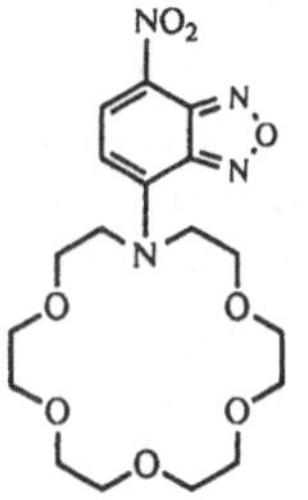

(54)

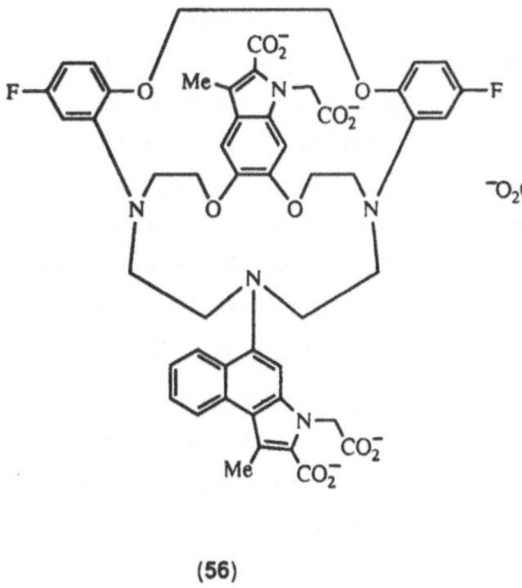

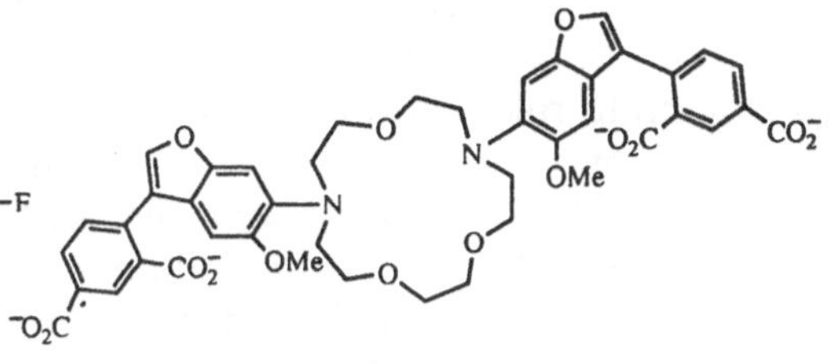

(55)

(56)

common occurrence with fluorescent pH sensors based on photoinduced proton transfer [50, 69]. A high point in this line of research is Tsien's sensor (55) [122] which selectively responded to physiological levels of sodium ions [122, 123]. The design of (55) allows the sodium induced deconjugation of fluorophore segments by a conformational change to allow additional *trans* diaxial ligation of the sodium ion held in the diazacrown ether moiety. Cryptand (56) [79], which is structurally more elaborate, also achieves the goal of selective measurement of sodium ions in physiological situations. The cryptand (26) [77] also possesses a dialkoxy coumarin fluorophore with an ICT excited state [65] and an overlapping [2,2,2] cryptand moiety. However, the similarity with the previous cases stops here. The nitrogen centres in the cryptand which are not conjugated with the fluorophore can act as the electron donors in a PET system. Related PET type behaviour concerning amines and coumarin fluorophores are known [124]. The experimental result of potassium ion induced fluorescence enhancement by a factor of ca. 3 in (26) without any significant wavelength shift is in line with PET sensory behaviour. Another cryptand (28c) [80] combines PET design principles with the idea of complexation-induced conformational changes previously used in this context by Tsien [78]. However, the conformational changes may be expected to be of a smaller magnitude due to the restrictions caused by the macrobicyclic framework. Nevertheless, the effect proves to be quite sufficient to cause ion-induced redox potential changes in the receptor module which give rise to order-of-magnitude enhancements of fluorescence with potassium and rubidium ions.

In contrast with integrated fluorophore-receptor system (51) [117, 118] the first-use of PET systems to employ benzocrown ether receptors arrived a decade later in the form of (14) [43]. The relatively large magnitude of the alkali cation induced-fluorescence changes seen with (14) which was detailed in Sect. 3 testifies to the usefulness of the PET approach. Earlier PET systems employed azacrown ether receptors (8) [15, 74, 125] and (12) [19]. (8) with its diazacrown ether receptor and an anthracene-containing alkane strap has found use as a constrained triple exciplex system [126] and in structural studies [127] besides its pioneering role as a fluorescent PET sensor for alkali cations. The ion-induced fluorescence enhancements are influenced by large conformational alterations which, in turn, are controlled by the length of the alkyl chains in the strap, by the hydrogen bonding ability of the solvent and, of course, by the nature of the cation itself. The structurally simpler (12) [19] allows extensive solvation of its electron transferred state and leads to low cation-free fluorescence quantum yields [128] which in turn sets the stage for large cation-induced fluorescence enhancements. The magnitude of these enhancements can be related to the electric field strength at the crucial lone electron pair on the nitrogen atom, in terms of the charge density of the guest ion and the centre-to-centre distance between the ion and the nitrogen atom. Another notable result [19, 43] is that the ion-binding constants measured fluorimetrically with these 'fluorophore-spacer-receptor' systems are virtually identical to the values found for the parent receptors by ground state techniques [30].

Alkali cations have also been involved in two interesting photophysical studies concerned with PET processes. In the first of these, sodium ion complexation to a diazacrown ether spacer between a PET donor–acceptor pair causes an acceleration of the electron transfer supporting a superexchange interaction [129]. The second case relies on sodium ion induced conformational changes in a calix[4]arene tetraester diagonally substituted with a PET donor-acceptor pair [130]. This results in a good fluorescent sensory action.

5.3 Systems Responsive to Alkaline Earth Cations

The dipositive nature of alkaline earth cations require hard multi-anionic receptors for optimum complexation [131]. (33) is an early example with some features suggestive of fluorescent PET sensory behaviour [86] (Sect. 5.1.2). The quenched fluorescence at high pH can be largely recovered with calcium ions. While a 'fluorophore-spacer-receptor' format can be discerned within the structure (33), the assistance of the phenoxide to the calcium complexation by the iminodiacetate moiety clouds this assignment. Tsien's elaboration of the iminodiacetate unit into a benzoannelated receptor (27a) led to adequate selectivity for calcium vis-a-vis magnesium ions and protons [78]. He was quick to follow up his discovery with practical fluorescent sensors designed with integrated fluorophore and receptor moieties, e.g. (57) [78, 132]. Calcium-induced conformational changes attenuate the electron delocalisation within the system giving rise to spectral hypsochromic shifts. Subsequently, we elaborated receptor (27a) into (27b) which possesses the 'fluorophore-spacer-receptor' format [81]. It is notable that the spacer unit in (27b) is also responsible for rigidising the π-electron system of the fluorophore to minimize nonradiative loss processes [133] other than the engineered electron transfer. The PET process is very efficient in quenching the fluorescence of (27b) when it is calcium-free. On the other hand, calcium ions decouple the iminodiacetate units from the phenoxy groups which raises the oxidation potential of the receptor module. Of course, the proximity of the dication exacerbates this situation which results in the supression of the PET process and the recovery of fluorescence. Such a combination of electrostatic and conformational influences produces very large fluorescence enhancement factors (FE) of ca. 100. Another attractive feature of (27b) and related PET systems is that their optical and ion-binding parameters are quantitatively predictable from those of the parent fluorophore and receptor modules respectively [81].

A similar series of events can give rise to fluorescent PET sensors for magnesium monitoring in physiological contexts. London's selective magnesium receptor (58a) [134] shares several features with (27a) but differs in the fact that (58a) would give rise to a smaller cavity in its binding conformation. Thus the PET sensor (58b) [135] is designed along lines identical to those used for (27b) and is expected to complement London's fluorescent sensor (59) [136]

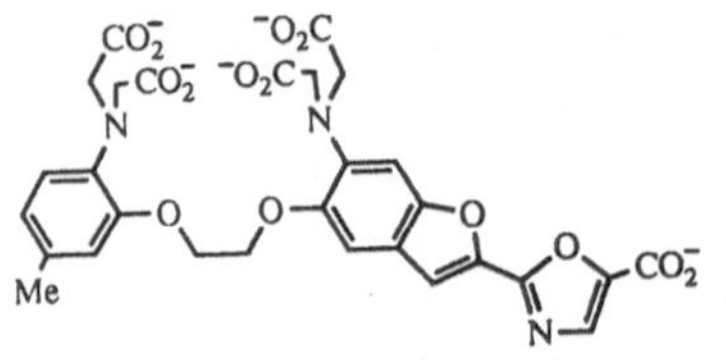

(57)

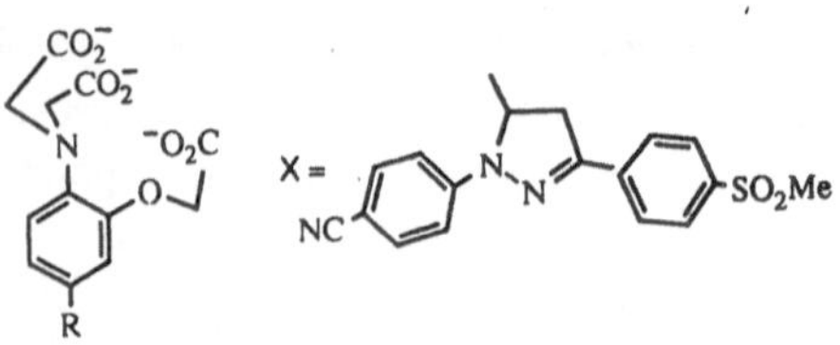

(58)a; R = F
b; R = X

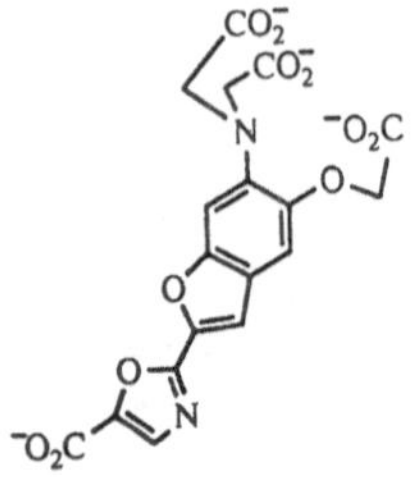

(59)

which is parallel to the calcium sensor (57). Indeed, (58b) does show FE values of ca. 40 with millimolar magnesium levels.

The discussion so far has centred on receptors based on the iminodiacetate 'element'. However, even crown ethers can serve as receptors in sensors where the dicationic alkaline earths can elicit a larger fluorescence response than the alkali monocations. Charge density effects of this sort are particularly noticeable in solvents of relatively low polarity. The best examples are to be found among fluoroionophores [137] with integrated fluorophore and receptor moieties such as (53) [120] and (54) [121]. No PET sensors of this type are currently available, though case (32) [85] is an approximate example of a 'fluorophore-spacer-receptor' system.

5.4 Systems Responsive to Other Species

The previous parts of Sect. 5 testify to the fact that development of sensors for a given chemical species is dependent on the availability of receptors with suitable selectivity, electroactivity and optical transparency. These criteria were most easily satisfied with amine derivatives with various structural modifications for the binding of protons, alkali and alkaline earth cations. Such is the ubiquity of the amine motif in fluorescent PET research that several of the examples to be discussed in this section also contain it within their receptors.

The ability of cyclam derivatives to bind heavy metal ions [131] has been exploited with (60) [75] which shows good fluorescence enhancements with zinc and cadmium ions at pH 10. In such polyamine receptors, it is necessary that all the nitrogen lone pairs occupy the first coordination sphere of the metal ion in order that the fluorescence is efficiently revived. The unusual spectral shape of the 'switched on' fluorescence in the case of (60), n = 4 with cadmium ions is ascribed to reversible cadmiation of the anthracene ring [76] and may be an effort by the pentacoordinated cadmium ion to obtain a sixth ligand. Distortions of the characteristic anthracenic spectral shape or position were also seen for (8) with thallium and silver ions [74]. The structural layout of (8) maximizes the interaction of the fluorophore with the guest metal ion. Mercuric ion with its higher redox activity, while closing off the designed PET channel in (60), opens up a new PET pathway directly involving the metal ion orbitals. Thus, no nett fluorescence is seen at pH 10. Mercuric ion-dependent fluorescence of (60) can however be achieved at lower pH values by displacing protons (which switch fluorescence on) by the quencher mercuric ion.

Pseudointramolecular PET type quenching of fluorescence by the guest within a host-guest complex (60)-Hg^{2+} is rather general and is an accentuated version of related intermolecular processes seen in (61)-Cl^-, for example [138]. Silver ions are similarly implicated in complexes with (8) [74] and (30) [83]. The PET process in the latter instance results in charge separation which lasts 5 μs. Such long-lived charge separation is useful for solar energy conversion schemes [139] but efficient fluorescent PET sensor designs prefer rapid back electron transfer to enhance photostability and to reduce the 'down time' in the non-sensory state. Other examples of pseudointramolecular fluorescence quenching within complexes may involve variable degree of electronic energy transfer to states involving the coordinated metal ion. (31)-Ni^{2+}[84c], (33)-Cu^{2+}[86] and iron-siderophore [140] complexes are among these. Two interesting variations on this general theme are the exploitation of cyclodextrin inclusion [141] and metalloporphyrin axial ligation [142] for the assembly of porphyrins and quinones as fluorophore-PET quencher pairs. The convenience and flexibility of such assembly strategies over conventional covalent syntheses is appealing. Fluorescence quenching schemes are quite usable in sensor designs, even though fluorescence enhancement schemes are preferable. However, cases such as (31)-Ni^{2+} exhibit slow kinetics of equilibration [84c], whereas real-time sensing requires fast equilibration of guest reception.

Organic cations have not received anything like the attention focussed upon their inorganic counterparts with fluorescent PET sensors. Ethyl ammonium ions cause a large fluorescence enhancement in (19) upon 2:1 complexation [59]. In contrast, putrescinium dications produce marginally larger FE values but upon 1:1 complexation at cation concentrations which are smaller by three orders of magnitude. (19) is interesting in that it sends out a visual signal only upon the successful selection of a bioactive guest of a predefined length. This action is distinct from the 'spectroscopic rulers' beloved of biochemists which measure distances according to energy transfer principles [143]. (62) which appeared earlier in the same year [144] is similar to (19) in signalling linear recognition via fluorescence. The difference however is that (62) relies on perturbation of a monomer-excimer equilibrium by the guest [145]. Organic zwitterions can also be included in fluorescent PET sensor designs. (63) is a non-ideal, but convenient, approach to the sensing of γ-aminobutyric acid (GABA)

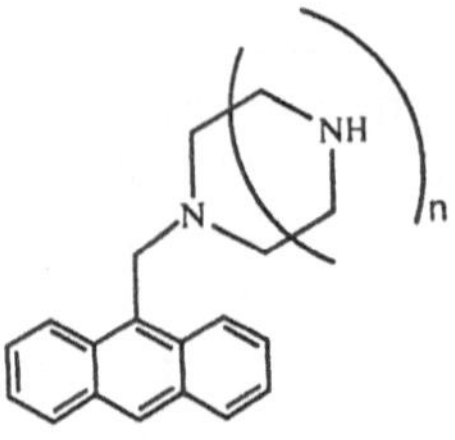

(60)

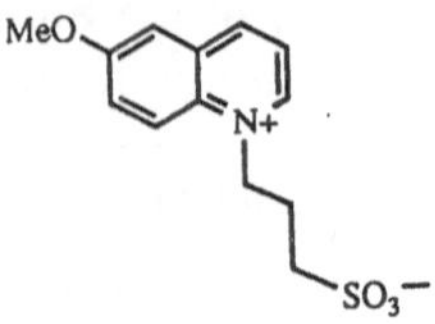

(61)

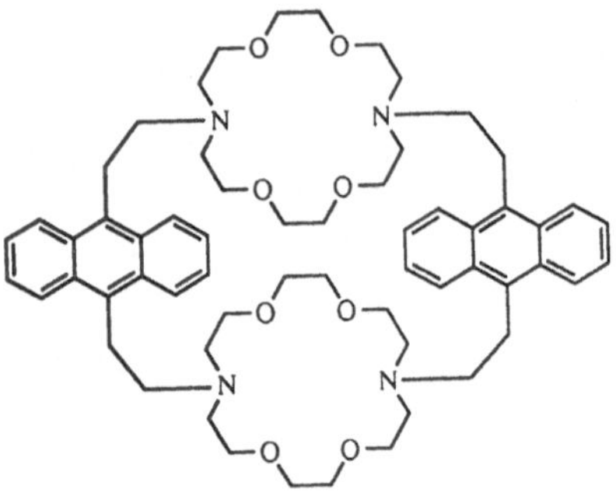

(62)

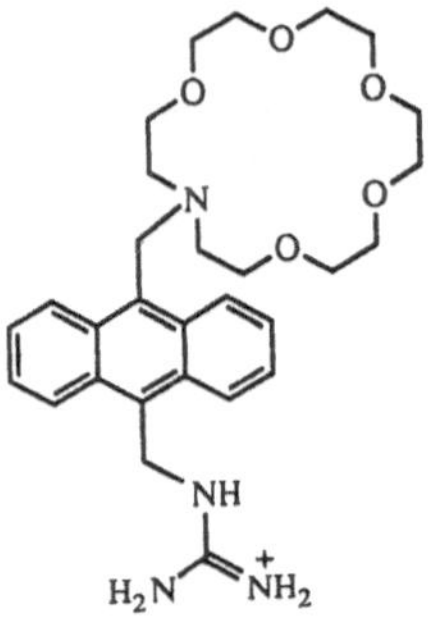

(63)

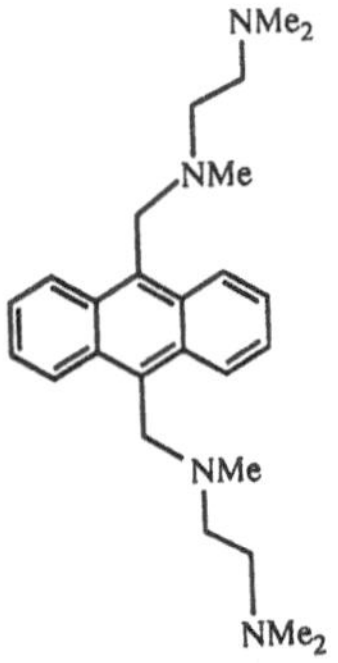

(64)

and related structures [135]. The importance of GABA within the central nervous system is the driving force behind the construction of (63). (13) is the only currently available fluorescent PET sensor for anions such as HPO_4^{2-} and citrate [20] and is discussed further in Sect. 7.

Electrically neutral species are less easy to accommodate within fluorescent PET sensor designs, especially if they are to operate under competitive solvent conditions. Strategies to overcome this hurdle are outlined in Sect. 6. However, the Lewis acid zinc chloride was employed to modulate the fluorescence of (40) by perturbing its amine moiety seventeen years ago [97a] (64). [64] is a modern version which produces a very large fluorescence enhancement in acetonitrile for which photographic evidence is available [54b]. Irreversible interactions involving neutral free radicals have also given rise to fluorescent probes and are discussed in Sect. 6.

5.5 Twisted Systems Akin to Fluorescent PET Sensors

The discussion so far has concentrated on 'fluorophore-spacer-receptor' systems where the terminal modules are largely independent of one another except for the crucial PET process. Spatially integrated fluorophore-receptor systems have also featured in a minor capacity for purposes of comparison and also because of their numerical prevalence in the fluorescent sensor literature. In between these two classes lies a third which is the subject of this section. This consists of geometrically twisted (or potentially twistable) fluorophore-receptor systems (Fig. 7), where the modular structure is largely maintained because of π-orbital orthogonality i.e. a virtual spacer is operational between the terminal components. It is conceptually worthwhile to consider the spectrum of coupling strength between fluorophore and receptor components in terms of the role played by the spacer. This has been done in the companion review [21]. Orthogonal fluorophore-receptor systems are of particular significance in the context of this review because of the twisted intramolecular charge transfer (TICT) nature of their lowest energy excited states [146]. The energetics of TICT excited state formation is described by an equation similar to the Weller equation, Eq. (1), and represents the separation of a full electronic charge between the two modules. While TICT excited states were first established for planar donor-acceptor systems capable of torsion e.g. (65), molecules which are twisted in their ground electronic states show a stronger propensity to produce TICT excited states. The cases to be discussed are of the latter type and belong to the structural category of twisted biaryls. In addition, it is interesting that all these cases except the first are best rationalized in terms of non-emissive TICT excited states.

In his second pioneering study in this general area (see Sect.1), Shizuka and his collaborators showed that protonation switched off the charge transfer emission from (66) while switching on its anthracenic fluorescence [148]. Both these fluorescence–pH profiles yield the same value for the acidity constant in the excited state, as does the analysis of the noticeable pH dependence of the electronic absorption spectrum. The latter is significantly larger than that seen in comparable 'fluorophore-spacer-receptor' systems suggesting that the virtual spacer allows some coupling. It could be inferred that a new protonation equilibrium was not established in the excited state from the fact that the two emissions from excited (66) decay as single exponential functions i.e. independently. This rationalizes the apparent identity of excited singlet and ground state

Fig. 7. The assembly of functional components in a twisted system related to fluorescent PET sensors

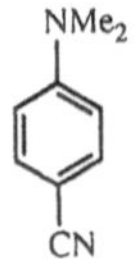

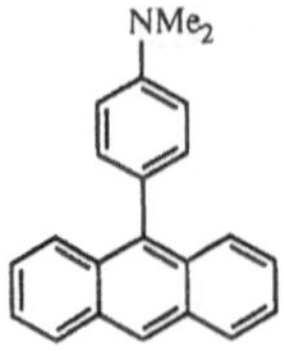

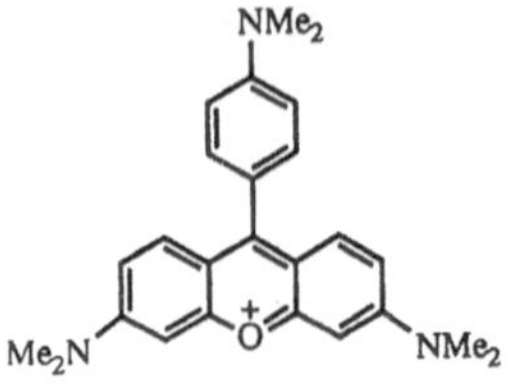

(65)

(66)

(67)

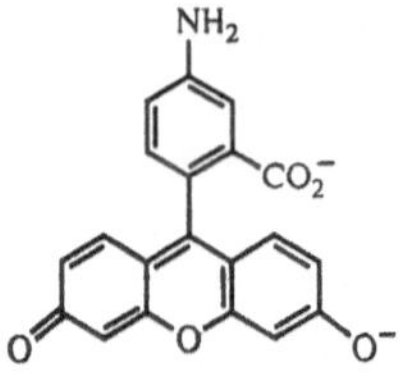

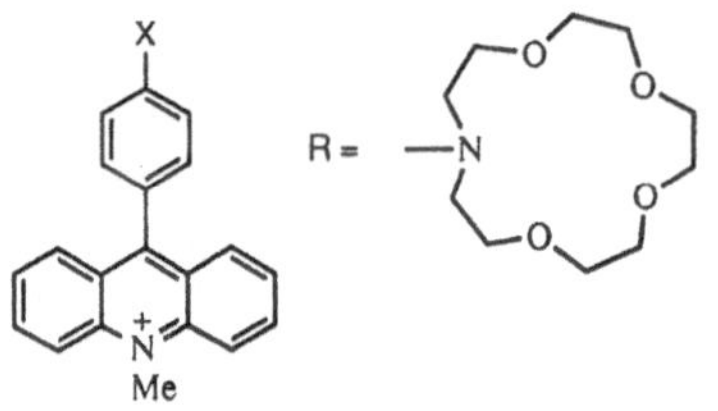

(68)

(69)a; X = NMe₂
b; X = R

R =

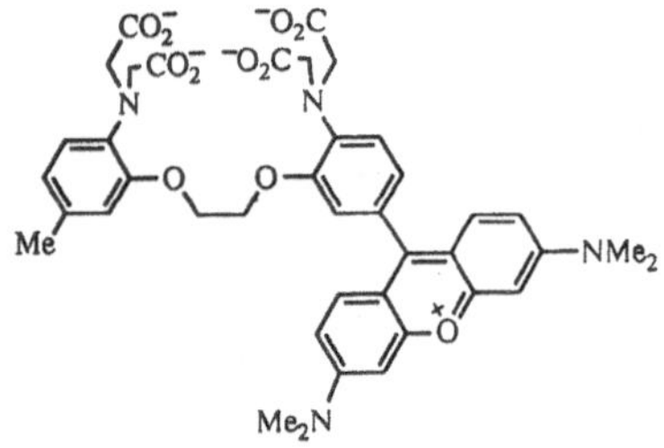

(70)

acidity constants. It was pointed out in Sect. 5.1.1 that this identity relation is common in modular systems with attenuated interactions between components. Thus, twisted systems share several features with fluorescent PET sensors at both the experimental and conceptual levels. Most importantly, protonation removes either PET or TICT processes from these systems.

There is an interesting structural similarity between (66) and the other subjects of this section – (67) [149], (68) [150], (69) [151] and (70) [152, 153] – in that they are all composed of 4-aminophenyl groups substituted at a meso position of a linearly fused tricyclic aromatic frame. The similarity continues even in their experimental behaviour in that protonation causes significant fluorescence enhancements in all cases. The structural elaboration of the aniline unit in (70) allows similar switching on with calcium ions. The case (70) is taken up further in Sect. 7. Additionally, (69) with its very electron deficient acridinium acceptor unit shows extensive perturbation of absorption spectra by several cations. Again, the acidity constants are experimentally identical in both the ground and excited singlet states. We close this final section of the survey chapter by noting that twisted systems involving TICT excited states considerably strengthen the concept and application base of fluorescent PET sensors.

6 Potentialities of the Fluorescent PET Sensing Principle for Molecular Association Research

The mechanism of fluorescent PET sensing required an association of the guest with the receptor module of the sensor supermolecule. From another viewpoint, the fluorophore-spacer combination acts as a reporter on the intimate details of the receptor-guest association. This means that the receptor has been fitted with an optically functional substituent which has only a limited influence on the guest-binding characteristics of the receptor. If we think laterally, we can see some parallels in the photoaffinity labelling reagents employed in enzymology [154]. An analogue to an enzyme substrate is created by substitution of a photochemically functional group at a remote location, so as to minimize the interference with the enzyme–substrate recognition and binding process. Most of our discussion has centred on cation–receptor associations with proton–Brønsted base interactions dominating. A reason for this bias is the easy availability of cation receptors in the wake of the growth of supramolecular chemistry [14] and the maturation of co-ordination chemistry [131]. In contrast, the development of anion receptors has lagged substantially behind. The special position of the proton reflects the fact that it underpinned many developments in solution chemistry for over a century [155]. A further reason is that an association involving electrically charged species naturally leads to significantly large oxidation (or reduction) potential modulation for controlling the PET process. Protons are extreme in this regard due to the covalent (though highly reversible) nature of their association with Brønsted bases. Indeed, for this reason, conceptual expansion of the fluorescent PET sensor design logic is usually best carried out with proton binding systems. A bonus for such work is that Brønsted bases are single point binders which are structurally simple and hence the corresponding sensor supermolecules are easily synthesized.

However, electric charges are not essential to create significant redox potential modulation in one component during a bimolecular association. Such modulation can be achieved if a) the microenvironment of the chosen component alters or b) a substantial redistribution of electron density occurs. Situation a) arises commonly during processes where a macromolecule engulfs a smaller molecule which, until then, enjoyed a micro-environment of bulk water. Therefore, powerful fluorescent signalling may be transduced during antibody–antigen, bioreceptor–hormone or enzyme–substrate interactions. Biochemists are the best equipped to take up these challenges. While situation b) is also expected to play some part in the above associations, it dominates interactions between small molecules which involve conformational changes, hydrogen bonds, charge transfer, dipolar forces or Lewis acid–base pairing. Conformational changes contribute to the large fluorescence enhancements observed in the calcium sensors (**27b**). Hydrogen bonds play a similar role in the sensors (**19**) and (**13**) for putrescinium and monohydrogen phosphate ions respectively. In these two instances, hydrogen bond arrays are additionally responsible for the binding between the two partners. Even hydrogen bonding solvents such as methanol retard PET processes arising from nitrogen lone pairs in inter- [156] and intramolecular [54b] fluorescence quenching situations. Charge transfer forces have been employed by Stoddart to produce molecular mechanochemical devices [27, 157]. The intriguing possibility exists that these forces can be harnessed to produce fluorescent switches. Lewis acid–base pairing is another growing reality for coupling to the fluorescent PET sensor logic [64, 97a]. In this regard, we can also take encouragement from the demonstration of BCl_3-induced enhancement of acridine fluorescence which is intrinsically low due to $n\pi^*–\pi\pi^*$ intersystem crossing [158]. Instances which can be understood as lithium ion-suppressed $n\pi^*–\pi\pi^*$ intersystem crossing can be seen in (**71**) and relatives [159].

Scheme 1 catalogues some associations which are available in the moecular sciences and are of interest to designers of fluorescent PET sensors. It must be emphasized that the fluorophore-spacer combination can be fixed to either partner of the association complex in the cases where a handle is available for attachment, thus widening the scope of the principle. Besides this means of shedding light on molecular associations, we must not forget that the present approach can afford fluorescent PET sensors for either partner. However, some of these cases involve covalent interactions and thus the binding would be irreversible. Even non-covalent interactions can be so strong that they lead to virtual irreversibility by decreasing dissociation rates. In addition, some of these

Proton	–	Brønsted base
Brønsted acid	–	Brønsted base
Metal ion	–	Ligand
Lewis acid	–	Lewis base
Anion	–	Anion receptor
Host	–	Guest
Electrophile	–	Nucleophile
Hydrogen bond donor	–	Hydrogen bond acceptor
Oxidant	–	Reductant
Charge transfer donor	–	Charge transfer acceptor
Ion	–	Counter ion in contact ion pair
Radical	–	Radical trap
Solute	–	Solvent
Adsorbent	–	Adsorbate
Enzyme	–	Substrate/inhibitor
Carrier (protein)	–	Ion/molecule
Receptor	–	Hormone/drug
Antibody	–	Antigen

Scheme 1. Molecular associative interactions as a resource for fluorescent PET sensor design

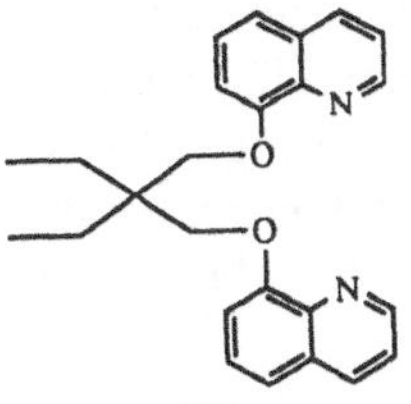

(71) 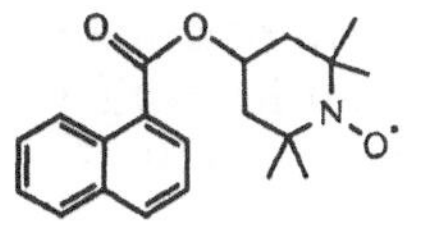   (72) 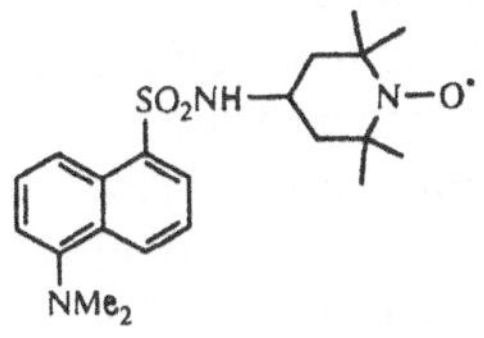   (73)

interactions can have substantial activation barriers in spite of overall ex-ergonicity. In all these situations one would be designing fluorescent PET reagents for single use rather than sensors for continuous monitoring. An extreme situation involving an association between electrically neutral species is that of radical–radical combination. A system constructed as a 'fluorophore-spacer-free radical trap' such as (72) [160] would have its fluorescence quenched by electron exchange interactions which can be viewed as two electron transfer acts in opposite directions. The primary electron transfer act in a very similar but earlier system (73) [161] proceeds essentially one-way. The binding of an external radical or reductant to the radical trap would suppress the quenching process. Thus, fluorescent reagents for free radical and reductant species are feasible [160].

7 Current Needs and Areas for Growth

While the principle of PET has been a guide to the design of many of the sensors described within these pages, no studies have been yet conducted, to the best of our knowledge, to establish the mechanism of action. The ion-induced recovery of luminescence and the successful application of the Weller equation, Eq. (1), are suggestive, but not proof, of PET in these sensors. In fact, ion-induced fluorescence enhancement is known to arise from other unrelated mechanisms such as the elimination of intersystem crossing from $n\pi^*$ to $\pi\pi^*$ excited states [159, 162]. Also, the Weller equation being a thermodynamic statement requires additional manipulation according to the Marcus formalism before ET rates can be extracted [67]. Photoinduced electron transfer was unequivocally estab-lished by, among other things, flash photolytic investigations which demon-strated the transient existence of radical anion–radical cation pairs following the photon absorption act. The studies with intermolecular systems were made easier due to the moderate nature of the rate of back ET [4]. Intramolecular system have required picosecond laser facilities due to the faster back ET [7]. Among the few exceptions are systems with long spacers, albeit with enhanced transmission, whose radical ion pairs evolve in the nanosecond regime [163]. The sensors discussed in this article are characterized by short or virtual spacers and hence fast back ET rates are expected [3d, 164]. In fact, the successful real-time monitoring function of these sensors can only be maintained if their 'down time' in the electron transferred radical ion pair state (where sensing activity is frozen) is minimized. Therefore, laboratories with picosecond laser facilities would be encouraged to examine fluorescent PET sensors for the possible presence of radical ion pairs.

The value of the Weller equation, Eq. (1), for the design of fluorescent PET sensors relies heavily on the availability of redox potential data for the various components. While extensive tables of redox potentials are available for simple structures [35], data is still lacking for many receptors and fluorophores of interest. Until such deficiencies are filled, resort must be made to linear free energy relationships [110] and crude modelizations [81] to obtain rough estimates. Sensor design solely on the basis of the Weller equation makes the naïve assumption that PET rates are controlled only by the thermodynamic driving force. Even in those situations where this is valid the inverted region at high exergonicities [67] has yet to be examined and possibly exploited by sensor designers. In more general situations however, it is by the construction and evaluation of new 'fluorophore-spacer-receptor' systems that we can investigate, and learn from, any regio-dependence of PET rates in sensory contexts. Recent developments regarding long range electron transfer [41, 165] show the value of research into the role of the spacer. Synthetic chemists have a crucial function in this regard as well as in the design and construction of receptors with new selectivity profiles and of fluoro/lumophores with relatively long communication wavelengths [166]. A further point is that, unlike in many other areas, the operating range of fluorescent PET sensors with regard to analyte concentration has been quantitatively predictable with some accuracy in several instances [19, 43, 81]. Therefore, more binding constant data are needed for known receptors [30] if these are to allow a quantitatively rational choice of components of PET sensors for successful use under a given set of conditions.

Fluorescent reagents and probes of various designs have long maintained a presence in the chemical and life sciences area. In fact, the classical protein probe (74) (1,8-ANS) can now be best understood as a TICT system with a virtual spacer (Sect. 5.5) [167] though water may exert a specific effect [168]. It is therefore natural to expect PET sensors to also play their part in chemical and biological research. Most of the systems described have not been directly targetted toward such applications. Such targetting involves the use of receptor modules of adequate selectivity and binding towards the chosen ion. For example, the calcium sensor (70), one of many successful fluorescent sensors

(74)

from Tsien's laboratory [152] has adequate selectivity versus magnesium and protons in the physiological range due to the well-proven (27a) receptor unit [78]. The sensing action of (70) is most simply understood as a twisted fluorophore-receptor system with a virtual spacer [153]. We have employed the (27a) receptor unit to build 'fluorophore-spacer-receptor' systems such as (27b) [81] which show very large calcium induced flourescence enhancements, while quantitatively preserving the excellent calcium binding properties of (27a). These systems with visible communication wavelengths should therefore complement the spatially integrated fluorophore-receptor systems like (57) which are popular with biologists [132]. However, the present systems lack the calcium induced spectral wavelength shift of (57) which allows fluorescence ratio imaging of calcium fields in single cells. Conceptual expansion [21] is needed if PET sensors are to contribute to this area, since first generation PET sensors are designed to give intensity on-off action at all wavelengths.

The potential for using PET systems in chemical assays may be illustrated with Czarnik's case (13) for phosphate [20]. The careful choice of pH is important in order to simultaneously provide a substantial fraction of phosphate species in the monoprotonated form and obtain triprotonation of the amine sites distal to the anthracene unit. The docking of these two complementary groups results in the hydrogen bonding of the hydroxyl group of monohydrogenphosphate to the benzylic nitrogen lone electron pair, causing the crucial oxidation potential increase leading to fluorescence enhancement. The participation of analytical chemists and sensor technologists could greatly aid the development of this line of research.

The suitability of fluorescent PET systems for use in the emerging field of molecular switching devices was pointed out in Sect. 4 due to their natural on-off action induced by ion input. Discussions in Sect. 6 have also illustrated the value of PET sensor ideas in the design of reagents and also of reporters on receptor–guest interactions. Such versatility of the fluorescent PET sensor logic makes this research worthwhile.

Acknowledgement. The contributions from our laboratory to this area were made possible by the support of The Science and Engineering Research Council, The Department of Education of Northern Ireland, The Nuffield Foundation, The University of Colombo and The Queen's University of Belfast. We value the continued encouragement of Dr J. Grimshaw, Dr J.T. Grimshaw, Professor R. Grigg, Dr B.J. Walker, Professor R.S. Ramakrishna and Professor A.M. Abeysekera over the years, as well as the typing support of Irene Campbell.

8 References

1. (a) Bender CJ (1986) Chem. Soc. Rev. 15: 475 (b) Mulliken RS, Person W (1969) Molecular complexes. Wiley, New York (c) Foster R (ed) (1975, 1979) Molecular association. vols 1, 2. Academic, New York
2. Gordon M, Ware WR (eds) (1975) The exciplex. Academic, New York

3. (a) Davidson RS (1983) Adv. Phys. Org. Chem. 19: 1 (b) Kavarnos GJ, Turro NJ (1986) Chem. Rev. 86: 401 (c) Mattes SL, Farid S (1983) Org. Photochem. 6: 233 (d) Fox MA (1986) Adv. Photochem. 13: 238 (e) Julliard M, Chanon M (1983) Chem. Rev. 83: 425 (f) Fox MA, Chanon M (eds) (1988) Photoinduced electron transfer. parts A-D. Elsevier, Amsterdam (g) Mattay J (ed) (1990) Top. Curr. Chem. 156; (1990) ibid. 158; (1991) ibid. 159; (1992) ibid. 163 (h) Wasielewski MR (1992) Chem. Rev. 92: 435
4. Weller A (1968) Pure Appl. Chem. 16: 115.
5. Reference 3(f). parts C, D
6. Selinger BK (1977) Aust. J. Chem. 30: 2087
7. (a) Migita M, Okada T, Mataga N, Sakata Y, Misumi S, Nakashima N, Yoshihara K (1981) Bull. Chem. Soc. Jpn. 54: 3304 (b) Wang Y, Crawford MC, Eisenthal KB (1982) J. Am. Chem. Soc. 104: 5874
8. Shizuka H, Nakamura M, Morita T (1979) J. Phys. Chem. 83: 2019
9. (a) Winnik MA (1981) Chem. Rev. 81: 491 (b) Winnik MA (1985) Acc. Chem. Res. 18: 173
10. Cox GS, Turro NJ, Yang NC, Chem. MJ (1984) J. Am. Chem. Soc. 106: 422
11. Tazuke S, Iwaya Y, Hayashi R (1982) Photochem. Photobiol. 35: 621
12. Balzani V, Scandola F (1991) Supramolecular photochemistry. Ellis-Horwood, Chichester
13. Mes GF, Van Ramesdonk HJ, Verhoeven JW (1984) J. Am. Chem. Soc. 106: 1335
14. (a) Pedersen CJ (1988) Angew. Chem. Int. Ed. Engl. 27: 1021; Angew. Chem. 100: 1053 (b) Lehn JM (1988) Angew. Chem. Int. Ed. Engl. 27: 89; Angew. Chem. 100: 91 (c) Cram DJ (1988) Angew. Chem. Int. Ed. Engl. 27: 1009; Angew. Chem. 100: 1041
15. Konopelski JP, Kotzyba-Hibert F, Lehn JM, Desvergne JP, Fages F. Castellan A, Bouas-Laurent H (1985) J. Chem. Soc. Chem. Commun. 433
16. Paris JP, Brandt WW (1959) J. Am. Chem. Soc. 81: 5001
17. Lieu VT, Handy CA (1974) Anal. Lett. 7: 267
18. (a) de Silva AP, Rupasinghe RADD, Peiris SLA (1982) Proc. Sri Lanka Assoc. Advmt. Sci. Abst. 38: 68 (b) de Silva AP, Rupasinghe RADD (1985) J. Chem. Soc. Chem. Commun. 1669
19. (a) de Silva AP, de Silva SA (1985) Proc. Sri Lanka Assoc. Advmt. Sci. Abst. 41: 83 (b) de Silva AP, de Silva SA (1986) J. Chem. Soc. Chem. Commun. 1709
20. Huston ME, Akkaya EU, Czarnik AW (1989) J. Am. Chem. Soc. 111: 8735
21. Bissell RA, de Silva AP, Gunaratne HQN, Lynch PM, Maguire GEM, Sandanayake KRAS (1992) Chem. Soc. Rev. 21: 187
22. Bryan AJ, de Silva AP, de Silva SA, Rupasinghe RADD, Sandanayake KRAS (1989) Biosensors 4: 169
23. Smith JN (1968) in: Patai S (ed) The chemistry of the amino group. Wiley, London, p 161
24. (a) Rebek J (1988) J. Mol. Recognit. 1: 1 (b) Rebek J (1990) Angew. Chem. Int. Ed. Engl. 29: 245; Angew. Chem. 102: 261
25. (a) Chang SK, Van Engen D, Fan E, Hamilton AD (1991) J Am. Chem. Soc. 113: 7640 (b) Murray TJ, Zimmerman SC (1992) J. Am. Chem. Soc. 114: 4010
26. Collier DA. Thuong NT, Helene C (1991) J. Am. Chem. Soc. 113: 1457
27. Anelli PL, Ashton PR, Ballardini R, Balzani V, Delgado M. Gandolfi MT, Goodnow TT, Kaifer AE, Philp D, Pietraszkiewicz M, Prodi L, Reddington MV, Slawin AMZ, Spencer N, Stoddart JF, Vicent C, Williams DJ (1992) J. Am. Chem. Soc. 114: 193
28. Iimori T, Still WC, Rheingold A, Staley DL (1989) J. Am. Chem. Soc. 111: 3439
29. (a) McLendon G (1988) Acc. Chem. Res. 21: 160 (b) Wasielewski MR (1998) in: ref 3(f), Part A, p 161 (c) Closs GL, Miller JR (1988) Science 240: 440
30. (a) Izatt RM, Pawlak K, Bradshaw JS, Bruening RL (1991) Chem. Rev. 91: 1721 (b) Izatt RM, Bradshaw JS, Nielsen SA, Lamb JD, Christensen JJ, Sen D (1985) Chem. Rev. 85: 271
31. Zweig A, Hodgson WG, Jura WH (1964) J. Am. Chem. Soc. 86: 4124
32. (a) Guilbault GG (1990) Practical fluorescence, 2nd edn. Dekker, New York (b) Lackowicz JR (1983) Principles of fluorescence spectroscopy. Plenum, New York (c) Turro NJ (1978) Modern molecular photochemistry. Benjamin, Menlo Park (d) Coyle JD (1986) Introduction to organic photochemistry. Wiley, Chichester
33. Rehm D, Weller A (1970) Isr. J. Chem. 8: 259
34. (a) Murov SL (1973) Handbook of photochemistry. Dekker, New York (b) Berlman IB (1971) Handbook of fluorescence spectra of aromatic molecules, 2nd edn. Academic, New York
35. (a) Mann CK, Barnes KK (1970) Electrochemical reactions in nonaqueous systems. Dekker, New York (b) Siegerman H (1975) in: Weinberg NL (ed) Technique of electroorganic synthesis. Part II. Wiley, New York, p 667

36. (a) Tada M, Suzuki A, Hirano H (1979) J. Chem. Soc. Chem. Commun. 1004 (b) Tada M, Hamazaki H, Hirano H (1980) Chem. Lett. 921
37. (a) Delgado M, Gustowski DA, Yoo HK, Gatto VJ, Gokel GW, Echegoyen L (1988) J. Am. Chem. Soc. 110: 119 (b) Beer PD (1989) Chem. Soc. Rev. 18: 409
38. Eriksen J, Foote CS (1978) J. Phys. Chem. 82: 2659
39. Kikuchi K, Hoshi M, Shiraishi, Y, Kokubun H (1987) J. Phys. Chem. 91: 574
40. Vogelmann E, Rauscher W, Traber R, Kramer HEA (1981) Z. Phys. Chem. Neue Folge 124: 13
41. (a) Jordan KD, Paddon-Row MN (1992) Chem. Rev. 92: 395 (b) Penfield KW, Miller JR, Paddon-Row MN, Cotsaris E, Olivier AM, Hush NS (1987) J. Am. Chem. Soc. 109: 5061
42. Hirayama F (1965) J. Chem. Phys. 42: 3163
43. de Silva AP, Sandanayake KRAS (1989) J. Chem. Soc. Chem. Commun 1183
44. Cherian AL, Pandit PY, Seshadri S (1972) Ind. J. Chem. 10: 267
45. Campaigne E, Archer WL (1952) J. Am. Chem. Soc. 75: 989
46. Rivett DE, Rosevear J, Wilshire JFK (1979) Aust. J. Chem. 32: 1601
47. Grimshaw J, Perera SD (1989) J. Electroanal. Chem. 265: 335
48. Takeda Y, Kumazawa T (1988) Bull. Chem. Soc. Jpn. 61: 655
49. de Silva AP, Sandanayake KRAS (1991) Tetrahedron Lett. 32: 421
50. Kirkbright GF (1972) in: Bishop E (ed) Indicators. Pergamon, Oxford, p 685
51. Michaelis L, Gyemant A (1920) Biochem. Ztschr. 109: 165
52. Vogel AI (1961) A textbook of quantitative inorganic analysis. 3rd edn. Longmans, London, p 59
53. Posch HE, Leiner MJP, Wolfbeis OS (1989) Fresenius Z. Anal. Chem. 334: 162
54. (a) Wolfbeis OS, Marhold H (1987) Fresenius Z. Anal. Chem. 327: 347 (b) Czarnik AW (1991) in: Schneider HJ, Durr H (eds) Frontiers in supramolecular organic chemistry and photochemistry. VCH, Weinheim, p 109
55. Fuh MRS, Burgess LW, Hirschfeld T, Christian GD, Wang F (1987) Analyst 112: 1159
56. de Silva AP, Goonesekere NCW, Lynch PLM, Patuwathavithana ST (1993) in preparation
57. (a) Shinkai S, Manabe O (1984) Top. Curr. Chem. 121: 67 (b) Adams SR, Kao JPY, Grynkiewicz G, Minta A, Tsien RY (1988) J. Am. Chem. Soc. 110: 3212 (c) Adams SR, Kao JPY, Tsien RY (1989) J. Am. Chem. Soc. 111: 7957 (d) Warmuth R, Grell E, Lehn JM, Bats JW, Quinkert G (1991) Helv. Chim. Acta 74: 671
58. (a) Fendler JH (1982) Membrane mimetic chemistry. Wiley, New York (b) Fendler JH, Fendler EJ (1975) Catalysis in micellar and macromolecular systems. Wiley, New York.
59. de Silva AP, Sandanayake KRAS (1990) Angew. Chem. Int. Ed. Engl. 29: 1173; Angew. Chem. 102: 1159
60. Gabor G, Walt DR (1991) Anal. Chem. 63: 793
61. Porter G, Topp MR (1978) Proc. Roy. Soc. A315: 163
62. (a) Pardo A, Poyato JML, Martin E, Camacho JJ, Reyman D, Brana MF, Castellano JM (1986) J. Photochem. Photobiol. A Chem. 46: 323 (b) Pardo A, Martin E, Poyato JML, Camacho JJ, Guerra JM, Weigand R, Brana MF, Castellano JM (1989) J. Photochem. Photobiol. A Chem. 48: 259 (c) Pardo A, Poyato JML, Martin E, Camacho JJ, Reyman D (1990) J. Lumin. 46: 381
63. Bissel RA, Calle E, de Silva AP, de Silva SA, Gunaratne HQN, Habib-Jiwan JL, Peiris SLA, Rupasinghe RADD, Samarasinghe TKSD, Sandanayake KRAS, Soumillion JP (1992) J. Chem. Soc. Perkin. Trans. 2: 1559
64. Huston ME, Haider KW, Czarnik AW (1988) J. Am. Chem. Soc. 110: 4460
65. de Silva AP, Gunaratne HQN, Lynch PLM, Spence GL (1993) in preparation
66. Grigg R, Norbert WDJA (1992) J. Chem. Soc. Chem. Commun. 1298
67. (a) Marcus RA (1982) Faraday Discuss. Chem. Soc. 74: 7 (b) Marcus RA (1986) J. Phys. Chem. 90: 3453
68. Grigg R, Norbert WDJA (1992) J. Chem. Soc. Chem. Commun. 1300
69. Ireland JF, Wyatt PAH (1976) Adv. Phys. Org. Chem. 12: 131
70. Scypinski S, Drake JM (1985) J. Phys. Chem. 89: 2432
71. (a) Degani Y, Willner I, Haas Y (1984) Chem. Phys. Lett. 104: 496 (b) Politzer IR, Crago KT, Hampton T, Joseph J, Boyer JH, Shah M (1989) Chem. Phys. Lett. 159: 258
72. (a) Turro NJ, Bolt JD, Kuroda Y, Tabushi I (1982) Photochem. Photobiol. 35: 69 (b) Turro NJ, Cox GS, Li X (1983) Photochem. Photobiol. 37: 149
73. Bissell RA, de Silva AP (1991) J. Chem. Soc. Chem. Commun. 1148

74. Fages F, Desvergne JP, Bousa-Laurent H, Marsau P, Lehn JM, Kotzyka-Hibert F, Albrecht-Gary AM, Al-Joubbeh M (1989) J. Am. Chem. Soc. 11: 8672
75. Akkaya EU, Huston ME, Czarnik AW (1990) J. Am. Chem. Soc. 112: 3590
76. Huston ME, Engleman C, Czarnik AW (1990) J. Am. Chem. Soc. 112: 7054
77. Golchini K, Mackovic-Basic M, Gharib SA, Masilamani D, Lucas M, Furtz I (1990) Am. J. Physiol. 285: F438
78. Tsien RY (1980) Biochemistry 19: 2396
79. Smith GA, Hesketh TR, Metcalfe JC (1988) Biochem. J. 250: 277
80. de Silva AP, Gunaratne HQN, Sandanayake KRAS (1990) Tetrahedron Lett. 31: 5193
81. de Silva AP, Gunaratne HQN (1990) J. Chem. Soc. Chem. Commun. 186
82. Bourson J, Borrel MN, Valeur B (1992) Anal. Chim Acta 257: 180
83. Gubelmann M, Harriman A, Lehn JM, Sessler JL (1988) J. Chem. Soc. Chem. Commun. 77
84. (a) Lehn JM, Ziessel R (1987) J. Chem. Soc. Chem. Commun. 1292 (b) Kimura E, Wada S, Shionoya, Takahashi T, Iitaka (1990) J. Chem. Soc. Commun. 397 (c) Rawle SC, Moore P, Alcock NW (1992) J. Chem. Soc. Chem. Commun. 684 (d) Fujita E, Milder SJ, Brunschwig BS (1992) Inorg. Chem. 31: 2079
85. Nishida H, Katayama Y, Katsuki H, Nakamura H, Takagi M, Ueno K (1982) Chem. Lett. 1853.
86. Wallach DFH, Steck TL (1963) Anal. Chem. 35: 1035
87. Van Arman SA, Carnik AW (1990) J. Am. Chem. Soc. 112: 5376
88. Kumar CV, Asuncion EH (1992) J. Chem. Soc. Chem. Commun. 470
89. Lee HC, Forte JG, Epel D (1982) in: Nuccitelli R, Deamer DW (eds) Intracellular pH: its measurement, regulation and utilization in cellular functions. Liss, New York, 1982, p 136
90. Martucci JD, Schulman SG (1975) Anal. Chim. Acta 77: 317
91. Van der Kooi G (1984) Photochem. Photobiol. 39: 755
92. Schulman SG, Threatte R, Capomacchia A, Paul W (1974) J. Pharm. Sci. 63: 876
93. Wilson DL, Wirz DR, Schenk GH (1973) Anal. Chem. 45: 1447
94. Baeyens W, De Moerloose P (1978) Analyst 103: 359
95. (a) Ibemesi JA, El-Bayoumi MA, Kinsinger JB (1978) Chem. Phys. Lett. 53: 270 (b) Ibemesi JA, El-Bayoumi (1979) Mol. Photochem. 9: 243
96. Beddard GS, Davidson RS, Whelan TD (1977) Chem. Phys. Lett. 56: 54
97. (a) Wang YC, Morawetz H (1976) J. Am. Chem. Soc. 98: 3611 (b) Goldenberg M, Emert J, Morawetz H (1978) J. Am. Chem. Soc. 100: 7171 (c) Liao TP, Okamoto Y, Morawetz H (1979) Macromolecules 12: 535
98. Saeva F, Luss H, Martic P (1989) J. Chem. Soc. Chem. Commun. 1477
99. (a) Brimage DRG, Davidson RS (1971) J. Chem. Soc. Commun. 1385 (b) Ide R, Sakata Y, Misumi S, Okada T, Mataga N (1972) J. Chem. Soc. Chem. Commun. 1009
100. Chandross EA, Thomas HT (1971) Chem. Phys. Lett. 9: 393
101. May EL, Mosettig E(1951) J. Am. Chem. Soc. 73: 1301
102. de Silva AP, Gunaratne HQN, Goonesekere NCW (1992) unpublished data
103. Camilleri P, Okafo GN (1992) J. Chem. Soc. Commun. 530
104. Hansch C, Leo A, Taft RW (1991) Chem. Rev. 91: 165
105. Meites L, Zuman P (1974) Electrochemical data, Part 1, Vol. 1A. Wiley, New York
106. Miwa T, Koizumi M (1963) Bull. Chem. Soc. Jpn. 36: 1619
107. Cooke NHC, Solomon BS (1972) J. Phys. Chem. 76: 3563
108. Royer J, Beugelmans-Verrier M, Biellmann JF (1981) Photochem. Photobiol 34: 667
109. Tournon J, El-Bayoumi MA (1971) J. Am. Chem. Soc. 93: 6396
110. de Silva AP, de Silva SA, Dissanayake AS, Sandanayake KRAS (1989) J. Chem. Soc. Chem. Commun. 1054
111. Pragst F, Weber FG (1976) J. Prakt. Chem. 318: 51
112. Rivett DE, Rosevear J, Wilshire JFK (1979) Aust. J. Chem. 32: 1601
113. Szucs L (1969) Chem. Zyestu. 23: 677
114. de Silva AP, Gunaratne HQN, Lynch PLM (1992) unpublished data
115. Pederson CJ (1967) J. Am. Chem. Soc. 89: 7017
116. (a) Sousa LR, Larson JM (1977) J. Am. Chem. Soc. 99: 307 (b) Larson JM, Sousa LR (1978) J. Am. Chem. Soc. 100: 1943
117. Shizuka H, Takada K, Morita T (1980) J. Phys. Chem. 84: 994
118. Wolfbeis OS, Offenbacher H (1984) Monatsh. Chem. 115: 647
119. Shirai M, Tanaka M (1988) J. Chem. Soc. Chem. Commun. 381

120. (a) Fery-Forgues S, Le Bris MT, Guette JP, Valeur B (1988) J. Phys. Chem. 92: 6223
     (b) Bourson J, Valeur B (1989) J. Phys. Chem. 93: 3871
121. Street KW, Krause SA (1986) Anal. Lett. 19: 735
122. Minta A, Tsien RY (1989) J. Biol. Chem. 264: 19449
123. Harootunian AT, Kao JPY, Eckert BK, Tsien RY (1989) J. Biol. Chem. 264: 19458
124. (a) Jones G, Griffin SF, Choi CY, Bergmark WR (1984) J. Org. Chem. 49: 2705
     (b) Priyadarshini KI, Mittal JP (1991) J. Photochem. Photobiol. A Chem. 61: 381
125. Bouas-Laurent H, Desvergne JP, Fages F, Marsau P (1991) in: Schneider HJ, Durr H (eds) Frontiers in supramolecular organic chemistry and photochemistry. VCH, Weinheim, p 265
126. Fages F, Desevergne JP, Bouas-Laurent H (1989) J. Am. Chem. Soc. 111: 96
127. (a) Marsau P, Bouas-Laurent H, Desevergne JP, Fages F, Lamotte M, Hinschberger J (1988) Mol. Cryst. Liq. Cryst. Inc. Nonlin. Opt. 156: 383 (b) Fages F, Desevergne JP, Bouas- Laurent H, Hinschberger J, Marsau P, Petraud M (1988) New J. Chem. 12: 95
128. Bissell RA, de Silva AP, Fernando WTLM, Patuwathavithana ST, Samarasinghe TKSD (1991) Tetrahedron Lett. 32: 425
129. Iyoda T, Morimoto M, Kasaki N, Shimidzu T (1991) J. Chem. Soc. Chem. Commun. 1480
130. Aoki I, Sakaki T, Shinkai S (1992) J. Chem. Soc. Chem. Commun. 730
131. Hancock RD, Martell AE (1989) Chem. Rev. 89: 1875
132. Grynkiewicz G, Poenie M, Tsien RY (1985) J. Biol. Chem. 260: 3440
133. (a) Dorlars A, Schellhammer CW, Schroeder J (1975) Angew. Chem. Int. Ed. Engl. 14: 665 Angew. Chem. 87: 693 (b) Krasovitskii BM, Bolotin BM (1989) Organic luminescent Materials. VCH, Weinheim, pp 78, 197
134. Levy LA, Murphy E, Raju B, London RE (1988) Biochemistry 27: 4041
135. de Silva AP, Gunaratne HQN, Maguire GEM (1992) unpublished results
136. Raju B, Murphy E, Levy LA, Hall RD, London RE (1989) Am. J. Physiol. 256: C540
137. Lohr HG, Vögtle F (1985) Acc. Chem. Res. 18: 65
138. Wolfbeis OS, Urbano E (1983) Fresenius Z. Anal. Chem. 314: 577
139. Gust A, Moore TA, Makings LR, Liddell PA, Nemeth GA, Moore AL (1986) J. Am. Chem. Soc. 108: 8028
140. Shanzer A (1991) Lecture at XVIth International Symposium on macrocyclic chemistry, Sheffield.
141. Gonzalez MC, McIntosh AR, Bolton JR, Weedon AC (1984) J. Chem. Soc. Chem. Commun. 1138
142. Hunter CA, Sanders JKM, Beddard GS, Evans S (1989) J. Chem. Soc. Chem. Commun. 1765
143. (a) Stryer L (1978) Ann Rev. Biochem. 47: 819 (b) Meares CF, Wensel TG (1984) Acc. Chem. Res. 17: 202
144. Fages F, Desvergne JP, Bouas-Laurent H, Lehn JM, Konopelski JP, Marsau P, Barrans Y (1990) J. Chem. Soc. Chem. Commun. 655
145. (a) Bouas-Laurent H, Castellan A, Daney M, Desvergne JP, Guinand G, Marsau P, Riffaud NH (1986) J. Am. Chem. Soc. 108: 315 (b) Jin T, Ichikawa K, Koyama T (1992) J. Chem. Soc. Commun. 499 (c) Aoki I, Kawabata H, Nakashima K, Shinkai S (1991) J. Chem. Soc. Chem. Commun. 1771
146. Rettig W (1986) Angew. Chem. Int. Ed. Engl. 25: 971; Angew. Chem. 98: 969
147. Grabowski ZR, Dobkowski J (1983) Pure Appl. Chem. 55: 245
148. Shizuka H, Ogiwara T, Kimura E (1985) J. Phys. Chem. 89: 4302
149. Vogel M, Rettig W, Sens R, Drexhage KH (1988) Chem. Phys. Lett. 147: 452
150. Munkholme C, Parkinson DR, Walt DR (1990) J. Am. Chem. Soc. 112: 2608
151. Jonker SA, Ariese F, Verhoeven JW (1989) Recl. Trav. Chim. Pays-Bas 108: 109
152. Minta A, Kao JPY, Tsien RY (1989) J. Biol. Chem. 164: 8171
153. de Silva AP, Gunaratne HQN, Kane ATM, Maguire GEM (1993) in preparation
154. Bayley. H, Knowles JR (1977) Meth. Enzymol. 46: 69
155. Bell RP (1960) The proton in chemistry. Methuen, London
156. Weller A (1977) Personal communication in reference 6
157. (a) Anelli PL, Spencer N, Stoddart JF (1991) J. Am. Chem. Soc. 113: 5131 (b) Reddington MV, Slawin AMZ, Spencer N, Stoddart JF, Vicent C, Williams DJ (1991) J. Chem. Soc. Chem. Commun. 630
158. Snyder R, Testa AC (1990) J. Lumin. 47: 35
159. (a) Hiratani K (1987) J. Chem. Soc. Chem. Commun. 960 (b) Hiratani K, Nomoto M,

Sugihara H, Okada T (1990) Chem. Lett. 43 (c) Hiratani K, Nomoto M, Ohuchi S, Taguchi K (1990) Bull. Chem. Soc. Jpn. 63: 1349
160. Blough NV, Simpson DJ (1988) J. Am. Chem. Soc. 110: 1915
161. Bystryak IM, Likhtenshtein GL, Kotel'nikov AI, Hankovskii HO, Hideg K (1986) Zh. Fiz. Khim. 60: 2796 (Chem. Abstr. 106: 75994t)
162. Kasama K, Kikuchi K, Yamamoto S, Uji-ie K, Nishida Y, Kokubun H (1981) J. Phys. Chem. 85: 1291
163. Finckh P, Heitele H, Volk M, Michel-Beyerle ME (1988) J. Phys. Chem. 92: 6584
164. Mauzerall DC (1988) in: reference 3(f). part A. p 228
165. Verhoeven JW (1990) Pure Appl. Chem. 62: 1585
166. (a) Imasaka T, Ishibashi N (1990) Anal. Chem. 62: 363A (b) Fabian J, Zahradnik R (1989) Angew. Chem. Int. Ed. Engl. 28: 677; Angew. Chem. 101: 693
167. Kosower EM, Huppert D (1986) Ann. Rev. Phys. Chem. 37: 127
168. Ebbesen TW, Ghiron CA (1989) J. Phys. Chem. 93: 7139

Note Added in Proof

The fluorescent sensing of electrically neutral species has received a boost with the report of the interaction of 2-anthryl boronic acid with sugars leading to fluorescence quenching [Yoon J, Czarnik AW (1992) J. Am. Chem. Soc. 114: 5874]. PET principles have also been employed in the design of 9-anthryl thiocarboxamide as a flourescent reagent for the redox active Hg^{2+} [Chae MY, Czarnik AW (1992) J. Am. Chem. Soc. 114: 9704]. The reagent molecule is converted to the strongly fluorescent 9-anthroate ion. This is to be compared with the (60)-Hg^{2+} interaction [75]. The twisted biaryl system (69b) has been followed up with a derivative possessing an additional methyl substituent ortho to the biaryl link [Jonker SA, Van Dijk SI, Goubitz K, Reiss CA, Schuddeboom W, Verhoeven JW (1990) Mol. Cryst. Liq. Cryst. 183: 273]. Such ortho substitution substantially increases the twist angle towards orthogonality which results in a fluorescence quantum yield of unity upon protonation. Remarkably, redox active Ag^+ causes a significant fluorescence enhancement in this case, in sharp contrast to the cases discussed in Sect. 5.4. The favourable soft–soft interaction between Ag^+ and the nitrogen centre in the azacrown ether moiety, which is observable by X-ray crystallography, is one reason for this result [Jonker SA, Verhoeven JW, Reiss CA, Goubitz K, Heijdendrijk D (1990) Recl. Trav. Chim. Pays-Bas 109: 154]. The strongly electron deficient nature of the acridinium fluorophore is another likely contributor to this result by disfavouring a new PET process from the fluorophore to the Ag^+ centre. The solvent dependence of the photophysical properties of (53) has been interpreted in terms of a fluorescent TICT excited state [Frey-Forgues S, Le Bris MT, Mialocq JC, Pouget J, Rettig W, Valeur B (1992) J. Phys. Chem. 96: 701].

Author Index Volumes 151–168

Author Index Vols. 26–50 see Vol. 50
Author Index Vols. 50–100 see Vol. 100
Author Index Vols. 101–150 see Vol. 150

The volume numbers are printed in italics